Natural Bioactives as Leading Molecules for Drug Development

Natural Bioactives as Leading Molecules for Drug Development

Guest Editors

Chang Liu
Hang Ma
Ang Cai

MDPI

Basel • Beijing • Wuhan • Barcelona • Belgrade • Novi Sad • Cluj • Manchester

Guest Editors

Chang Liu
Department of Biomedical
and Pharmaceutical Sciences
University of Rhode Island
Kingston
USA

Hang Ma
Department of Biomedical
and Pharmaceutical Sciences
University of Rhode Island
Kingston
USA

Ang Cai
College of Pharmacy
University of Rhode Island
Kingston
USA

Editorial Office
MDPI AG
Grosspeteranlage 5
4052 Basel, Switzerland

This is a reprint of the Special Issue, published open access by the journal *Biomolecules* (ISSN 2218-273X), freely accessible at: https://www.mdpi.com/journal/biomolecules/special_issues/019EC076F7.

For citation purposes, cite each article independently as indicated on the article page online and as indicated below:

Lastname, A.A.; Lastname, B.B. Article Title. *Journal Name* **Year**, *Volume Number*, Page Range.

ISBN 978-3-7258-6884-1 (Hbk)
ISBN 978-3-7258-6885-8 (PDF)
https://doi.org/10.3390/books978-3-7258-6885-8

Contents

About the Editors

Chang Liu

Chang Liu earned his Ph.D. in pharmaceutical sciences from the University of Rhode Island, an M.S. in pharmacology from Yangzhou University, and a B.E. in pharmaceutical engineering from Yangzhou University. He is currently a research assistant professor and the director of the Proteomics Core Facility at the University of Rhode Island College of Pharmacy. Dr. Liu's research centers on drug discovery and the pharmacological evaluation of small molecules, with a particular emphasis on immune checkpoint inhibitors, natural product bioactives, and proteomics-driven approaches. His work integrates biophysical and omics technologies to elucidate the molecular mechanisms underlying inflammation, oxidative stress, and cancer immunotherapy. His research has been supported by the NIH, NSF, and the Rhode Island Foundation.

Hang Ma

Hang Ma, Ph.D., is a research assistant professor in the Department of Biomedical and Pharmaceutical Sciences at the University of Rhode Island (URI) College of Pharmacy. He has over 20 years of research experience in natural product chemistry and translational drug discovery. Dr. Ma directs the Bioactive Botanical Research Laboratory at URI, where his group integrates phytochemistry, medicinal chemistry, and mechanistic biology to identify, characterize, and optimize bioactive molecules from medicinal plants and functional foods. His research emphasizes structure–activity relationships, molecular target identification, and mechanism-of-action studies relevant to inflammation, immune regulation, oxidative stress, ferroptosis, cancer, skin disorders, and age-associated diseases. Leveraging advanced analytical platforms and multi-omics approaches, his work bridges natural product discovery with pharmacological validation and translational applications. Dr. Ma has authored over 120 peer-reviewed research articles and received both his M.S. and Ph.D. degrees in pharmaceutical sciences from the URI.

Ang Cai

Ang Cai is a research assistant professor and director of the RI-INBRE Centralized Research Core Facility at the University of Rhode Island. His interests center on drug discovery and development, biophysical and biochemical characterization of molecular interactions, and the application of advanced analytical technologies in biomedical research. He oversees the operation, maintenance, and strategic development of a multidisciplinary core facility supporting chemistry and molecular biology research, with expertise in mass spectrometry, surface plasmon resonance, flow cytometry, spectroscopy, and related platforms. His research projects address cancer therapeutics, infectious diseases, and small-molecule screening, including published work on DNA damage–induced replication heterogeneity and the identification of inhibitors targeting the SARS-CoV-2 main protease. He has organized numerous technology workshops and instrument demonstrations and has contributed to strengthening shared research infrastructure and collaborative research across the INBRE network.

biomolecules

MDPI

Review

Natural Alternatives in the Treatment of Colorectal Cancer: A Mechanisms Perspective

Karen Vanessa Fernandez-Muñoz [1,2], Cuauhtémoc Ángel Sánchez-Barrera [1], Marco Meraz-Ríos [2], Jose Luis Reyes [1], Eloy Andrés Pérez-Yépez [3], Maria Teresa Ortiz-Melo [1], Luis I. Terrazas [1,4,*] and Monica Graciela Mendoza-Rodriguez [1,*]

1 Unidad de Biomedicina, Facultad de Estudios Superiores Iztacala, Universidad Nacional Autónoma de México, Avenida de los Barrios 1, Los Reyes Iztacala, Tlalnepantla 54090, Mexico; karen.fernandez@cinvestav.mx (K.V.F.-M.); angel.casb94@gmail.com (C.Á.S.-B.); teresaortiz@iztacala.unam.mx (M.T.O.-M.)

2 Departamento de Biomedicina Molecular, Centro de Investigación y de Estudios Avanzados del Instituto Politécnico Nacional, Avenida Instituto Politécnico Nacional 2508, Ciudad de México 07360, Mexico

3 Laboratorio de Genómica, Instituto Nacional de Cancerología, Tlalpan 14080, Mexico

4 Laboratorio Nacional en Salud, Facultad de Estudios Superiores Iztacala, Universidad Nacional Autónoma de México, Avenida de los Barrios 1, Los Reyes Iztacala, Tlalnepantla 54090, Mexico

* Correspondence: literrazas@unam.mx (L.I.T.); monica.mendoza@iztacala.unam.mx (M.G.M.-R.)

Abstract: Colorectal cancer (CRC) is one of the deadliest neoplasia. Intrinsic or acquired resistance is the main cause of failure of therapy regimens that leads to relapse and death in CRC patients. The widely used chemotherapeutic agent 5-fluorouracil (5-FU) remains the mainstay for therapeutic combinations. Unfortunately, chemotherapeutic resistance and side effects are frequent events that compromise the success of these therapies; the dysregulation of enzymes that regulate 5-FU metabolism increases the expression and activity of efflux pumps. Additional tumor cell adaptations such as epithelial–mesenchymal transition (EMT), autophagy shaping of the tumor microenvironment, and inflammation contribute to chemoresistance. Finding new strategies and alternatives to enhance conventional chemotherapies has become necessary. Recently, the study of natural compounds has been gaining strength as an alternative to chemotherapeutics in different cancers. Curcumin, trimethylglycine, resveratrol, artemisinin, and some helminth-derived molecules, among others, are some natural compounds studied in the context of CRC. This review discusses the main benefits, mechanisms, advances, and dark side of conventional chemotherapeutics currently evaluated in CRC treatment. We also analyzed the landscape of alternative non-conventional compounds and their underlying mechanisms of action, which could, in the short term, provide fundamental knowledge to harness their anti-tumor effects and allow them to be used as alternative adjuvant therapies.

Keywords: colorectal cancer; chemoresistance; natural compounds; bioactive compounds; adjuvants; chemosensitivity

1. Introduction

Colorectal cancer (CRC) is one of the main concerns worldwide due to its increasing incidence and mortality rate in the last few years. Currently, CRC is the third most common cancer (1.9 million new cases) and the second leading cause of cancer-related death (930,000 deaths) [1]. Furthermore, the World Health Organization (WHO) estimates that by 2030 there will be 2.2 million new cases of CRC and 1.09 million deaths caused by CRC [2].

The etiological origin of CRC can be classified into two principal groups: hereditary, around 20%, and sporadic, representing 80% of total cases [3]. Sporadic CRC emerges from the accumulation of acquired somatic mutations and epigenetic alterations related to several modifiable lifestyle risk factors, such as obesity, sedentary behavior, high alcohol and tobacco intake, low fiber consumption, high-fat diets, and processed meats [4]. Three main pathways have been described for the formation of sporadic CRC tumors: (I) the adenoma–carcinoma sequence, influenced by chromosomal instability (CIN), microsatellite instability (MSI), KRAS, APC, and Tp53 mutations; (II) the serrated pathway, governed by the methylation of CpG island methylator phenotype (CIMP), MSI, BRAF, and CDKN2A mutations; and (III) the inflammatory pathway, related to the activation of inflammatory signaling pathways such as NFκB, IL-6/STAT3, COX-2/PGE2 and IL-23/Th17 [5,6].

The heterogeneity of tumor origins influences the treatment and prognosis of CRC. The classification and identification of tumor subtypes can improve disease management and predictions [7]. The WHO classifies CRC tumors according to their histological characteristics, including (I) adenocarcinoma, which accounts for about 85%; (II) mucinous carcinoma, which represents around 5–20% of the total cases; (III) medullary carcinoma (4%); and (IV) Signet-ring carcinoma, with less than 2% [8].

The staging-based classification of CRC tumors is supported by the tumor–node–metastasis (TNM) system, where T represents tumor invasion into the layers of the intestinal wall subdivided into T1: submucosa; T2: muscularis propria; T3: mesocolon or mesorectal fat; and T4: perforation of serosa or invasion into other organs. The N refers to the number of lymph nodes with metastases, where N0: no lymph nodes; N1: 1–3; N2: 4 or more. The M indicates the presence of distant metastasis, including M0, which has no distant metastasis, and M1, with metastasis beyond regional lymph nodes. In this context, colorectal tumors are classified as early stage or stage I, where T1 or T2 indicate tumors without metastases, stage II is T3 or T4 tumors without metastases, stage III is tumors with lymph node metastases, and stage IV is indicative of tumors with distant metastases [9]. Since detecting CRC in the pre-cancerous and early stages reduces mortality, periodic screening is recommended for adults aged 50–70. However, due to the increasing incidence of CRC in young patients, recent guidelines recommend screening at age 45 [10]. In this context, the treatment choice and its success depend on the tumor stage at diagnosis.

This is a narrative review structured as shown in Figure 1. First, the keywords related to the research question were identified, and recognized databases were utilized for the search (PubMed, Scopus, and Google Scholar). The inclusion criteria focused on all the works on colorectal cancer treatment and natural compounds and excluded the articles that did not have experimental validation and relation to colorectal cancer treatment with 5-Fluorouracil (5-FU). All the publications selected were thoroughly analyzed in their full-text forms to ensure alignment with the inclusion criteria. This system allowed for the development of the following topics of the review.

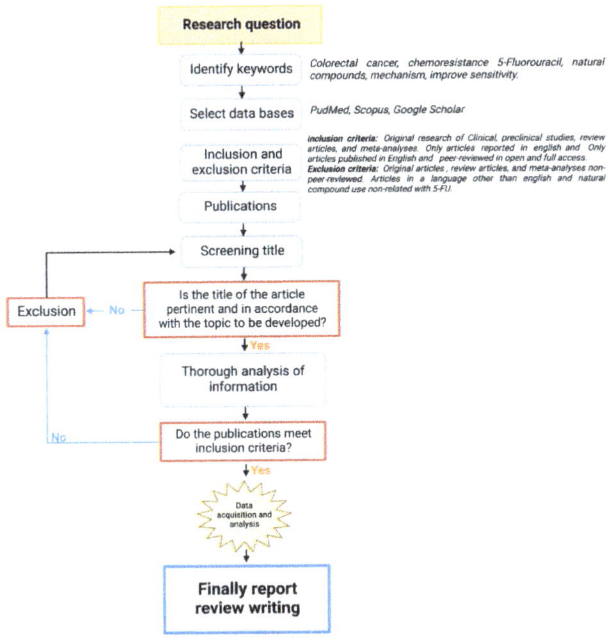

Figure 1. A flowchart of the methodological approach that was followed for the selection/exclusion of the articles.

2. CRC Treatment

The treatment management of patients with CRC mostly depends on the stage of their diagnosis. Surgery is the primary treatment used in the early stages (I and II), and in the advanced stages, surgery accompanied by systemic therapy is the option for patients with positive lymph nodes and patients with a low-burden spread of tumors [11]. Unfortunately, it is well known that more than 50% of CRC patients will develop metastasis, mainly to the liver; in this situation, most tumors are unresectable, and systemic therapy becomes the primary treatment for metastatic CRC (mCRC) [12,13]. Systemic therapy consists of chemotherapy, targeted therapy, and immunotherapy. Chemotherapy is the primary treatment for mCRC inducing cytotoxicity against cancer cells; targeted therapy recognizes specific molecules that are involved in signaling pathways related to proliferation and survival, such as epidermal growth factor receptor (EGFR) and vascular endothelial growth factor (VEGF); and immunotherapy blocks immune checkpoint molecules such as PD-1, PD-L1, and CTLA-4 which restore the activation of adaptative immune cells [14].

5-Fluorouracil (5-FU) is the central axis for therapeutic combinations in advanced CRC treatment. However, individual therapy's function and success rates are limited (10–15%). Therefore, it is frequently administered together with other drugs, such as leucovorin, oxaliplatin, or irinotecan, increasing the response rates by 40–50% [15]. Nonetheless, it has been determined that 40–50% of all patients with CRC will develop metastasis either at the time of diagnosis or as a recurrent disease after therapy [13,16]. mCRC remains incurable, and the five-year relative survival rate is only about 14% [14].

It is well known that current pharmacological chemotherapeutic drugs are closely related to severe side effects in cancer patients [17]. The adverse effects of chemotherapy in CRC patients include typical symptoms such as asthenia, nausea, vomiting, hair loss, diarrhea, and immunosuppression [18]. However, other symptoms with more significant repercussions on the patient's health may occur. For instance, the pharmacological scheme of 5-FU plus irinotecan causes intestinal mucositis in 5–15% of patients [19]. Cardiotoxicity and thrombosis have been reported in 5-FU treatment, and oxaliplatin administration can provoke peripheral neuropathy in about 89% of patients [20,21]. Usually, these side effects are intolerable for patients who are forced to reduce chemotherapeutic drug doses, therefore limiting the therapeutic response, or even requiring that they drop the treatment scheme [22], favoring the acquisition of chemoresistance, which is one of the main problems clinicians face in treating and managing CRC patients.

3. Chemoresistance in CRC

Chemoresistance is associated with a lack of response to therapy and relapse in CRC patients, representing a significant problem in medical oncology. Drug resistance occurs in about 90% of patients with mCRC, and high mortality rates can be attributed to a deficient response to chemotherapeutics [23,24].

Chemoresistance can be classified into two main categories: intrinsic or acquired. Intrinsic chemoresistance exists before drug treatment, implicating pre-existing genetic and cellular alterations. These mechanisms are critical for the result of the initial response to therapies and may influence subsequent outcomes that lead to acquired resistance [25]. Thus, acquired chemoresistance is the adaptation of cancer cells after treatment exposure, in which these cells display the ability to self-renew and re-initiate the cancer-like parental tumor as well as the expression of a distinctive set of surface biomarkers; this hypothesis proposes the existence of cancer stem cells [25]. Both mechanisms lead to central hallmarks such as the evasion of tumoral cell death by increasing anti-apoptotic proteins, inflammation, increasing potential drug targets, alteration in the functions of efflux pumps, the acquisition of EMT, autophagy, the presence of cancer stem cells (CSCs), and the tumor microenvironment as the primary weapons of tumoral cells in resistance (Figure 2).

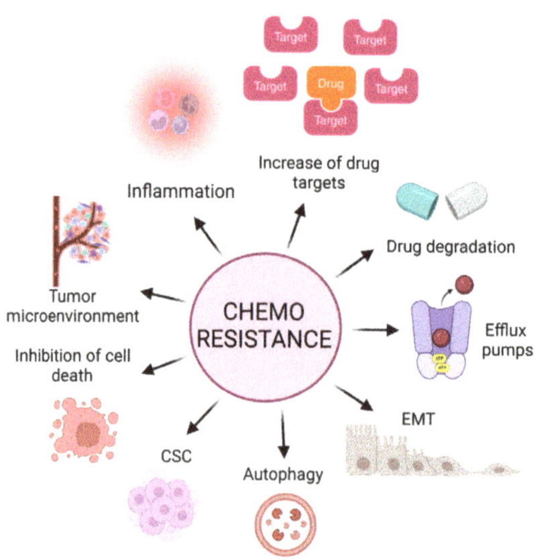

Figure 2. Main mechanisms and alterations in tumors that lead to evasion of cell death and, consequently, chemoresistance and tumor relapse. Inflammation, increment of drug targets, alteration in functions of efflux pumps, acquisition of epithelial–mesenchymal transition (EMT) markers, autophagy, presence of cancer stem cells (CSCs), increment of anti-apoptotic proteins, and tumor microenvironment.

3.1. Enzymes Involved in the 5-FU Metabolism Processing

One of the main features of 5-FU is its functioning as an analog of uracil with a fluorine atom bound at the C-5 position in place of hydrogen that favors cell entry through the exact same facilitated transport mechanism as uracil [26]. Once inside the cells, this compound is converted to 5-fluorodeoxyuridine monophosphate (FdUMP), a potent thymidylate synthase (TS) inhibitor. TS is the main enzyme required for DNA replication, acting as a dimer. It contains a nucleotide-binding site for 5,10-methylenetetrahydrofolate (CH2THF) as the methyl donor necessary for the conversion of deoxyuridine monophosphate (dUMP) to deoxythymidine monophosphate (dTMP), supplying the sole de novo source of thymidylate, which is required for DNA replication and repair [15]. This interaction of 5-FU and TS is performed when FdUMP binds to the nucleotide-binding site of TS, forming a ternary complex with the enzyme and CH2THF, thus blocking the binding site of the normal substrate dUMP and blocking dTMP synthesis and DNA replication [27] (Figure 3). Therefore, the alteration and increase in TS expression or mutation in this protein that disables the binding of FdUMP are the most well-established mechanisms of chemoresistance to 5-FU and imply poor prognosis in cancer patients [28]. As observed in breast and prostate cancer patients, the high expression of TS has been associated with a poor prognosis and survival [29,30]. In primary stomach adenocarcinoma, the expression of TS was associated with poor response and survival in patients who received 5-FU as a therapy [31,32]. In CRC, the main neoplasia treated with 5-FU, the landscape is too similar, and the high expression of TS is frequently observed in refractory patients to 5-FU treatment; in contrast, diminished TS expression correlates with a better response to chemotherapeutic treatments [32]. Indeed, when TS is found in low concentrations, CRC patients have a higher 3-year survival rate [33]. On the other hand, when TS is overexpressed, CRC patients usually do not respond to 5-FU-based schemes, such as the FOLFOX regimen (5-FU, oxaliplatin, and leucovorin) [34].

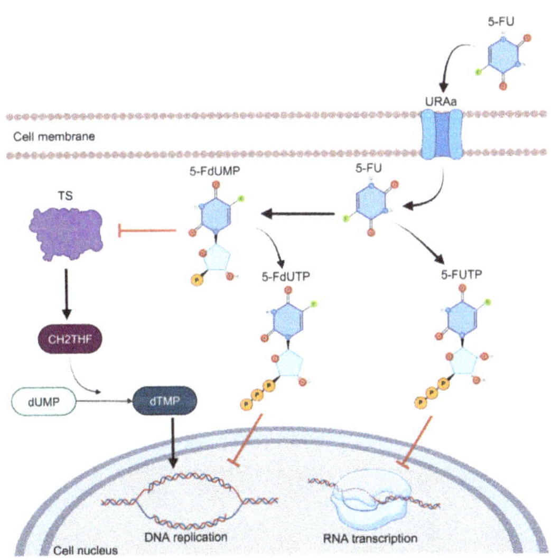

Figure 3. The 5-FU metabolism and inhibition of thymidylate synthase activity.

Dihydropyrimidine dehydrogenase (DPD) is an enzyme involved in 5-FU catabolism. In the intravenous administration of 5-FU, a significant percentage of this drug (~80%) is metabolically degraded in the liver by the DPD enzyme, converting 5-FU to dihydro fluorouracil (DHFU), an inactive metabolite unable to stop DNA replication and synthesis due to its inability to join to TS [35]. Therefore, there is an inverse correlation between DPD expression and 5-FU-based chemotherapeutic response; in addition, DPD overexpression is associated with an unfavorable prognosis in different types of cancers, including CRC. The low expression and activity of DPD have been related to a better response to 5-FU in advanced head and neck cancer, and the same has been observed in colorectal cancer, in which the inhibition of DPD levels reverses 5-FU resistance [36].

5-FU is an analog to the nucleotide uracil; thus, its entry into cells is through facilitated transport, and it is then converted to 5-fluorouridine monophosphate (FdUMP) and may act as an inhibitor of DNA replication through the inhibition of thymidylate synthase (TS) activity, binding to the nucleotide-specific site in TS for the attachment of methylenetetrahydrofolate (CH2THF), the methyl donor necessary for the conversion of deoxyuridine monophosphate (dUMP) to deoxythymidine monophosphate (dTMP). This is how it blocks the binding of the normal substrate dUMP as 5-FdUTP. Owing to this, the sole source of thymidylate is blocked; therefore, DNA replication is inhibited. The analogous structure of 5-FU to the nucleotide uracil allows it to disrupt the transcription of RNA.

3.2. *Efflux Pumps*

The activity of efflux pumps also contributes to chemoresistance, and efflux pumps are mainly members of the ATP-binding cassette (ABC) family. These drug transporters increase chemotherapeutic efflux, reducing its intracellular accumulation in cancer cells [37]. There are three main proteins involved in chemoresistance: P-glycoprotein ABCB1 (MDR1), multidrug resistance protein ABCC1 (MRP1), and breast cancer resistance protein ABCG2 [38]. Recently, compelling evidence has demonstrated that the inhibition of these transporters can improve the efficacy of chemotherapy [39]. In CRC, it has been shown that, as a secondary effect, 5-FU treatment positively regulates the expression of ABCB1, ABCC1, and ABCG2a through the activation of the IRE1α-XBP1 pathway; thus, favoring the chemoresistance and inhibition of this protein decreased the ABC transporter expression and in turn restored the efficacy of 5-FU treatment in colon cancer cells [38,40].

3.3. *Epithelial–Mesenchymal Transition*

One of the diverse hallmarks of malignancy in tumor cells is the acquisition of a phenotype of epithelial–mesenchymal transition (EMT). This phenotype is characterized by the loss of cell–cell adhesion and apical–basal polarity and undergoing cytoskeleton reorganization, contributing to increased migration and invasion capabilities being displayed by tumor cells [41]. These changes allow cancer cells to gain the properties of mesenchymal cells, such as a greater capacity for migration and invasion, favoring metastasis [42]. The relationship between EMT and chemoresistance to different drugs, including 5-FU in CRC and other cancers, has been broadly studied [43]. Previous reports showed that markers associated with EMT, such as Twist, Zeb1, and Zeb2, after 5-FU treatments, were overexpressed in resistant colon cancer cells (HT29) compared to non-resistant cells. Hence, they display characteristic features of loose cell–cell interaction: a spindle shape, intercellular spaces, and scattering [44]. In CRC patients receiving 5-FU, leucovorin, and oxaliplatin (FOLFOX), nuclear ZEB2 expression was a marker of a poor response to chemotherapy, favoring metastasis, early recurrence, and reduced survival [45]. Thus, the pharmacological modulation of molecules re-emerging during the EMT represents a potential target for CRC therapy [46].

3.4. *Autophagy*

Autophagy is a fundamental biological process where lysosomes degrade cytoplasmic materials. It is contained in double-membrane vesicles named autophagosomes. Although it is also a type of cell death, it has been shown that tumor cells exhibit increased autophagy as a metabolic adaptation to the harsh conditions present in their surrounding microenvironment. Therefore, this process can be a double-edged sword in cancer therapy because of its tumor suppressor and pro-oncogenic properties [47]. In CRC, the dual role of autophagy in 5-FU-based therapy resistance is no exception. CRC patients overexpressing the autophagy-related gene *HSPB8* have shorter overall survival and disease-free survival, and higher tumor stages and lymph node invasion [48]. In some in vitro models, autophagy activation contributes to 5-FU resistance [49]. However, many studies have established that autophagy induction leads to cancer cell death [50,51]. These controversial roles require further studies to be clarified since several early and late (i.e., 3-MA and chloroquine) autophagy inhibitors exist. However, the lack of consensus impairs the pharmacological targeting of autophagy as a therapy.

3.5. *Chemoresistance Induced by Components of the Tumor Microenvironment*

The tumor microenvironment is a complex interaction of different cell types of the immune system, such as T cells, B cells, tumor-associated macrophages (TAMs), and myeloid-derived suppressor cells (MDSCs), with stromal cells like cancer-associated fibroblasts (CAFs), endothelial cells, and soluble extracellular components as growth factors, as well as cytokines, chemokines, hormones, extracellular matrix, etc. This tumor microenvironment can influence cancer patients' therapeutic response and clinical outcomes in several ways [52,53].

In CRC, there is evidence that CAFs can promote chemoresistance to 5-FU and oxaliplatin by different mechanisms, promoting cell stemness and EMT [54,55]. Recent reports indicate an intricate regulation of CAFs and malignancy in colon cancer cells through regulating the axis molecules, STATs, involved in the EMT process and the regulation of the malignant tumor microenvironment [56,57]. CAFs secrete the pro-inflammatory cytokine IL-6 that induces STAT3 activation through its tyrosine phosphorylation, leading to translocation to the nucleus and activating pro-survival and anti-apoptotic genes [58]. Previous studies have demonstrated that the inhibition of STAT3 activity directly impacts the levels of anti-apoptotic molecules and renders colon tumor cells susceptible to 5-FU treatment [59]. Another STAT molecule involved in the pathogenesis of CRC is STAT6. Previous reports have indicated the critical role of the signaling of this molecule in tumor growth in colonic tissue in a pre-clinical model of CRC. Interestingly, the absence of STAT6 (STAT6-KO mice) and the pharmacological inhibition of STAT6 phosphorylation significantly reduced the number and size of colonic tumors while diminishing the expression of markers related to EMT, such as β-catenin nuclear expression, and increasing the response to 5-FU [60–62]. The regulation of these molecules is closely related to the tumor microenvironment and susceptibility to chemotherapy in CRC, making them an excellent therapeutic target that needs further study. On the other hand, the effect of TAMs can also induce oxaliplatin and 5-FU resistance by maintaining high levels of Cathepsin B, increasing the levels of efflux pump proteins and suppressing caspase-mediated apoptosis, to name a few of its effects [63,64]. The interplay between the tumor and microenvironment impacts the biological characteristics of tumoral cells frequently related to the malignancy of tumors, in which specific factors and signaling mechanisms are dysregulated.

3.6. *Inflammation*

It is well known that inflammation plays a crucial role in cancer initiation, progression, and malignancy [65]. Chronic intestinal inflammation is, without a doubt, a significant risk factor for CRC development and can also be involved in a patient's prognosis and response to therapy [66]. Inflammation contributes to the drug resistance of tumoral cells by increasing the expression of efflux pumps and altering the levels of drug-metabolizing enzymes. In addition, inflammation can protect cancer cells from drug-mediated cell death by regulating DNA damage repair, activating oncogenic pathways, and modulating the influence of the tumor microenvironment [67,68]. The high expression of inflammatory molecules is associated with the nuclear factor kappa-light-chain-enhancer of activated B cells (NFκB) that is considered a master regulator of the transcriptional activation of a great variety of inflammatory molecules, such as IL-1β, IL-6, COX2, and STAT3, that are closely related to malignancy and drug resistance [69]. Different studies indicate that high expression levels of NFκB are associated with perineural invasion, lymph node metastasis, and pathologic tumor node metastasis in around 62% of analyzed CRC patients [70]. Therefore, targeting inflammation is a great option to help increase therapy effectiveness. Currently, there is much interest in the use of non-steroidal anti-inflammatory drugs (NSAIDs) as possible adjuvants to conventional therapy with promising results. However, the side effects and toxicity of NSAIDs limit the establishment of their use in the clinic [71,72].

All these mechanisms involved in chemoresistance prevent therapy success. Thus, it is necessary to find new alternatives that help to increase the response to chemotherapy by counteracting chemoresistance and avoiding greater side effects and toxicity.

4. Natural Compounds as a Promising Therapeutic Option for CRC

The search for new cancer treatment strategies to improve patients' therapeutic response has led to a broader vision of non-synthetic compounds. In this regard, natural products are a rich reservoir of bioactive compounds with therapeutic potential [73]. Moreover, around 50% of anti-cancer drugs were obtained directly or indirectly from natural products, such as alkaloids, polysaccharides, polyphenols, diterpenoids, and unsaturated fatty acids possessing various structures, among others [74]. The search for molecules of a natural origin has become relevant. It has been suggested that some of these compounds can sensitize cancer cells to chemotherapy, enhancing the effect of drugs and making them more specific to acting only on cancerous cells. Furthermore, natural compounds tend to be multi-target, so they can interfere with different signaling pathways and thus prevent the appearance of chemoresistance [75]. The origin of natural products can be diverse: microbes, plants, and other living organisms, such as protozoa and helminths, all present anti-cancer properties [73,76].

In vitro and in vivo assays have demonstrated the potential use of natural products in inhibiting cancer cell proliferation, cell cycle arrest, and tumoral cell death [77,78]. This does not mean that natural compounds (NCs) alone are the cure for all types of neoplasia. Previous reports have highlighted that the use of these compounds as a single therapy does not seem to be a viable option to cure a disease as aggressive as cancer; nonetheless, their use as adjuvants to conventional chemotherapy could be a promising therapeutic option to increase anti-cancer therapy responses [79]. Curcumin, resveratrol, artemisinin, trimethylglycine, and helminth-derived molecules are common as new NCs with potential uses as adjuvants to improve 5-FU-based therapy in CRC models. The mechanisms underlying their roles in enhancing the effect of chemotherapeutics and their abilities to revert chemoresistance are areas of intense activity.

4.1. Curcumin

Curcumin is an NC derived from the root of the turmeric plant (*Curcuma longa*); it has been used for many centuries as a spice and in traditional medicine [79]. This compound has anti-inflammatory, antioxidant, and anti-cancerogenic properties associated with the regulation of different signaling pathways involved in cell proliferation and survival, which make it a promising therapeutic option for being used as an adjuvant with gemcitabine, docetaxel, oxaliplatin, erlotinib, 5-FU, and celecoxib [80–82].

The mechanism of action of curcumin against cancer cells involves the negative regulation of NF-κB, EGFR, and PI3K/AKT signaling pathways that lead to the diminished activity and expression of proteins involved in anti-apoptotic processes, such as Bcl-2 and Bcl-xL, while inducing Bax, caspase-3, and p53 pro-apoptotic molecules. These effects have been observed in many types of cancer, mainly in breast, prostate, gastric, pancreatic, lung, and colorectal cancer [83,84]. In CRC, curcumin has been combined with 5-FU-based therapies in different models where it has been demonstrated to potentiate the chemotherapy effect and reverse chemoresistance through diverse pathways, including TS regulation in colon cancer cell lines, due to its inhibition of the nuclear translocation of NFκB and, consequently, the decrease in E2F1 transcription factor, a promoter of cell proliferation [85,86]. Table 1 summarizes the discovered mechanisms of the synergistic effects of curcumin with 5-FU-based therapies in different in vitro and in vivo models of CRC, including apoptosis inhibition, the dysregulation of survival pathways associated with inflammation such as NFκB and COX2, and the inhibition of stemness and EMT properties.

Table 1. Effects of administration of combined curcumin and 5-FU-based therapies in different CRC models.

		Cell Lines	
Author	Model	Condition	Mechanism
[87]	HT-29	20 µM Cur + 50 µM 5-FU	Synergistic effect ↓ COX-2
[88]	HCT-116	10 µM Cur + 200 µM 5-FU + 5 µM OXA	↑ sensitivity to chemotherapeutics and apoptosis ↓ p-EGFR, p-HER-2, p-HER-3, p-IGF-1R, p-AKT, and COX-2
[89]	HCT-116	20 µM Cur + 50 µM 5-FU + 1.12 µM OXA	↓ stemness (↓ CD166, CD44), colony formation, and EGFR. ↑ EGFR promoter methylation
[90]	HCT-116	25 µM Cur + 50 µM 5-FU + 1.25 µM OXA	↓ cell survival, p-EGFR, p-HER-2, p-IGF-1R, p-AKT, COX-2
[91]	HCT-116	5 µM Cur + 1 mM 5-FU	Cell cycle arrest in S-phase ↑ apoptosis ↓ NF-κB, Src, PI3K
[92]	HCT-8/VCR	Different concentrations	Enhanced 5-FU sensitivity
[93]	HCT-116/5-FUR	20 µM Cur + 5 µM 5-FU	↑ sensitivity to 5-FU, apoptosis, and colonosphere formation, and ↓ stemness (↓ CD133, CD44, ALDH1)
[94]	HCT-116/5-FUR and SW480/5-FUR	10 µM Cur + 10 µM 5-FU	↑ sensitivity to 5-FU, apoptosis ↓colony number, stemness (↓ polycomb repressive complex), EMT, and cell cycle arrest.
[95]	Co-culture of HCT-115 and MRC-5	5 µM Cur + different concentrations of 5-FU	↓ colony number, stemness (↓ CD133, ALDH1, CD44), MMP13, NF-κB-p65, p50, TGF-β3, TGF-βR, Smad2, and mesenchymal properties (↓ vimentin, Slug, ↑ E-cad)
[96]	HCT-116/5-FUR in alginate-based 3D culture	5 µM Cur + 0.01 nM 5-FU	↑ sensitivity to 5-FU, apoptosis ↓ cell viability, migration, CXCR4, MMP9, p-NF-κB-p65
[97]	HCT-8/5-FUR	10 µM Cur + 10 mM 5-FU	↑ sensitivity to 5-FU, apoptosis ↓ Nrf2 signaling
[98]	HCT-8/5-FUR	Different concentrations	↑ sensitivity to 5-FU, apoptosis ↓ P-gp and HSP27 G0/G1 cell cycle arrest

Table 1. *Cont.*

Cell Lines			
Author	**Model**	**Condition**	**Mechanism**
[99]	HT-29	30 μM Cur + 20 mg/L 5-FU	Cell cycle arrest in G1/S and G2/M, ↑ p21, ↓cyclinB1, CDK1, cyclin A2, CDK2, p-Rb, cyclin D1, CDK4
[99]	SW480 NNMT overexpression	10 μM Cur + 10 mg/L 5-FU	Cell cycle arrest in G1/S and G2/M ↓ cyclin B1, CDK1, cyclin A2, CDK2, p-Rb, CDK4
[100]	SW620	10 μM Cur + 5 μM 5-FU	↑ apoptosis ↓ cell viability, pERK, pSTAT1, L1
[101]	HT-29	Drug-loaded hydrogels with 60 mg/mL Cur + 5 mg/mL 5-FU	Synergism effect, cell cycle arrest ↑ apoptosis
[102]	HT-29	Nanoparticles GO-Alb-Cur-5FU-FA with 32 μg/mL Cur + 200 μg/mL 5-FU	Potentiates cytotoxicity ↑ apoptosis
[103]	SW480 and SW620	Hybrid 6a, 6d, 6e from 0.625 to 40 μM	↓ cell viability ↑ apoptosis cell cycle arrest
Animal Models			
Author	Model	Condition	Mechanism
[104]	Orthotopic transplant of HCT-116	60 mg/kg, gavage, twice/week of 5-FU (capecitabine) + 1 g/kg, gavage, daily, of cur, for 28 days	↓ tumor volume; ascites; metastasis to liver, intestine, spleen, and rectum; proliferation index; angiogenesis; NF-κB; COX-2; cyclin D1; c-Myc; ICAM-1; MMP-9; CXCR4; VEGF ↑ apoptosis
[94]	Xenotransplant-HCT-116/5-FUR	20 mg/kg once every 2 days of 5-FU + 50 mg/kg daily of cur, ip. for 40 days	↓tumor volume and tumor weight
[100]	Xenotransplant-SW620	Not specified	↑ cell death and survival rates ↓ tumor volume and proliferation
[105]	Titanium dioxide with dimethylhydrazine-induced CRC	50 mg/kg 5-FU coated with pectin + 200 mg/kg cur coated with pectin	Ameliorate histopathology

Cur: curcumin; 5-FU: 5-Fluorouracil; OXA: oxaliplatin; VCR: vincristine resistant; NNMT: Nicotinamide N-methyltransferase; 5-FUR: 5-Fluorouracil-resistant; GO: graphene oxide; Alb: albumin; FA: folic acid; ip: intraperitoneal.

Several studies in CRC cell lines have shown that induced 5-FU resistance is reverted with exposure to curcumin through cell cycle arrest, decreased ABC-transporter protein levels, the inhibition of stemness, and a reduction in EMT; the mechanisms related to the activity of curcumin in revert 5-FU resistance (Figure 4) [93,94]. In oral cancer, using curcumin as a treatment has been demonstrated to significantly diminish EMT features via c-Met blockade and the inhibition of the ERK effector pathway [106].

Recently, different target drug delivery systems of curcumin and 5-FU have been tested to increase their specificity and cytotoxicity and reduce side effects. Moreno-Quintero et al. (2023) designed 5-FU–curcumin hybrids, tested in the malignant colorectal cancer cells SW-480 and SW-620, and showed that combinatory therapy was more effective in tumoral cells than individual treatment [103]. New alternatives for drug delivery have been proposed to increase the effects of curcumin together with 5-FU, folic acid (FA)-modified nanoparticles, or micelle-crosslinked hydrogel with 5-FU, and curcumin has been tested in HT-29 CRC cells, showing a more significant inhibitory effect than free drugs [102].

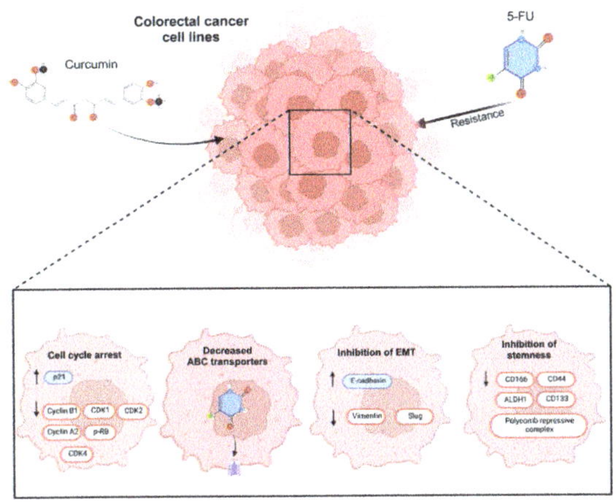

Figure 4. The main mechanisms of action related to the effect of curcumin in 5-Fluorouracil (5-FU) resistance colorectal cancer cells. Previous studies have verified the use of curcumin in 5-FU and its effects as regulator of cell cycle arrest, increasing P21 protein expression and the negative regulation of cyclin B1, cyclin-dependent kinases 1, 2, and 4 (CDK1, CDK2, CDK4), cyclin A2, and the phosphorylation of retinoblastoma, decreasing the expression of ABC transporters, regulating EMT markers such as E-cadherin positively, and reducing Vimentin and Slung proteins. Finally, curcumin has an important effect on regulating stemness markers such as CD116, CD44, ALDH1, CD33, and the polycomb repressive complex.

The combination of curcumin and 5-FU in animal models demonstrated that this combinatory therapy decreases the tumor volume and weight in CRC xenotransplant models [94,100]. In addition, this combination significantly reduced metastasis to different organs, such as the liver, intestine, spleen, and rectum, in an orthotopic transplant by decreasing angiogenesis and the number of NFκB, COX-2, cyclin D1, c-Myc, ICAM-1, MMP-9, CXCR4, and VEGF molecules [104].

In humans, a randomized Phase II trial evaluated the combination of curcumin with FOLFOX (folinic acid, oxaliplatin, and 5-FU) in a cohort of patients with metastatic CRC, showing a better survival in the FOLFOX plus curcumin treatment in comparison to patients receiving FOLFOX as a single therapy. This combined therapy was safe and well-tolerated, reducing neuropathic pain and neuronal functional abnormalities; however, no conclusive results were obtained regarding effectiveness; hence, more evidence is needed in clinical trials [99].

All these studies indicate that curcumin strongly enhances the effects of 5-FU-based chemotherapy. It avoids chemoresistance mainly by reducing inflammation and tumor cell stemness and enhancing apoptosis cell death, making curcumin a valuable candidate for use as a natural adjuvant in CRC therapy.

4.2. *Resveratrol*

Resveratrol is a polyphenol abundantly present in grape skin and seeds but also in peanuts, berries, tea, and wine [107]. Resveratrol has been reported to have anti-inflammatory and antioxidant activities that confer anti-aging and anti-cancer properties [108]. Its anti-cancer properties have been demonstrated against many types of cancers, such as breast, prostate, liver, and CRC, among others [109,110].

The administration of resveratrol as monotherapy induces apoptosis and autophagy and inhibits tumor growth in in vitro and in vivo models of CRC [111,112]. It is essential to point out that autophagy displays dual roles in cancer treatments, being both related to resistance and an inductor of death in tumoral cells. Resveratrol has demonstrated potential use in inhibiting the progression of colorectal cancer through autophagy-related apoptosis using a SIRT1/FOXQ1/ATG16L pathway axis [112]. In cisplatin-resistant human oral cancer CAR (a tongue squamous cell carcinoma cell line derived from the CAL 27 cell line), the resveratrol treatment induces cell death by an autophagic pathway, which ultimately converges with apoptotic cell death [113]. The dual role of autophagy in response to chemotherapies appears to be dependent on the origin of the neoplasia and used drugs, as observed in p53-null CRC cell lines, in which irinotecan treatment increases the autophagy favoring resistance to tumoral cell death, while data observed in hepatocellular carcinoma, colon, and pancreatic cancer showed that the restoration of autophagy favored 5-FU activity [114,115]. Therefore, the evidence suggests that resveratrol could be an inductor of autophagy and could act synergistically with 5-FU to improve the efficacy of the drug.

Several in vitro and in vivo studies reveal alternative and different mechanisms used by resveratrol to enhance 5-FU activity and cell cycle arrest and increase apoptotic proteins, mitochondrial oxidative stress, and proteins related to the integrity of cell junctions, increasing E-cadherin and claudin-2, among others [116–119]. The combination of 5-FU plus resveratrol leads to caspase-6 cleavage, increasing cancerous cell death and centrosome amplification, making CRC cells more sensitive to resveratrol and DNA damage [116,117]. Furthermore, the use of resveratrol in induced chemoresistance to the 5-FU cell line (HCT-116R) diminished the hypoxia-inducible factor (HIF) that is associated with poor prognosis in CRC patients [118]. Meanwhile, in a N-methyl nitrosourea-induced CRC model in rats, the administration of 5-FU together with resveratrol ameliorated the histopathological lesions of colon tissues and attenuated inflammation and epithelial damage, increased antioxidant molecules as superoxide dismutase (SOD) and oxidant marker malondialdehyde (MDA), and advanced oxidation protein products (AOPP) in the colon, favoring the anti-tumor response [119]. All these results and the most relevant mechanisms are presented in Table 2.

Table 2. Effects of resveratrol and 5-FU administration in combination therapy in different colorectal cancer models.

Cell Lines			
Author	**Model**	**Condition**	**Mechanism**
[116]	HCT-116 p53$^{+/+}$ and HCT-116 p53$^{-/-}$	5-FU IC_{50} + RSV 25–200 μM	↑ cleaved caspase-6 and apoptosis G1 cell cycle arrest
[117]	HCT-116 p53$^{+/+}$ and HCT-116 p53$^{-/-}$	5-FU 50 μM + RSV 200 μM	↑ cleaved caspase-6, apoptosis, centrosome amplification
[120]	HCT-116	5-FU 0.5 μM + RSV 15 μM	↓ cell viability, cell proliferation ↑ apoptosis (↓ Bcl-xL, ↑ Bax, cleaved caspase-9, -3, -8, and PARP), DNA damage, p-JNK, p38 Cell cycle arrest at S-phase
[121]	HT-29 and SW620	5-FU 10 μM + RSV 100 μM	Sensitivity to 5-FU ↑ mitochondrial oxidative stress, ROS, lipid peroxidation ↓ catalase GPx, SOD, p-AKT, p-STAT3
[122]	HCT-116/HCT-116-5-FUR in a 3D-alginate tumor microenvironment	5-FU 0.01, 0.1 and 1 nM + RSV 5 μM	↓ proliferation ↑ sensitivity to 5-FU, colony number, intercellular junctions (↑ E-cadherin, claudin-2), apoptosis, and cleaved caspase-3 ↓EMT (↓ vimentin, Slug), p-NF-κB-p65, p-NF-κB-p50, IκBα
[123]	SW620	5-FU 10 μM + RSV 10 μM	↓ cell viability ↑ H_2O_2 production

Table 2. *Cont.*

		Cell Lines	
Author	**Model**	**Condition**	**Mechanism**
[124]	HCT-116/HCT-116-5-FUR in a 3D-alginate tumor microenvironment stimulated by TNF-β	5-FU 1 nM + RSV 5 μM	↑ apoptosis ↓ colony number, stemness (↓ ALDH1, CD44, CD133), p-NF-κB-p65, CXCR4, MMP9, EMT (↓ vimentin, Slug, ↑ E-cadherin)
[125]	HCT-116 and DLD1	5-FU 10 μM + RSV 25 μM	Cell cycle arrest in S-phase ↑ apoptosis ↓ EMT (↓ vimentin, Slug), migration, stemness (↓ CD51, CD44), pSTAT3, pAkt, and telomerase activity
[126]	SW480-CD133^{+} and LoVo-CD133^{+}	5-FU 15 μM + RSV 80 μM	↑ sensitivity to 5-FU, apoptosis (↑ Bax)
[127]	HCT-116 and HCT-116-5-FUR	5-FU 2 nM + RSV 5 μM	↑5-FU sensitivity by β1-Integrin receptors ↓ cell viability, cell colony formation, migration and invasion, mesenchymal phenotype, apoptosis, HIF-α expression, and inflammation (↓ p-NF-κB) ↓ vascularization (↓ VEGF), stemness (↓ CD44, CD133)
		Animal Models	
Author	Model	Condition	Mechanism
[128]	N-Nitroso-N-methyl urea induced colorectal cancer in rats	i.p. treatment 5-FU 6.25 mg/kg on days 1, 3, and 5, with cycle being repeated every four weeks for 2 months, and orally treated with RSV dissolved in DMSO: 100μ/kg daily for 2 months.	More intact surface epithelium ↑ normal colon cells, integrity of mucosal architecture ↓ inflammation, severity of colon injury, epithelial loss, inflammatory cell infiltrate, epithelial hyperplasia, irregular crypts, goblet cell, proliferation, COX-2
[119]	N-methyl nitrosourea-induced colon cancer in rats	Colon cancer rats orally treated with RSV (10 mg/kg/day) and i.p. injected with 5-FU (12.5 mg/kg, on days 1, 3, and 5 with the cycle being repeated every 4 weeks) over 4 months	Ameliorate liver function (↓ AST, ↓ ALT), kidney function (↓ urea, ↓ creatinine), AOPP, MDA ↑ SOD, p53 ↓ NF-κB, COX-2 Ameliorate histopathology

5-FU: 5-Fluorouracil, RSV: resveratrol, 5-FUR: 5-Fluorouracil-resistant.

In summary, the combinatory treatment of resveratrol and 5-FU can potentially enhance the apoptotic effect of 5-FU, regulate molecules involved in oxidative stress and autophagy, and reduce EMT markers, stemness, cell proliferation, and inflammation.

4.3. *Artemisinin*

Artemisinin is a sesquiterpene lactone isolated by the Chinese Professor Tu Youyou from sweet wormwood leaves (*Artemisia annua* L.). Initially, artemisinin and its derivates were used as antimalarial drugs. However, these compounds demonstrated immunosuppressive effects primarily associated with the suppression of T cell proliferation and the downregulation of pro-inflammatory cytokines such as IL-2 and IFN-γ [129,130]. Additionally, these NCs reduced the symptoms of autoimmune diseases in murine models of atopic dermatitis, experimental encephalomyelitis, systemic lupus erythematosus, and ulcerative colitis [131,132]. The immunosuppressive effects of artemisinin and its derivates have been tested in murine models of ulcerative colitis and colitis-associated colorectal cancer. First, in dextran sodium sulfate (DSS)-induced colitis, artemisinin and its derivates decreased the disease activity index by downregulating pro-inflammatory cytokines such as IL-12 and TNF-α, recruiting alternatively activated CD206+ macrophages, and blocking the TLR4 signaling pathway and NFκB signaling pathway, which consequently decreased the number of pro-inflammatory cytokines such as IL-1β, IL-6, and TNF-α. In recent reports, artemisinin and its derivates directly affected the human colorectal cancer cell lines HCT-116 and RKO by inducing apoptotic processes [133,134].

Artemisinin may also function as a chemosensitizer, as reported in esophageal cancer EC109 cells, where exposure to artemisinin reduced the activity of the Wnt/β-catenin signaling pathway, one of the main drivers associated with resistance to chemotherapies, favoring the effect of oxaliplatin [134], and, as with resveratrol, because of its additional immunosuppressive and anti-inflammatory effects, artemisinin can be considered as a potential drug repositioning candidate due to its targeting of autophagy to induce cancer cell death [135].

However, there is limited evidence for the role of artemisinin in regulating chemoresistance processes. Still, its potential beneficial activities, such as having fewer side effects than traditional cancer treatments and its beneficial use in early cancer development, need more studies to establish direct participation in these processes and their interaction with conventional therapies.

4.4. *Helminth-Derived Molecules*

Infections with parasites are more prevalent in developing countries than industrialized ones, correlating with the hygiene hypothesis [136]. The relationship between parasites and CRC has been controversial in the last ten years [137]. Clinical observations suggest that patients infected with *Echinococcus granulossus* have a lower probability of colon cancer development [138]. In concordance with this, it has been shown that early treatment using molecules excreted/secreted by *Taenia crassiceps* (TcES) in a colitis-associated colon cancer (CAC) mouse model leads to reduced pro-inflammatory and proliferative signaling pathways related to cancer establishment and progression, such as STAT3/NFκB, indicating that these molecules may be an excellent early adjuvant therapy in experimental CRC treatment [76]. Using other helminth antigens in vitro, such as those from *Heligmosomoides po-lygyrus*, increased the levels of P53 and P21 proteins in CRC cell lines from mice and humans [139]. These remarkable effects of helminth-derived molecules on CRC were reinforced when used as adjuvant therapy with 5-FU in a mouse model of CAC and the human colorectal cancer cell lines HCT-116 and RKO. Mice receiving the combination of TcES plus 5-FU developed a significantly lower colon tumor load than mice receiving single 5-FU. Moreover, this combination improved the recruitment and activity of NK cells and P53 proteins, inducing pro-apoptotic and anti-proliferative effects [140]. Previously, TcES had been demonstrated to inhibit STAT1 phosphorylation in vitro by activating protein tyrosine phosphatases, such as SHP1, in human and mouse cells [141]. Further research must determine whether this mechanism happens when TcES inhibits STAT3 phosphorylation as an adjuvant to 5-FU during CAC.

On the other hand, human CRC lines HCT-116 and RKO exposed simultaneously to TcES and 5-FU displayed lower proliferative rates and significantly reduced their migration activity [139]. Collectively, these studies show evidence of the potential use of some helminth-derived molecules as an alternative adjuvant to enhance 5-FU activity in the therapy of colorectal cancer. These molecules may negatively regulate both STAT3 and NFκB signaling, altering the interaction between tumoral cells and their microenvironment, reducing the acquisition of malignant phenotypes in tumoral cells, favoring response to therapies, and improving survival.

4.5. *Trimethylglycine*

Trimethylglycine (TMG), or betaine, is a glycine-methyl derivative with a zwitterionic quaternary ammonium structure. It was first discovered in sugar beets (*Beta vulgaris*). It is also found in high quantities in spinach, wheat germ, wheat bran, and seafood, and is part of the metabolism of microorganisms, plants, and animals. TMG has two main physiological functions: the first is to act as an osmolyte to protect cells from stress conditions, and the second is to act as a methyl group donor [142]. In humans, TMG osmolyte function is mainly executed in the kidneys, which protects renal cells from high levels of urea and electrolytes. TMG acts as a methyl group donor mainly in the liver, where it helps to convert homocysteine to methionine via betaine–homocysteine methyl transferase (BHMT) (Figure 5) [143]. Besides its physiological functions, TMG has antioxidant and anti-inflammatory properties, making it a great option for cancer chemoprevention and cancer treatment in neoplasia with inflammatory components [144].

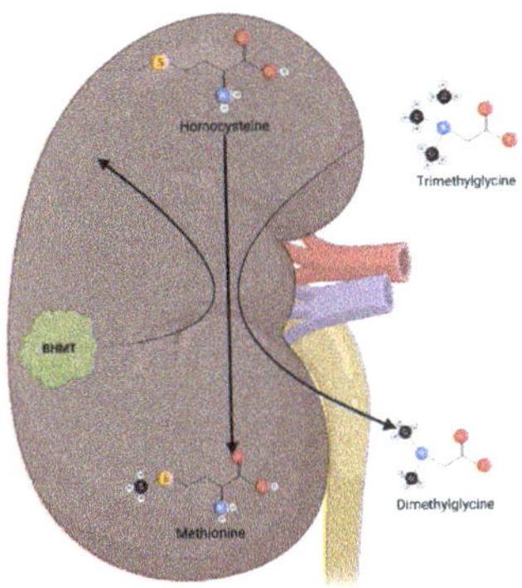

Figure 5. The mechanisms of action and metabolism of TMG. TMG is an osmolyte that functions mainly in the kidney, protecting renal cells from high urea levels and electrolytes. TMG acts as a methyl group donor, switching from trimethylglycine to dimethylglycine, which helps to convert homocysteine to methionine via betaine–homocysteine methyl transferase (BHMT).

Some reports have indicated the potential role of TMG as a cancer chemopreventive. A meta-analysis study found a relationship between TMG levels ingested in the diet and the reduced incidence of different types of cancer, especially CRC [144]. An in vivo model of CAC indicated that the administration of TMG in the early stages of the disease significantly reduced the inflammation and development of tumors in the colon [145]. As observed in prostate and lung cancer cell lines, treatment with different doses of TMG reduces oxidative stress and inflammation, favoring the apoptosis process through the negative regulation of NFκB [146,147]. Therefore, it is possible to suggest that TMG, an NFκB activity inhibitor, could regulate mechanisms in resistance to 5-FU, such as TS expression, similarly to curcumin and resveratrol. There is evidence, shown in an in vivo CAC model, that using TMG in combination with 5-FU increases the effectiveness of 5-FU by suppressing the indirect markers of NFκB activity, such as SNAIL1 and β-catenin nuclear translocation, and recovering E-cadherin protein levels; additionally, the use of TMG in this model was shown to significantly reduce the number and size of tumors, as a potential mechanism of the downregulation of STAT6 activity [60]. Thus, TMG is an attractive molecule to be used as an adjuvant for cancer therapy; additionally, previous advantages have been reported in its apoptosis induction of tumoral cells and its anti-inflammatory effect. TMG is considered a safe compound that ameliorates the secondary effects of chemotherapy, such as nephrotoxicity and hepatic damage caused by oxaliplatin treatment [148,149]. Currently, there is limited research using TMG together with conventional chemotherapeutics. Thus, further in vitro and in vivo studies are necessary to translate these results into benefits for colon cancer patients. Nevertheless, there is strong evidence suggesting that TMG could be a great candidate as adjuvant therapy for different types of cancer.

5. Conclusions

Colorectal cancer has become a concerning disease in recent years; the high mortality rates of nearly 50% of the diagnosticated patients reflect mainly two problems: (1) a delayed time of diagnosis leading to advanced stages of CRC; (2) the failure of conventional schemes of therapeutic regimens to overcome acquired or intrinsic resistance that contributes to relapses, metastasis, and, finally, mortality in the patients. The current chemotherapeutics—5-FU, oxaliplatin, and folinic acid—are clinicians' best weapons to fight advanced CRC. Improving these chemotherapeutic treatments and avoiding their most significant chemoresistance problems would benefit patients' health [150].

Natural compounds have recently attracted attention, given their functions as cancer preventives and potential adjuvants in chemotherapy. NCs have displayed direct and indirect effects in regulating frequent events characteristic of intrinsic and acquired chemoresistance, such as increased drug metabolism, regulated efflux pumps, autophagy, and the acquisition of EMT markers induced by inflammation and the tumor microenvironment [74,75]. Curcumin, resveratrol, artemisinin, TMG, and some molecules derived from ancient parasites like helminths, are natural compounds that have demonstrated their safe use in combinatory therapy together with 5-FU, improving the anti-tumoral response or restoring sensitivity to the drug (Figure 5). All these data support the idea that most NCs, where safe doses for use have been proven, could be postulated as potential compounds for drug repositioning.

Interestingly, many of these natural alternatives share similar mechanisms of action, such as STAT signaling modulation and NFκB inhibition, and may have less toxicity than synthetic drugs. These characteristics make NCs an attractive field of research to support their use as adjuvant therapies against colorectal cancer in more preclinical and clinical studies (Figure 6).

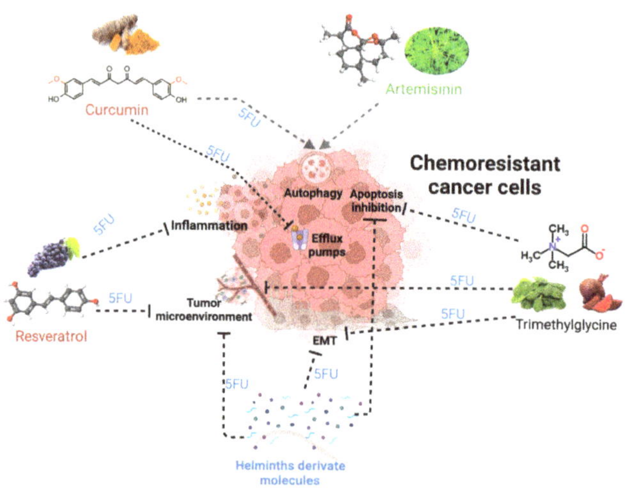

Figure 6. The mechanisms and alterations in tumors that lead to chemoresistance and its potential regulation by natural compounds, such as trimethylglycine, curcumin, resveratrol, artemisinin, and, recently, helminth-derived molecules together with 5FU. The black dotted lines indicate the blocking effect of a natural compound, while the gray lines with arrowheads show the induction of the mechanism related to better responses to treatments.

Author Contributions: Conceptualization, K.V.F.-M., M.G.M.-R. and L.I.T.; formal analysis, C.Á.S.-B., M.M.-R. and E.A.P.-Y.; investigation, J.L.R. and M.T.O.-M.; writing—original draft preparation,

K.V.F.-M. and C.Á.S.-B.; writing—review and editing, M.G.M.-R., L.I.T. and J.L.R.; funding acquisition, M.G.M.-R. and L.I.T. All authors have read and agreed to the published version of the manuscript.

Funding: This work was supported by grants from the Consejo Nacional de Humanidades, Ciencia y Tecnología (CF-2023-I-563), Consejo Mexiquense de Ciencia y Tecnología (FICDTEM-2023-132), Programa de Apoyo a Proyectos de Investigación e Innovación Tecnológica (PAPIIT-UNAM IA208424 and IV200425) de la Dirección General de Asuntos de Personal Académico, and the Programa de Apoyo a Profesores de Carrera (PAPCA), (FESI-PAPCA 2021-2022-38). Karen V. Fernandez-Muñoz would like to thank the Consejo Nacional de Humanidades, Ciencia y Tecnología (CONAHCYT) for PhD scholarship number 1008708.

Institutional Review Board Statement: Not applicable.

Informed Consent Statement: Not applicable.

Data Availability Statement: Not applicable.

Conflicts of Interest: The authors declare no conflicts of interest.

Abbreviations

The following abbreviations are used in this manuscript:

3-MA	3-methyladenine
5-FU	5-fluorouracil
5-FUR	5-fluorouracil resistance
ABC	ATP-binding cassette
ABCB1	ATP-binding cassette subfamily B member 1
ABCC1	ATP-binding cassette subfamily C member 1
ABCG2	ATP-binding cassette subfamily G member 2
Alb	Albumin
ALDH1	Aldehyde dehydrogenase 1
ALT	Alanine aminotransferase
AOM	Azoxymethane
AOPP	Advanced oxidation protein products
APC	Adenomatous polyposis coli
AST	Aspartate transferase
ATG16L	Autophagy related 16-like
Bcl-2	B-cell lymphoma 2
Bcl-xL	B-cell lymphoma-extra large
BHMT	Betaine-homocysteine methyl transferase
BRAF	B-Raf proto-oncogene
CAC	Colitis-associated colon cancer
CAFs	Cancer-associated fibroblasts
CDK1/2/4	Cyclin-dependent kinase 1/2/4
CDKN2A	Cyclin-dependent kinase inhibitor 2A
CH2THF	5,10-methylenetetrahydrofolate
CIMP	CpG island methylator phenotype
CIN	Chromosomal instability
COX-2	Cyclooxygenase 2
CRC	Colorectal cancer
CSC	Cancer stem cells
CTLA-4	Cytotoxic T-lymphocyte-associated protein 4
Cur	Curcumin
CXCR2/4	Chemokine receptor type 2/4
DHFU	Dihydro fluorouracil

DMSO	Dimethyl sulfoxide
DPD	Dihydropyrimidine dehydrogenase
DSS	Dextran sodium sulfate
dTMP	Deoxythymidine monophosphate
dUMP	Deoxyuridine monophosphate
E2F1	E2F transcription factor 1
EGFR	Epidermal growth factor receptor
EMT	Epithelial–mesenchymal transition
ERCC1	Excision repair cross-complementing rodent repair deficiency, complementation group 1
ERK	Extracellular signal-regulated kinase
FA	Folinic acid
FdUMP	5-fluorodeoxyuridine monophosphate
FOLFOX	5-fluorouracil, oxaliplatin, and leucovorin
FOXQ1	Forkhead box Q1
GO	Graphene oxide
GPx	Glutathione peroxidase
HER-2/3	Human epidermal growth factor receptor 2/3
HIF	Hypoxia-inducible factor
HSP27	Heat shock protein 27
HSPB8	Heat shock protein family B member 8
ICAM-1	Intercellular adhesion molecule 1
IFN-γ	Interferon gamma
IGF-1R	Insulin-like growth factor-1 receptor
IL-1β/2/6/ 10/12/23	Interleukin 1β/2/6/10/12/23
Ip.	Intraperitoneal
IRE1α	Inositol-requiring transmembrane kinase endoribonuclease-1α
IκBα	NFκB inhibitor alpha
JNK	c-Jun N-terminal kinase
KRAS	Kirsten rat sarcoma viral proto-oncogene
mCRC	Metastatic colorectal cancer
MDA	Malondialdehyde
MDR1	Multidrug resistance protein 1
MDR1	Multidrug resistance protein 1
MDSC	Myeloid-derived suppressor cells
MMP9/13	Matrix metalloproteinases 9/13
MRP1	Multidrug resistance-associated protein 1
MSI	Microsatellite instability
NC	Natural compound
NFκB	Nuclear factor kappa-light-chain-enhancer of activated B cells
NK	Natural killer cell
NNMT	Nicotinamide N-methyltransferase
NRF2	Nuclear factor erythroid 2-related factor 2
NSAIDs	Non-steroidal anti-inflammatory drugs
OXA	Oxaliplatin
PARP	Poly (ADP-ribose) polymerase
PD-1	Programmed cell death protein 1
PD-L1	Programmed death-ligand 1
PGE2	Prostaglandin E2
P-gp	P-glycoprotein 1
PI3K	Phosphatidylinositol 3-kinase
pRb	Retinoblastoma protein

ROS	Reactive oxygen species
RSV	Resveratrol
SHP1	Src homology 2 domain-containing protein tyrosine phosphatase 1
SIRT1	Sirtuin 1
SOD	Superoxide dismutase
STAT1/3/6	Signal transducer and activator of transcription 1/3/6
TAMs	Tumor-associated macrophages
TcES	Molecules excreted/secreted by *Taenia crassiceps*
TGF-β	Transforming growth factor beta
TGFβ-R	Transforming growth factor-β receptor
TLR4	Toll-like receptor 4
TMG	Trimethylglycine
TNF-α/β	Tumor necrosis factor alpha/beta
TNM	Tumor–node–metastasis
Tp53	Tumor protein p53
TS	Thymidylate synthase
VCR	Vincristine-resistant
VEGF	Vascular endothelial growth factor
WHO	World Health Organization
XBP1	X-box binding protein 1
Zeb1/2	Zinc finger E-box binding homeobox ½

References

1. Morgan, E.; Arnold, M.; Gini, A.; Lorenzoni, V.; Cabasag, C.J.; Laversanne, M.; Vignat, J.; Ferlay, J.; Murphy, N.; Bray, F. Global Burden of Colorectal Cancer in 2020 and 2040: Incidence and Mortality Estimates from GLOBOCAN. *Gut* **2022**, *72*, 338–344. [CrossRef] [PubMed]
2. Rawla, P.; Sunkara, T.; Barsouk, A. Epidemiology of Colorectal Cancer: Incidence, Mortality, Survival, and Risk Factors. *Przegląd Gastroenterol.* **2019**, *14*, 89–103. [CrossRef] [PubMed]
3. Yamagishi, H.; Kuroda, H.; Imai, Y.; Hiraishi, H. Molecular Pathogenesis of Sporadic Colorectal Cancers. *Chin. J. Cancer* **2016**, *35*, 4. [CrossRef]
4. Lewandowska, A.; Rudzki, G.; Lewandowski, T.; Stryjkowska-Góra, A.; Rudzki, S. Risk Factors for the Diagnosis of Colorectal Cancer. *Cancer Control* **2022**, *29*. [CrossRef]
5. Ahmad, R.; Singh, J.K.; Wunnava, A.; Al-obeed, O.; Abdulla, M.; Srivastava, S.K. Emerging Trends in Colorectal Cancer: Dysregulated Signaling Pathways. *Int. J. Mol. Med.* **2021**, *47*, 14. [CrossRef]
6. Khalyfa, A.A.; Punatar, S.; Aslam, R.; Yarbrough, A. Exploring the Inflammatory Pathogenesis of Colorectal Cancer. *Diseases* **2021**, *9*, 79. [CrossRef]
7. Sánchez-Barrera, C.Á.; Olguin, J.E.; Terrazas, L. *A Dual Role for Neutrophils and Myeloid-Derived Suppressor Cells in the Development of Colorectal Cancer*; Morales-Montor, J., Ed.; Serie: Immunology and Immune System Disorders; Nova Science Publishers, Inc.: Hauppauge, NY, USA, 2022; pp. 145–185.
8. Alzahrani, S.M.; Al Doghaither, H.A.; Al-Ghafar, A.B. General Insight into Cancer: An Overview of Colorectal Cancer. *Mol. Clin. Oncol.* **2021**, *15*, 271. [CrossRef]
9. IARC Working Group on the Evaluation of Cancer-Preventive Interventions. *Colorectal Cancer Screening [Internet]*; World Health Organization: Geneva, Switzerland, 2019; ISBN 9789283230236.
10. Wilson, L.A.M.; Browne, S.; Barnes, J.; El Hoyek, N.; Helmueller, L.; Janeiro, M.J.; Wendt, B. Opportunities and Challenges in Screening for Colorectal Cancer. *Popul. Health Manag.* **2023**, *26*, 246–253. [CrossRef]
11. Buccafusca, G.; Proserpio, I.; Tralongo, A.C.; Rametta Giuliano, S.; Tralongo, P. Early Colorectal Cancer: Diagnosis, Treatment and Survivorship Care. *Crit. Rev. Oncol. Hematol.* **2019**, *136*, 20–30. [CrossRef]
12. Mattar, R.E.; Al-Alem, F.; Simoneau, E.; Hassanain, M. Preoperative Selection of Patients with Colorectal Cancer Liver Metastasis for Hepatic Resection. *World J. Gastroenterol.* **2016**, *22*, 567–581. [CrossRef]
13. Biller, L.H.; Schrag, D. Diagnosis and Treatment of Metastatic Colorectal Cancer: A Review. *JAMA* **2021**, *325*, 669–685. [CrossRef] [PubMed]
14. Shin, A.E.; Giancotti, F.G.; Rustgi, A.K. Metastatic Colorectal Cancer: Mechanisms and Emerging Therapeutics. *Trends Pharmacol. Sci.* **2023**, *44*, 222–236. [CrossRef] [PubMed]

15. Longley, D.B.; Harkin, D.P.; Johnston, P.G. 5-Fluorouracil: Mechanisms of Action and Clinical Strategies. *Nat. Rev. Cancer* **2003**, *3*, 330–338. [CrossRef] [PubMed]
16. Cao, H.; Xu, E.; Liu, H.; Wan, L.; Lai, M. Epithelial-Mesenchymal Transition in Colorectal Cancer Metastasis: A System Review. *Pathol. Res. Pract.* **2015**, *211*, 557–569. [CrossRef]
17. van den Boogaard, W.M.C.; Komninos, D.S.J.; Vermeij, W.P. Chemotherapy Side-Effects: Not All DNA Damage Is Equal. *Cancers* **2022**, *14*, 627. [CrossRef]
18. Wigmore, P.M.; Mustafa, S.; El-Beltagy, M.; Lyons, L.; Umka, J.; Bennett, G. Effects of 5-FU. In *Chemo Fog. Advances in Experimental Medicine and Biology*; Springer: New York, NY, USA, 2010; Volume 678. [CrossRef]
19. Ribeiro, R.A.; Wanderley, C.W.S.; Wong, D.V.T.; Mota, J.M.S.C.; Leite, C.A.V.G.; Souza, M.H.L.P.; Cunha, F.Q.; Lima-Júnior, R.C.P. Irinotecan- and 5-Fluorouracil-Induced Intestinal Mucositis: Insights into Pathogenesis and Therapeutic Perspectives. *Cancer Chemother. Pharmacol.* **2016**, *78*, 881–893. [CrossRef]
20. Anaka, M.; Abdel-Rahman, O. Managing 5FU Cardiotoxicity in Colorectal Cancer Treatment. *Cancer Manag. Res.* **2022**, *14*, 273–285. [CrossRef]
21. Kang, L.; Tian, Y.; Xu, S.; Chen, H. Oxaliplatin-Induced Peripheral Neuropathy: Clinical Features, Mechanisms, Prevention and Treatment. *J. Neurol.* **2021**, *268*, 3269–3282. [CrossRef]
22. Palmieri, L.J.; Dubreuil, O.; Bachet, J.B.; Trouilloud, I.; Locher, C.; Coriat, R.; Moryoussef, F.; Landi, B.; Perkins, G.; Hautefeuille, V.; et al. Reasons for Chemotherapy Discontinuation and End-of-Life in Patients with Gastrointestinal Cancer: A Multicenter Prospective AGEO Study. *Clin. Res. Hepatol. Gastroenterol.* **2021**, *45*, 101431. [CrossRef]
23. Ramos, A.; Sadeghi, S.; Tabatabaeian, H. Battling Chemoresistance in Cancer: Root Causes and Strategies to Uproot Them. *Int. J. Mol. Sci.* **2021**, *22*, 9451. [CrossRef]
24. Hammond, W.A.; Swaika, A.; Mody, K. Pharmacologic Resistance in Colorectal Cancer: A Review. *Ther. Adv. Med. Oncol.* **2016**, *8*, 57–84. [CrossRef] [PubMed]
25. Cornelison, R.; Llaneza, D.C.; Landen, C.N. Emerging Therapeutics to Overcome Chemoresistance in Epithelial Ovarian Cancer: A Mini-Review. *Int. J. Mol. Sci.* **2017**, *18*, 2171. [CrossRef] [PubMed]
26. Wohlhueter, R.M.; McIvor, R.S.; Plagemann, P.G. Facilitated Transport of Uracil and 5-Fluorouracil, and Permeation of Orotic Acid into Cultured Mammalian Cells. *J. Cell. Physiol.* **1980**, *104*, 309–319. [CrossRef] [PubMed]
27. Santi, D.V.; McHenry, C.S.; Sommer, H. Mechanism of Interaction of Thymidylate Synthetase with 5-Fluorodeoxyuridylate. *Biochemistry* **1974**, *13*, 471–481. [CrossRef]
28. Kumar, A.; Singh, A.K.; Singh, H.; Thareja, S.; Kumar, P. Regulation of Thymidylate Synthase: An Approach to Overcome 5-FU Resistance in Colorectal Cancer. *Med. Oncol.* **2023**, *40*, 3. [CrossRef]
29. Song, S.; Tian, B.; Zhang, M.; Gao, X.; Jie, L.; Liu, P.; Li, J. Diagnostic and Prognostic Value of Thymidylate Synthase Expression in Breast Cancer. *Clin. Exp. Pharmacol. Physiol.* **2021**, *48*, 279–287. [CrossRef]
30. Li, Y.; Mizutani, Y.; Shiraishi, T.; Okihara, K.; Ukimura, O.; Kawauchi, A.; Nonomura, N.; Fukushima, M.; Sakai, T.; Miki, T. Prognostic Significance of Thymidylate Synthase Expression in Patients with Prostate Cancer Undergoing Radical Prostatectomy. *Urology* **2007**, *69*, 988–995. [CrossRef]
31. Lenz, H.J.; Leichman, C.G.; Danenberg, K.D.; Danenberg, P.V.; Groshen, S.; Cohen, H.; Laine, L.; Crookes, P.; Silberman, H.; Baranda, J.; et al. Thymidylate Synthase MRNA Level in Adenocarcinoma of the Stomach: A Predictor for Primary Tumor Response and Overall Survival. *J. Clin. Oncol.* **1996**, *14*, 176–182. [CrossRef]
32. Dotor, E.; Cuatrecases, M.; Martínez-Iniesta, M.; Navarro, M.; Vilardell, F.; Guinó, E.; Pareja, L.; Figueras, A.; Mollevi, D.G.; Serrano, T.; et al. Tumor Thymidylate Synthase 1494del6 Genotype as a Prognostic Factor in Colorectal Cancer Patients Receiving Fluorouracil-Based Adjuvant Treatment. *J. Clin. Oncol.* **2006**, *24*, 1603–1611. [CrossRef]
33. Jiang, H.; Li, B.; Wang, F.; Ma, C.; Hao, T. Expression of ERCC1 and TYMS in Colorectal Cancer Patients and the Predictive Value of Chemotherapy Efficacy. *Oncol. Lett.* **2019**, *18*, 1157–1162. [CrossRef]
34. Badary, D.M.; Elkabsh, M.M.; Mady, H.H.; Gabr, A.; Kroosh, S.S. Prognostic and Predictive Role of Excision Repair Cross-Complementation Group 1 and Thymidylate Synthase in Colorectal Carcinoma Patients Received FOLFOX Chemotherapy: An Immunohistochemical Study. *Appl. Immunohistochem. Mol. Morphol.* **2020**, *28*, 741–747. [CrossRef] [PubMed]
35. Avendaño, C.; Menéndez, J.C. Antimetabolites. In *Medicinal Chemistry of Anticancer Drugs*; Elsevier: Amsterdam, The Netherlands, 2008; pp. 9–52.
36. Zhang, Y.H.; Luo, D.D.; Wan, S.B.; Qu, X.J. S1PR2 Inhibitors Potently Reverse 5-FU Resistance by Downregulating DPD Expression in Colorectal Cancer. *Pharmacol. Res.* **2020**, *155*, 104717. [CrossRef] [PubMed]
37. Ughachukwu, P.; Unekwe, P. Efflux Pump-Mediated Resistance in Chemotherapy. *Ann. Med. Health Sci. Res.* **2012**, *2*, 191–198. [CrossRef] [PubMed]

38. Alketbi, L.; Al-Ali, A.; Talaat, I.M.; Hamid, Q.; Bajbouj, K. The Role of ATP-Binding Cassette Subfamily A in Colorectal Cancer Progression and Resistance. *Int. J. Mol. Sci.* **2023**, *24*, 1344. [CrossRef]
39. Chen, N.; Kong, Y.; Wu, Y.; Gao, Q.; Fu, J.; Sun, X.; Geng, Q. CAC1 Knockdown Reverses Drug Resistance through the Downregulation of P-Gp and MRP-1 Expression in Colorectal Cancer. *PLoS ONE* **2019**, *14*, e0222035. [CrossRef]
40. Gao, Q.; Li, X.X.; Xu, Y.M.; Zhang, J.Z.; Rong, S.D.; Qin, Y.Q.; Fang, J. IRE1α-Targeting Downregulates ABC Transporters and Overcomes Drug Resistance of Colon Cancer Cells. *Cancer Lett.* **2020**, *476*, 67–74. [CrossRef]
41. Zhang, N.; Ng, A.S.; Cai, S.; Li, Q.; Yang, L.; Kerr, D. Novel Therapeutic Strategies: Targeting Epithelial–Mesenchymal Transition in Colorectal Cancer. *Lancet Oncol.* **2021**, *22*, e358–e368. [CrossRef]
42. Du, F.; Liu, H.; Lu, Y.; Zhao, X.; Fan, D. Epithelial-to-Mesenchymal Transition: Liaison between Cancer Metastasis and Drug Resistance. *Crit. Rev. Oncog.* **2017**, *22*, 275–282. [CrossRef]
43. Ebrahimi, N.; Manavi, M.S.; Faghihkhorasani, F.; Fakhr, S.S.; Baei, F.J.; Khorasani, F.F.; Zare, M.M.; Far, N.P.; Rezaei-Tazangi, F.; Ren, J.; et al. Harnessing Function of EMT in Cancer Drug Resistance: A Metastasis Regulator Determines Chemotherapy Response. *Cancer Metastasis Rev.* **2024**, *43*, 457–479. [CrossRef]
44. Kim, A.Y.; Kwak, J.H.; Je, N.K.; Lee, Y.H.; Jung, Y.S. Epithelial-Mesenchymal Transition Is Associated with Acquired Resistance to 5-Fluorocuracil in HT-29 Colon Cancer Cells. *Toxicol. Res.* **2015**, *31*, 151–156. [CrossRef]
45. Sreekumar, R.; Al-Saihati, H.; Emaduddin, M.; Moutasim, K.; Mellone, M.; Patel, A.; Kilic, S.; Cetin, M.; Erdemir, S.; Navio, M.S.; et al. The ZEB2-Dependent EMT Transcriptional Programme Drives Therapy Resistance by Activating Nucleotide Excision Repair Genes ERCC1 and ERCC4 in Colorectal Cancer. *Mol. Oncol.* **2021**, *15*, 2065–2083. [CrossRef] [PubMed]
46. Tan, T.Z.; Miow, Q.H.; Miki, Y.; Noda, T.; Mori, S.; Huang, R.Y.; Thiery, J.P. Epithelial-mesenchymal Transition Spectrum Quantification and Its Efficacy in Deciphering Survival and Drug Responses of Cancer Patients. *EMBO Mol. Med.* **2014**, *6*, 1279–1293. [CrossRef] [PubMed]
47. Kocaturk, N.M.; Akkoc, Y.; Kig, C.; Bayraktar, O.; Gozuacik, D.; Kutlu, O. Autophagy as a Molecular Target for Cancer Treatment. *Eur. J. Pharm. Sci.* **2019**, *134*, 116–137. [CrossRef]
48. Gao, T.; Yuan, D.; He, B.; Gao, Y.; Liu, C.; Sun, H.; Nie, J.; Wang, S.; Nie, Z. Identification of Autophagy Related Genes in Predicting the Prognosis and Aiding 5- Fluorouracil Therapy of Colorectal Cancer. *Heliyon* **2022**, *8*, e09033. [CrossRef] [PubMed]
49. Sun, H.; Wang, R.; Liu, Y.; Mei, H.; Liu, X.; Peng, Z. USP11 Induce Resistance to 5-Fluorouracil in Colorectal Cancer through Activating Autophagy by Stabilizing VCP. *J. Cancer* **2021**, *12*, 2317–2325. [CrossRef] [PubMed]
50. Chen, H.; Huang, J.; Chen, C.; Jiang, Y.; Feng, X.; Liao, Y.; Yang, Z. NGFR Increases the Chemosensitivity of Colorectal Cancer Cells by Enhancing the Apoptotic and Autophagic Effects of 5-Fluorouracil via the Activation of S100A9. *Front. Oncol.* **2021**, *11*, 652081. [CrossRef]
51. Bai, C.; Zhang, Z.; Zhou, L.; Zhang, H.Y.; Chen, Y.; Tang, Y. Repurposing Ziyuglycoside II Against Colorectal Cancer via Orchestrating Apoptosis and Autophagy. *Front. Pharmacol.* **2020**, *11*, 576547. [CrossRef]
52. Wu, T.; Dai, Y. Tumor Microenvironment and Therapeutic Response. *Cancer Lett.* **2017**, *387*, 61–68. [CrossRef]
53. Blondy, S.; David, V.; Verdier, M.; Mathonnet, M.; Perraud, A.; Christou, N. 5-Fluorouracil Resistance Mechanisms in Colorectal Cancer: From Classical Pathways to Promising Processes. *Cancer Sci.* **2020**, *111*, 3142–3154. [CrossRef]
54. Hu, J.L.; Wang, W.; Lan, X.L.; Zeng, Z.C.; Liang, Y.S.; Yan, Y.R.; Song, F.Y.; Wang, F.F.; Zhu, X.H.; Liao, W.J.; et al. CAFs Secreted Exosomes Promote Metastasis and Chemotherapy Resistance by Enhancing Cell Stemness and Epithelial-Mesenchymal Transition in Colorectal Cancer. *Mol. Cancer* **2019**, *18*, 91. [CrossRef]
55. Linares, J.; Sallent-Aragay, A.; Badia-Ramentol, J.; Recort-Bascuas, A.; Méndez, A.; Manero-Rupérez, N.; Re, D.L.; Rivas, E.I.; Guiu, M.; Zwick, M.; et al. Long-Term Platinum-Based Drug Accumulation in Cancer-Associated Fibroblasts Promotes Colorectal Cancer Progression and Resistance to Therapy. *Nat. Commun.* **2023**, *14*, 746. [CrossRef] [PubMed]
56. Jin, W. Role of JAK/STAT3 Signaling in the Regulation of Metastasis, the Transition of Cancer Stem Cells, and Chemoresistance of Cancer by Epithelial– Mesenchymal Transition. *Cells* **2020**, *9*, 217. [CrossRef] [PubMed]
57. Sánchez-Ramírez, D.; Mendoza-Rodríguez, M.G.; Alemán, O.R.; Candanedo-González, F.A.; Rodríguez-Sosa, M.; Montesinos-Montesinos, J.J.; Salcedo, M.; Brito-Toledo, I.; Vaca-Paniagua, F.; Terrazas, L.I. Impact of STAT-Signaling Pathway on Cancer-Associated Fibroblasts in Colorectal Cancer and Its Role in Immunosuppression. *World J. Gastrointest. Oncol.* **2024**, *16*, 1705–1724. [CrossRef] [PubMed]
58. Tošić, I.; Frank, D.A. STAT3 as a Mediator of Oncogenic Cellular Metabolism: Pathogenic and Therapeutic Implications. *Neoplasia* **2021**, *23*, 1167–1178. [CrossRef]
59. Qin, A.; Yu, Q.; Gao, Y.; Tan, J.; Huang, H.; Qiao, Z.; Qian, W. Inhibition of STAT3/CyclinD1 Pathway Promotes Chemotherapeutic Sensitivity of Colorectal Caner. *Biochem. Biophys. Res. Commun.* **2015**, *457*, 681–687. [CrossRef]
60. Mendoza-Rodríguez, M.G.; Sánchez-Barrera, C.Á.; Callejas, B.E.; García-Castillo, V.; Beristain-Terrazas, D.L.; Delgado-Buenrostro, N.L.; Chirino, Y.I.; León-Cabrera, S.A.; Rodríguez-Sosa, M.; Gutierrez-Cirlos, E.B.; et al. Use of STAT6 Phosphorylation Inhibitor

and Trimethylglycine as New Adjuvant Therapies for 5-Fluorouracil in Colitis-Associated Tumorigenesis. *Int. J. Mol. Sci.* **2020**, *21*, 2130. [CrossRef]
61. Leon-Cabrera, S.A.; Molina-Guzman, E.; Delgado-Ramirez, Y.G.; Vázquez-Sandoval, A.; Ledesma-Soto, Y.; Pérez-Plasencia, C.G.; Chirino, Y.I.; Delgado-Buenrostro, N.L.; Rodríguez-Sosa, M.; Vaca-Paniagua, F.; et al. Lack of STAT6 Attenuates Inflammation and Drives Protection against Early Steps of Colitis-Associated Colon Cancer. *Cancer Immunol. Res.* **2017**, *5*, 385–396. [CrossRef]
62. Jayakumar, A.; Bothwell, A.L.M. Stat6 Promotes Intestinal Tumorigenesis in a Mouse Model of Adenomatous Polyposis by Expansion of MDSCs and Inhibition of Cytotoxic CD8 Response. *Neoplasia* **2017**, *19*, 595–605. [CrossRef]
63. Zhang, L.; Lu, X.; Xu, Y.; La, X.; Tian, J.; Li, A.; Li, H.; Wu, C.; Xi, Y.; Song, G.; et al. Tumor-Associated Macrophages Confer Colorectal Cancer 5-Fluorouracil Resistance by Promoting MRP1 Membrane Translocation via an Intercellular CXCL17/CXCL22–CCR4–ATF6–GRP78 Axis. *Cell Death Dis.* **2023**, *14*, 582. [CrossRef]
64. Shi, Q.; Shen, Q.; Liu, Y.; Shi, Y.; Huang, W.; Wang, X.; Li, Z.; Chai, Y.; Wang, H.; Hu, X.; et al. Increased Glucose Metabolism in TAMs Fuels O-GlcNAcylation of Lysosomal Cathepsin B to Promote Cancer Metastasis and Chemoresistance. *Cancer Cell* **2022**, *40*, 1207–1222.e10. [CrossRef]
65. Niu, T.; Zhou, F. Inflammation and Tumor Microenvironment. *J. Cent. South Univ. (Med. Sci.)* **2023**, *48*, 1899–1913. [CrossRef]
66. Tuomisto, A.E.; Mäkinen, M.J.; Väyrynen, J.P. Systemic Inflammation in Colorectal Cancer: Underlying Factors, Effects, and Prognostic Significance. *World J. Gastroenterol.* **2019**, *25*, 4383–4404. [CrossRef] [PubMed]
67. Lu, S.; Li, Y.; Zhu, C.; Wang, W.; Zhou, Y. Managing Cancer Drug Resistance from the Perspective of Inflammation. *J. Oncol.* **2022**, *2022*, 3426407. [CrossRef] [PubMed]
68. Mirzaei, S.A.; Dinmohammadi, F.; Alizadeh, A.; Elahian, F. Inflammatory Pathway Interactions and Cancer Multidrug Resistance Regulation. *Life Sci.* **2019**, *235*, 116825. [CrossRef]
69. Li, Y.; Zhao, B.; Peng, J.; Tang, H.; Wang, S.; Peng, S.; Ye, F.; Wang, J.; Ouyang, K.; Li, J.; et al. Inhibition of NF-KB Signaling Unveils Novel Strategies to Overcome Drug Resistance in Cancers. *Drug Resist. Updat.* **2024**, *73*, 101042. [CrossRef]
70. Pyo, J.S.; Kim, E.K. Clinicopathological Significance and Prognostic Implication of Nuclear Factor-KB Activation in Colorectal Cancer. *Pathol. Res. Pract.* **2019**, *215*, 152469. [CrossRef]
71. Thiruchenthooran, V.; Sánchez-López, E.; Gliszczyńska, A. Perspectives of the Application of Non-Steroidal Anti-Inflammatory Drugs in Cancer Therapy: Attempts to Overcome Their Unfavorable Side Effects. *Cancers* **2023**, *15*, 475. [CrossRef]
72. Hua, X.; Phipps, A.I.; Burnett-Hartman, A.N.; Adams, S.V.; Hardikar, S.; Cohen, S.A.; Kocarnik, J.M.; Ahnen, D.J.; Lindor, N.M.; Baron, J.A.; et al. Timing of Aspirin and Other Nonsteroidal Anti-Inflammatory Drug Use among Patients with Colorectal Cancer in Relation to Tumor Markers and Survival. *J. Clin. Oncol.* **2017**, *35*, 2806–2813. [CrossRef]
73. Huang, M.; Lu, J.J.; Ding, J. Natural Products in Cancer Therapy: Past, Present and Future. *Nat. Prod. Bioprospect.* **2021**, *11*, 5–13. [CrossRef]
74. Naeem, A.; Hu, P.; Yang, M.; Zhang, J.; Liu, Y.; Zhu, W.; Zheng, Q. Natural Products as Anticancer Agents: Current Status and Future Perspectives. *Molecules* **2022**, *27*, 8367. [CrossRef]
75. Rejhová, A.; Opattová, A.; Čumová, A.; Slíva, D.; Vodička, P. Natural Compounds and Combination Therapy in Colorectal Cancer Treatment. *Eur. J. Med. Chem.* **2018**, *144*, 582–594. [CrossRef] [PubMed]
76. Callejas, B.E.; Mendoza-Rodríguez, M.G.; Villamar-Cruz, O.; Reyes-Martínez, S.; Sánchez-Barrera, C.A.; Rodríguez-Sosa, M.; Delgado-Buenrostro, N.L.; Martínez-Saucedo, D.; Chirino, Y.I.; León-Cabrera, S.A.; et al. Helminth-Derived Molecules Inhibit Colitis-Associated Colon Cancer Development through NF-KB and STAT3 Regulation. *Int. J. Cancer* **2019**, *145*, 3126–3139. [CrossRef] [PubMed]
77. Calibasi-Kocal, G.; Pakdemirli, A.; Bayrak, S.; Ozupek, N.M.; Sever, T.; Basbinar, Y.; Ellidokuz, H.; Yigitbasi, T. Curcumin Effects on Cell Proliferation, Angiogenesis and Metastasis in Colorectal Cancer. *J. BUON* **2019**, *24*, 1482–1487.
78. Liang, Z.; Xie, H.; Shen, W.; Shao, L.; Zeng, L.; Huang, X.; Zhu, Q.; Zhai, X.; Li, K.; Qiu, Z.; et al. The Synergism of Natural Compounds and Conventional Therapeutics against Colorectal Cancer Progression and Metastasis. *Front. Biosci.* **2022**, *27*, 263. [CrossRef]
79. Kotha, R.R.; Luthria, D.L. Curcumin: Biological, Pharmaceutical, Nutraceutical, and Analytical Aspects. *Molecules* **2019**, *24*, 2930. [CrossRef]
80. Lin, Y.G.; Kunnumakkara, A.B.; Nair, A.; Merritt, W.M.; Han, L.Y.; Armaiz-Pena, G.N.; Kamat, A.A.; Spannuth, W.A.; Gershenson, D.M.; Lutgendorf, S.K.; et al. Curcumin Inhibits Tumor Growth and Angiogenesis in Ovarian Carcinoma by Targeting the Nuclear Factor-KB Pathway. *Clin. Cancer Res.* **2007**, *13*, 3423–3430. [CrossRef]
81. Yoysungnoen, P.; Wirachwong, P.; Changtam, C.; Suksamram, A.; Patumraj, S. Anti-Cancer and Anti-Angiogenic Effects of Curcumin and Tetrahydrocurcumin on Implanted Hepatocellular Carcinoma in Nude Mice. *World J. Gastroenterol.* **2008**, *14*, 2003–2009. [CrossRef]

82. Li, L.; Ahmed, B.; Mehta, K.; Kurzrock, R. Liposomal Curcumin with and without Oxaliplatin: Effects on Cell Growth, Apoptosis, and Angiogenesis in Colorectal Cancer. *Mol. Cancer Ther.* **2007**, *6*, 1276–1282. [CrossRef]
83. Pricci, M.; Girardi, B.; Giorgio, F.; Losurdo, G.; Ierardi, E.; Di Leo, A. Curcumin and Colorectal Cancer: From Basic to Clinical Evidences. *Int. J. Mol. Sci.* **2020**, *21*, 2364. [CrossRef]
84. Giordano, A.; Tommonaro, G. Curcumin and Cancer. *Nutrients* **2019**, *11*, 2376. [CrossRef]
85. Rajitha, B.; Belalcazar, A.; Nagaraju, G.P.; Shaib, W.L.; Snyder, J.P.; Shoji, M.; Pattnaik, S.; Alam, A.; El-Rayes, B.F. Inhibition of NF-κB Translocation by Curcumin Analogs Induces G0/G1 Arrest and Downregulates Thymidylate Synthase in Colorectal Cancer. *Cancer Lett.* **2016**, *373*, 227–233. [CrossRef] [PubMed]
86. Yang, H.; Huang, S.; Wei, Y.; Cao, S.; Pi, C.; Feng, T.; Liang, J.; Zhao, L.; Ren, G. Curcumin Enhances the Anticancer Effect of 5-Fluorouracil against Gastric Cancer through down-Regulation of COX-2 and NF-KB Signaling Pathways. *J. Cancer* **2017**, *8*, 3697–3706. [CrossRef] [PubMed]
87. Du, B.; Jiang, L.; Xia, Q.; Zhong, L. Synergistic Inhibitory Effects of Curcumin and 5-Fluorouracil on the Growth of the Human Colon Cancer Cell Line HT-29. *Chemotherapy* **2005**, *52*, 23–28. [CrossRef]
88. Patel, B.B.; Sengupta, R.; Qazi, S.; Vachhani, H.; Yu, Y.; Rishi, A.K.; Majumdar, A.P.N. Curcumin Enhances the Effects of 5-Fluorouracil and Oxaliplatin in Mediating Growth Inhibition of Colon Cancer Cells by Modulating EGFR and IGF-1R. *Int. J. Cancer* **2008**, *122*, 267–273. [CrossRef]
89. Yu, Y.; Kanwar, S.S.; Patel, B.B.; Nautiyal, J.; Sarkar, F.H.; Majumdar, A.P.N. Elimination of Colon Cancer Stem-like Cells by the Combination of Curcumin and FOLFOX. *Transl. Oncol.* **2009**, *2*, 321–328. [CrossRef]
90. Patel, B.B.; Gupta, D.; Elliott, A.A.; Sengupta, V.; Yu, Y.; Majumdar, A.P.N. Curcumin Targets FOLFOX-Surviving Colon Cancer Cells via Inhibition of EGFRs and IGF-1R. *Anticancer Res.* **2010**, *30*, 319–325.
91. Shakibaei, M.; Mobasheri, A.; Lueders, C.; Busch, F.; Shayan, P.; Goel, A. Curcumin Enhances the Effect of Chemotherapy against Colorectal Cancer Cells by Inhibition of NF-KB and Src Protein Kinase Signaling Pathways. *PLoS ONE* **2013**, *8*, e57218. [CrossRef]
92. Lu, W.D.; Qin, Y.; Yang, C.; Li, L.; Fu, Z.X. Effect of Curcumin on Human Colon Cancer Multidrug Resistance in Vitro and in Vivo. *Clinics* **2013**, *68*, 694–701. [CrossRef]
93. Shakibaei, M.; Buhrmann, C.; Kraehe, P.; Shayan, P.; Lueders, C.; Goel, A. Curcumin Chemosensitizes 5-Fluorouracil Resistant MMR-Deficient Human Colon Cancer Cells in High Density Cultures. *PLoS ONE* **2014**, *9*, e85397. [CrossRef]
94. Toden, S.; Okugawa, Y.; Jascur, T.; Wodarz, D.; Komarova, N.L.; Buhrmann, C.; Shakibaei, M.; Richard Boland, C.; Goel, A. Curcumin Mediates Chemosensitization to 5-Fluorouracil through MiRNA-Induced Suppression of Epithelialto-Mesenchymal Transition in Chemoresistant Colorectal Cancer. *Carcinogenesis* **2014**, *36*, 355–367. [CrossRef]
95. Buhrmann, C.; Kraehe, P.; Lueders, C.; Shayan, P.; Goel, A.; Shakibaei, M. Curcumin Suppresses Crosstalk between Colon Cancer Stem Cells and Stromal Fibroblasts in the Tumor Microenvironment: Potential Role of EMT. *PLoS ONE* **2014**, *9*, e107514. [CrossRef] [PubMed]
96. Shakibaei, M.; Kraehe, P.; Popper, B.; Shayan, P.; Goel, A.; Buhrmann, C. Curcumin Potentiates Antitumor Activity of 5-Fluorouracil in a 3D Alginate Tumor Microenvironment of Colorectal Cancer. *BMC Cancer* **2015**, *15*, 250. [CrossRef] [PubMed]
97. Zhang, C.; He, L.-J.; Ye, H.-Z.; Liu, D.-F.; Zhu, Y.-B.; Miao, D.-D.; Zhang, S.-P.; Chen, Y.-Y.; Jua, Y.-W.; Shen, J.; et al. Nrf2 Is a Key Factor in the Reversal Effect of Curcumin on Multidrug Resistance in the HCT-8/5-Fu Human Colorectal Cancer Cell Line. *Mol. Med. Rep.* **2018**, *18*, 5409–5416. [CrossRef] [PubMed]
98. He, W.-T.; Zhu, Y.-H.; Zhang, T.; Abulimiti, P.; Zeng, F.-Y.; Zhang, L.-P.; Luo, L.-J.; Xie, X.-M.; Zhang, H.-L. Curcumin Reverses 5-Fluorouracil Resistance by Promoting Human Colon Cancer HCT-8/5-FU Cell Apoptosis and Down-Regulating Heat Shock Protein 27 and P-Glycoprotein. *Chin. J. Integr. Med.* **2019**, *25*, 416–424. [CrossRef]
99. Li, G.; Fang, S.; Shao, X.; Li, Y.; Tong, Q.; Kong, B.; Chen, L.; Wang, Y.; Yang, J.; Yu, H.; et al. Curcumin Reverses Nnmt-Induced 5-Fluorouracil Resistance via Increasing Ros and Cell Cycle Arrest in Colorectal Cancer Cells. *Biomolecules* **2021**, *11*, 1295. [CrossRef]
100. Zheng, X.; Yang, X.; Lin, J.; Song, F.; Shao, Y. Low Curcumin Concentration Enhances the Anticancer Effect of 5-Fluorouracil against Colorectal Cancer. *Phytomedicine* **2021**, *85*, 153547. [CrossRef]
101. Sadeghi-Abandansari, H.; Pakian, S.; Nabid, M.R.; Ebrahimi, M.; Rezalotfi, A. Local Co-Delivery of 5-Fluorouracil and Curcumin Using Schiff's Base Cross-Linked Injectable Hydrogels for Colorectal Cancer Combination Therapy. *Eur. Polym. J.* **2021**, *157*, 110646. [CrossRef]
102. Bardania, H.; Jafari, F.; Baneshi, M.; Mahmoudi, R.; Ardakani, M.T.; Safari, F.; Barmak, M.J. Folic Acid-Functionalized Albumin/Graphene Oxide Nanocomposite to Simultaneously Deliver Curcumin and 5-Fluorouracil into Human Colorectal Cancer Cells: An in Vitro Study. *BioMed Res. Int.* **2023**, *2023*, 8334102. [CrossRef]
103. Moreno-Quintero, G.; Betancur-Zapata, E.; Herrera-Ramírez, A.; Cardona-Galeano, W. New Hybrid Scaffolds Based on 5-FU/Curcumin: Synthesis, Cytotoxic, Antiproliferative and Pro-Apoptotic Effect. *Pharmaceutics* **2023**, *15*, 1221. [CrossRef]

104. Kunnumakkara, A.B.; Diagaradjane, P.; Anand, P.; Kuzhuvelil, H.B.; Deorukhkar, A.; Gelovani, J.; Guha, S.; Krishnan, S.; Aggarwal, B.B. Curcumin Sensitizes Human Colorectal Cancer to Capecitabine by Modulation of Cyclin D1, COX-2, MMP-9, VEGF and CXCR4 Expression in an Orthotopic Mouse Model. *Int. J. Cancer* **2009**, *125*, 2187–2197. [CrossRef]
105. Karthika, C.; Sureshkumar, R.; Sajini, D.V.; Ashraf, G.M.; Rahman, M.H. 5-Fluorouracil and Curcumin with Pectin Coating as a Treatment Regimen for Titanium Dioxide with Dimethylhydrazine-Induced Colon Cancer Model. *Environ. Sci. Pollut. Res.* **2022**, *29*, 63202–63215. [CrossRef] [PubMed]
106. Ohnishi, Y.; Sakamoto, T.; Li, Z.; Yasui, H.; Hamada, H.; Kubo, H.; Nakajima, M. Curcumin Inhibits Epithelial-Mesenchymal Transition in Oral Cancer Cells via c-Met Blockade. *Oncol. Lett.* **2020**, *19*, 4177–4182. [CrossRef]
107. Galiniak, S.; Aebisher, D.; Bartusik-Aebisher, D. Health Benefits of Resveratrol Administration. *Acta Biochim. Pol.* **2019**, *66*, 13–21. [CrossRef]
108. Meng, T.; Xiao, D.; Muhammed, A.; Deng, J.; Chen, L.; He, J. Anti-Inflammatory Action and Mechanisms of Resveratrol. *Molecules* **2021**, *26*, 229. [CrossRef] [PubMed]
109. Rauf, A.; Imran, M.; Butt, M.S.; Nadeem, M.; Peters, D.G.; Mubarak, M.S. Resveratrol as an Anti-Cancer Agent: A Review. *Crit. Rev. Food Sci. Nutr.* **2018**, *58*, 1428–1447. [CrossRef] [PubMed]
110. Carter, L.G.; D'Orazio, J.A.; Pearson, K.J. Resveratrol and Cancer: Focus on in Vivo Evidence. *Endocr. Relat. Cancer* **2014**, *21*, R209–R225. [CrossRef] [PubMed]
111. Honari, M.; Shafabakhsh, R.; Reiter, R.J.; Mirzaei, H.; Asemi, Z. Resveratrol Is a Promising Agent for Colorectal Cancer Prevention and Treatment: Focus on Molecular Mechanisms. *Cancer Cell Int.* **2019**, *19*, 180. [CrossRef]
112. Zhou, M.; Niu, H.; Cui, D.; Huang, G.; Li, J.; Tian, H.; Xu, X.; Liang, F.; Chen, R. Resveratrol Triggers Autophagy-related Apoptosis to Inhibit the Progression of Colorectal Cancer via Inhibition of FOXQ1. *Phytother. Res.* **2024**, *38*, 3218–3239. [CrossRef]
113. Chang, C.H.; Lee, C.Y.; Lu, C.C.; Tsai, F.J.; Hsu, Y.M.; Tsao, J.W.; Juan, Y.N.; Chiu, H.Y.; Yang, J.S.; Wang, C.C. Resveratrol-Induced Autophagy and Apoptosis in Cisplatin-Resistant Human Oral Cancer CAR Cells: A Key Role of AMPK and Akt/MTOR Signaling. *Int. J. Oncol.* **2017**, *50*, 873–882. [CrossRef]
114. Wu, J.; Hu, D.; Zhang, R. Depletion of Bmi-1 Enhances 5-Fluorouracil-Induced Apoptosisand Autophagy in Hepatocellular Carcinoma Cells. *Oncol. Lett.* **2012**, *4*, 723–726. [CrossRef]
115. He, X.X.; Huang, C.K.; Xie, B.S. Autophagy Inhibition Enhanced 5-FU-Induced Cell Death in Human Gastric Carcinoma BGC-823 Cells. *Mol. Med. Rep.* **2018**, *17*, 6768–6776. [CrossRef]
116. Chan, J.Y.; Meng, S.P.; Clement, M.V.; Pervaiz, S.; Shao, C.L. Resveratrol Displays Converse Dose-Related Effects on 5-Fluorouracilevoked Colon Cancer Cell Apoptosis: The Roles of Caspase-6 and P53. *Cancer Biol. Ther.* **2008**, *7*, 1305–1312. [CrossRef]
117. Lee, S.C.; Chan, J.Y.; Pervaiz, S. Spontaneous and 5-Fluorouracil-Induced Centrosome Amplification Lowers the Threshold to Resveratrol-Evoked Apoptosis in Colon Cancer Cells. *Cancer Lett.* **2010**, *288*, 36–41. [CrossRef]
118. Baba, Y.; Nosho, K.; Shima, K.; Irahara, N.; Chan, A.T.; Meyerhardt, J.A.; Chung, D.C.; Giovannucci, E.L.; Fuchs, C.S.; Ogino, S. HIF1A Overexpression Is Associated with Poor Prognosis in a Cohort of 731 Colorectal Cancers. *Am. J. Pathol.* **2010**, *176*, 2292–2301. [CrossRef] [PubMed]
119. Abdel Latif, Y.; El-Bana, M.; Hussein, J.; El-Khayat, Z.; Farrag, A.R. Effects of Resveratrol in Combination with 5-Fluorouracil on N-Methylnitrosourea-Induced Colon Cancer in Rats. *Comp. Clin. Pathol.* **2019**, *28*, 1351–1362. [CrossRef]
120. Mohapatra, P.; Preet, R.; Choudhuri, M.; Choudhuri, T.; Kundu, C.N. 5-Fluorouracil Increases the Chemopreventive Potentials of Resveratrol through DNA Damage and MAPK Signaling Pathway in Human Colorectal Cancer Cells. *Oncol. Res.* **2011**, *19*, 311–321. [CrossRef]
121. Santandreu, F.M.; Valle, A.; Oliver, J.; Roca, P. Resveratrol Potentiates the Cytotoxic Oxidative Stress Induced by Chemotherapy in Human Colon Cancer Cells. *Cell. Physiol. Biochem.* **2011**, *28*, 219–228. [CrossRef] [PubMed]
122. Buhrmann, C.; Shayan, P.; Kraehe, P.; Popper, B.; Goel, A.; Shakibaei, M. Resveratrol Induces Chemosensitization to 5-Fluorouracil through up-Regulation of Intercellular Junctions, Epithelial-to-Mesenchymal Transition and Apoptosis in Colorectal Cancer. *Biochem. Pharmacol.* **2015**, *98*, 51–68. [CrossRef] [PubMed]
123. Blanquer-Rosselló, M. del M.; Hernández-López, R.; Roca, P.; Oliver, J.; Valle, A. Resveratrol Induces Mitochondrial Respiration and Apoptosis in SW620 Colon Cancer Cells. *Biochim. Biophys. Acta (BBA)-Gen. Subj.* **2017**, *1861*, 431–440. [CrossRef]
124. Buhrmann, C.; Yazdi, M.; Popper, B.; Shayan, P.; Goel, A.; Aggarwal, B.B.; Shakibaei, M. Resveratrol Chemosensitizes TNF-β-Induced Survival of 5-FU-Treated Colorectal Cancer Cells. *Nutrients* **2018**, *10*, 888. [CrossRef]
125. Chung, S.S.; Dutta, P.; Austin, D.; Wang, P.; Awad, A.; Vadgama, J.V. Combination of Resveratrol and 5-Flurouracil Enhanced Antitelomerase Activity and Apoptosis by Inhibiting STAT3 and Akt Signaling Pathways in Human Colorectal Cancer Cells. *Oncotarget* **2018**, *9*, 32943–32957. [CrossRef]

126. Huang, L.; Zhang, S.; Zhou, J.; Li, X. Effect of Resveratrol on Drug Resistance in Colon Cancer Chemotherapy. *RSC Adv.* **2019**, *9*, 2572–2580. [CrossRef]
127. Brockmueller, A.; Girisa, S.; Kunnumakkara, A.B.; Shakibaei, M. Resveratrol Modulates Chemosensitisation to 5-FU via B1-Integrin/HIF-1α Axis in CRC Tumor Microenvironment. *Int. J. Mol. Sci.* **2023**, *24*, 4988. [CrossRef]
128. Soliman, B.A.; Farrag, A.R.H.; Khaled, H.A.S. Combinational Effect of 5-Flourouracil and Resveratrol against N-Nitroso-N-Methyl Urea Induced Colorectal Cancer. *Egypt. J. Hosp. Med.* **2018**, *70*, 994–1006. [CrossRef]
129. Shi, C.; Li, H.; Yang, Y.; Hou, L. Anti-Inflammatory and Immunoregulatory Functions of Artemisinin and Its Derivatives. *Mediat. Inflamm.* **2015**, *2015*, 435713. [CrossRef] [PubMed]
130. Pras, N.; Visser, J.F.; Batterman, S.; Woerdenbag, H.J.; Malingré, T.M.; Lugt, C.B. Laboratory Selection of *Artemisia annua* L. for High Artemisinin Yielding Types. *Phytochem. Anal.* **1991**, *2*, 80–83. [CrossRef]
131. Xie, K.; Li, Z.; Zhang, Y.; Wu, H.; Zhang, T.; Wang, W. Artemisinin and Its Derivatives as Promising Therapies for Autoimmune Diseases. *Heliyon* **2024**, *10*, e27972. [CrossRef]
132. Khakzad, M.R.; Ganji, A.; Ariabod, V.; Farahani, I. Artemisinin Therapeutic Efficacy in the Experimental Model of Multiple Sclerosis. *Immunopharmacol. Immunotoxicol.* **2017**, *39*, 348–353. [CrossRef]
133. Wang, T.; Wang, J.; Ren, W.; Liu, Z.L.; Cheng, Y.F.; Zhang, X.M. Combination Treatment with Artemisinin and Oxaliplatin Inhibits Tumorigenesis in Esophageal Cancer EC109 Cell through Wnt/β-Catenin Signaling Pathway. *Thorac. Cancer* **2020**, *11*, 2316–2324. [CrossRef]
134. Jung, E.J.; Paramanantham, A.; Kim, H.J.; Shin, S.C.; Kim, G.S.; Jung, J.M.; Ryu, C.H.; Hong, S.C.; Chung, K.H.; Kim, C.W.; et al. *Artemisia annua* L. Polyphenol-Induced Cell Death Is Ros-Independently Enhanced by Inhibition of Jnk in Hct116 Colorectal Cancer Cells. *Int. J. Mol. Sci.* **2021**, *22*, 1366. [CrossRef]
135. Ma, Z.; Woon, C.Y.N.; Liu, C.G.; Cheng, J.T.; You, M.; Sethi, G.; Wong, A.L.A.; Ho, P.C.L.; Zhang, D.; Ong, P.; et al. Repurposing Artemisinin and Its Derivatives as Anticancer Drugs: A Chance or Challenge? *Front. Pharmacol.* **2021**, *12*, 828856. [CrossRef]
136. Rastogi, T.; Devesa, S.; Mangtani, P.; Mathew, A.; Cooper, N.; Kao, R.; Sinha, R. Cancer Incidence Rates among South Asians in Four Geographic Regions: India, Singapore, UK and US. *Int. J. Epidemiol.* **2008**, *37*, 147–160. [CrossRef]
137. Callejas, B.E.; Martínez-Saucedo, D.; Terrazas, L.I. Parasites as Negative Regulators of Cancer. *Biosci. Rep.* **2018**, *38*, BSR20180935. [CrossRef]
138. Rostamirad, S.; Daneshpour, S.; Mofid, M.R.; Andalib, A.; Eskandariyan, A.; Mousavi, S.; Yousofi Darani, H. Inhibition of Mouse Colon Cancer Growth Following Immunotherapy with a Fraction of Hydatid Cyst Fluid. *Exp. Parasitol.* **2023**, *249*, 108501. [CrossRef] [PubMed]
139. Jacobs, B.A.; Prince, S.; Smith, K.A. Gastrointestinal Nematode-Derived Antigens Alter Colorectal Cancer Cell Proliferation and Migration through Regulation of Cell Cycle and Epithelial-Mesenchymal Transition Proteins. *Int. J. Mol. Sci.* **2020**, *21*, 7845. [CrossRef] [PubMed]
140. Mendoza-Rodríguez, M.G.; Medina-Reyes, D.; Sánchez-Barrera, C.A.; Fernández-Muñoz, K.V.; García-Castillo, V.; Ledesma-Torres, J.L.; González-González, M.I.; Reyes, J.L.; Pérez-Plascencia, C.; Rodríguez-Sosa, M.; et al. Helminth-Derived Molecules Improve 5-Fluorouracil Treatment on Experimental Colon Tumorigenesis. *Biomed. Pharmacother.* **2024**, *175*, 116628. [CrossRef] [PubMed]
141. Becerra-Díaz, M.; Terrazas, L.I. Taenia Crassiceps Infection and Its Excreted/Secreted Products Inhibit STAT1 Activation in Response to IFN-γ. *Int. J. Parasitol.* **2014**, *44*, 613–623. [CrossRef]
142. Dobrijević, D.; Pastor, K.; Nastić, N.; Özogul, F.; Krulj, J.; Kokić, B.; Bartkiene, E.; Rocha, J.M.; Kojić, J. Betaine as a Functional Ingredient: Metabolism, Health-Promoting Attributes, Food Sources, Applications and Analysis Methods. *Molecules* **2023**, *28*, 4824. [CrossRef]
143. Zhao, G.; He, F.; Wu, C.; Li, P.; Li, N.; Deng, J.; Zhu, G.; Ren, W.; Peng, Y. Betaine in Inflammation: Mechanistic Aspects and Applications. *Front. Immunol.* **2018**, *9*, 1070. [CrossRef]
144. Youn, J.; Cho, E.; Lee, J.E. Association of Choline and Betaine Levels with Cancer Incidence and Survival: A Meta-Analysis. *Clin. Nutr.* **2019**, *38*, 100–109. [CrossRef]
145. Kim, D.H.; Sung, B.; Kang, Y.J.; Jang, J.Y.; Hwang, S.Y.; Lee, Y.; Kim, M.; Im, E.; Yoon, J.-H.; Kim, C.M.; et al. Anti-Inflammatory Effects of Betaine on AOM/DSS-induced Colon Tumorigenesis in ICR Male Mice. *Int. J. Oncol.* **2014**, *45*, 1250–1256. [CrossRef]
146. Bingula, R.; Dupuis, C.; Pichon, C.; Berthon, J.Y.; Filaire, M.; Pigeon, L.; Filaire, E. Study of the Effects of Betaine and/or C-Phycocyanin on the Growth of Lung Cancer A549 Cells In Vitro and In Vivo. *J. Oncol.* **2016**, *2016*, 8162952. [CrossRef]
147. Kar, F.; Hacioglu, C.; Kacar, S.; Sahinturk, V.; Kanbak, G. Betaine Suppresses Cell Proliferation by Increasing Oxidative Stress–Mediated Apoptosis and Inflammation in DU-145 Human Prostate Cancer Cell Line. *Cell Stress Chaperones* **2019**, *24*, 871–881. [CrossRef]

148. Hagar, H.; Husain, S.; Fadda, L.M.; Attia, N.M.; Attia, M.M.A.; Ali, H.M. Inhibition of NF-KB and the Oxidative Stress -Dependent Caspase-3 Apoptotic Pathway by Betaine Supplementation Attenuates Hepatic Injury Mediated by Cisplatin in Rats. *Pharmacol. Rep.* **2019**, *71*, 1025–1033. [CrossRef] [PubMed]
149. Al Za'abi, M.; Ali, H.; Al Sabahi, M.; Ali, B.H. The Salutary Action of Melatonin and Betaine, given Singly or Concomitantly, on Cisplatin-Induced Nephrotoxicity in Mice. *Naunyn-Schmiedeberg's Arch. Pharmacol.* **2021**, *394*, 1693–1701. [CrossRef] [PubMed]
150. Wang, X.; Zhang, H.; Chen, X. Drug Resistance and Combating Drug Resistance in Cancer. *Cancer Drug Resist.* **2019**, *2*, 141–160. [CrossRef] [PubMed]

biomolecules

MDPI

Review

Pharmacological and Therapeutic Potential of *Chrysopogon zizanioides* (Vetiver): A Comprehensive Review of Its Medicinal Applications and Future Prospects

Conjeevaram J. Gunasekar [1], Amin F. Majdalawieh [2,3,*], Imad A. Abu-Yousef [2] and Sham A. Al Refaai [2]

1 Loyola College, Madras University, Chennai 600034, Tamil Nadu, India; drsekhar62@gmail.com
2 Department of Biology, Chemistry and Environmental Sciences, College of Arts and Sciences, American University of Sharjah, Sharjah P.O. Box 26666, United Arab Emirates; iabuyousef@aus.edu (I.A.A.-Y.); g00092818@alumni.aus.edu (S.A.A.R.)
3 Advanced Biosciences and Bioengineering Research Center, American University of Sharjah, Sharjah P.O. Box 26666, United Arab Emirates
* Correspondence: amajdalawieh@aus.edu

Abstract: *Chrysopogon zizanioides* (Linn.) Nash, commonly known as vetiver, has been an integral component of traditional medicinal systems across India and Asia for centuries. The roots and essential oils of this aromatic grass have been widely utilized for their anti-inflammatory, analgesic, anticancer, antioxidant, antimicrobial, and wound-healing properties. Recent scientific investigations have provided substantial evidence supporting these traditional claims, revealing a diverse array of bioactive phytochemicals with significant pharmacological potential. Preclinical studies have demonstrated the efficacy of *C. zizanioides* extracts in mitigating inflammation, alleviating pain, combating microbial infections, and even exhibiting anticancer and antidiabetic effects. This review provides a comprehensive analysis of the current literature on the therapeutic properties of *C. zizanioides*, summarizing findings from in vitro assays, cell line studies, animal models, and available clinical studies. The bioactive constituents responsible for these pharmacological effects, including essential oil components and isolated fractions, are discussed, along with their proposed mechanisms of action. These mechanisms involve modulation of oxidative stress, inflammatory pathways, microbial proliferation, and pain perception. Additionally, current research limitations, gaps in knowledge, and future directions for investigating medicinal applications of *C. zizanioides* are explored. Emerging scientific evidence increasingly validates traditional claims regarding the healing properties of this versatile medicinal grass.

Keywords: *C. zizanioides*; vetiver; anticancer; anti-inflammatory; analgesic; antioxidant; antimicrobial

1. Introduction

Chrysopogon zizanioides (Linn.) Nash, commonly known as vetiver, is a densely tufted, perennial grass belonging to the Poaceae family. Native to India, it is widely distributed across tropical and subtropical regions of South and Southeast Asia, including Malaysia, Sri Lanka, the Philippines, Pakistan, and Thailand [1]. The plant is primarily valued for its fragrant root system, which produces essential oils and bioactive compounds with diverse medicinal applications [1]. *C. zizanioides* has been extensively utilized in traditional Indian medical systems such as Unani, Siddha, and Ayurveda for centuries. In Sanskrit, it is referred to as *Ushira* and is known for its cooling, astringent, and therapeutic properties [2]. Historically, it has been employed in the treatment of heat stroke, inflammation, headaches, digestive disorders, skin conditions, and wound healing [1,2]. Ancient medical texts, including the Ayurveda and ancient Tamil literature, documented its medicinal use for treating fever, rheumatism, and urinary tract infections [3]. These historical applications suggest a broad pharmacological potential, which has been recently explored through scientific research.

Recent pharmacological research has provided substantial scientific validation for many of *C. zizanioides*' traditional medicinal applications, linking its therapeutic efficacy to its diverse bioactive constituents. Extracts from *C. zizanioides* roots and essential oils have exhibited a broad spectrum of biological activities, including antimicrobial, antioxidant, anti-inflammatory, analgesic, and anticancer properties, in both in vitro and in vivo models [4]. The complexity of *C. zizanioides* essential oil is remarkable, with over 300 identified compounds, among which sesquiterpenes, flavonoids, and phenolic derivatives are the most biologically active [5]. Notably, key sesquiterpenoids such as cedr-8-en-13-ol display strong antimicrobial properties [6], while tannins display antifungal activity [7], and phenolic compounds contribute significantly to its anti-inflammatory effects [8]. As a result, *C. zizanioides*-based therapeutic applications have garnered increasing scientific interest and have been extensively explored for their potential in managing inflammatory disorders, infections, cancer, oxidative stress, and metabolic diseases such as diabetes and hypertension [9].

Despite promising preclinical findings, significant gaps remain in understanding the pharmacokinetics, bioavailability, and molecular mechanisms of action associated with *C. zizanioides*. Current research is limited by the lack of clinical trials to substantiate its therapeutic efficacy. This review provides a comprehensive synthesis of existing research on the bioactive compounds and pharmacological effects of *C. zizanioides*, integrating in vitro, in vivo, and clinical studies. Additionally, this review aims to explore proposed mechanisms of action, safety profiles, and key research gaps to guide future studies aimed at its development. Despite promising preclinical findings, significant gaps remain in understanding the pharmacokinetics and bioavailability of *C. zizanioides* and the molecular mechanisms underlying its medicinal applications.

2. Anti-Inflammatory Effects

Inflammation is a complex immunological response to harmful stimuli, including tissue injury, pathogen invasion, or environmental stressors [10]. While acute inflammation is an essential protective mechanism, chronic inflammation is implicated in the pathogenesis of numerous diseases, such as diabetes, arthritis, atherosclerosis, and neurodegenerative disorders [10,11]. Given the increasing prevalence of inflammatory diseases worldwide, natural sources of anti-inflammatory agents have garnered significant scientific interest [12–14]. *C. zizanioides* has been studied for such properties. However, most evidence available so far remains at the in vitro or animal model stage, and clinical trials in humans are currently lacking. Preclinical studies indicate that extracts and phytochemical fractions derived from *C. zizanioides* exhibit significant anti-inflammatory properties [15]. Both essential oils and solvent-based extracts, such as ethanolic and aqueous formulations, have demonstrated the ability to downregulate inflammation in cell culture assays and animal models [8,16]. These extracts effectively decrease the expression of key inflammatory cytokines and mediators, including tumor necrosis factor alpha (TNFα), interleukin-1 (IL-1), interleukin-6 (IL-6), cyclooxygenase-2 (COX-2), and nitric oxide (NO), while also inhibiting nuclear factor kappa B (NF-κB), a critical regulator of inflammatory pathways [8,16,17]. A detailed overview of these studies is presented in Table 1, summarizing both the in vitro and in vivo evidence supporting the anti-inflammatory properties of *C. zizanioides*.

Table 1. Summary of the preclinical studies on the anti-inflammatory effects of *C. zizanioides*.

Study Design	Experimental Model	Extract Type and Bioactive Compounds Involved	Dosage	Main Observed Effects	References
In vitro	RAW 264.7 macrophages (LPS-induced)	Essential oil; major constituents: cedr-8-en-13-ol (12.4%), α-amorphene (7.8%), β-vatirenene (5.94%), α-gurjunene (5.91%)	5–12.5 μg/mL	NO ↓ (to 26.5% at 12.5 μg/mL), iNOS ↓ (~67%), COX-2 ↓ (~75%), TNF-α ↓ (≥23%), IL-1β ↓ (up to 81%), IFN-β ↓; HO-1 ↑ (10–25%); superoxide anion ↓ 12–20%; lipid peroxidation ↓; apoptosis ↓	[16]
In vivo	CFA-induced rheumatoid arthritis in Wistar albino rats	*C. zizanioides* aerial (CA, phenolic-rich: flavonoids, lignans, flavolignans) and root (CR, triterpene- and sesquiterpene-rich) ethanolic extracts	200 mg/kg	ACPA ↓, IL-6 ↓, TNF-α ↓, IL-10 ↑ (CA > CR); JAK2/STAT3 ↓, SOCs3 ↑; ERK1/ERK2 ↓; TRAF6/c-FOS/NFATC1 ↓ (CA stronger), NF-κB ↓; RANKL ↓; synovial inflammation, pannus formation, cartilage destruction ↓	[8]
In vivo	Carrageenan-induced paw edema and cotton pellet-induced granuloma in Wistar rats	Root methanol extract	300–600 mg/kg	Acute: paw edema ↓ 66.17% at 600 mg/kg (6 h); chronic: granuloma formation ↓ 53.69% at 600 mg/kg; dose-dependent inhibition of early (histamine, serotonin, kinins) and late (prostaglandin) mediators; fibroblast proliferation and collagen synthesis ↓	[18]
In vitro	Albumin denaturation assay	Root ethanol extract; bioactives: vetiverol, vetivones	0–240 μg/mL	Protein denaturation ↓ dose-dependently; IC50 = 157.63 μg/mL (~2.8 × less potent than diclofenac)	[19]

↑ indicates an increase (upregulation), and ↓ indicates a decrease (downregulation) in the corresponding molecule or effect.

Several in vitro studies have provided strong evidence of these effects. A study evaluated the essential oil extracted from roots in lipopolysaccharide (LPS)-stimulated RAW 264.7 murine macrophages, identifying 25 chemical constituents, of which cedr-8-en-13-ol (12.4%), α-amorphene (7.80%), β-vatirenene (5.94%), and α-gurjunene (5.91%) predominated. At non-cytotoxic concentrations of 5–12.5 μg/mL, *C. zizanioides* essential oil markedly reduced nitric oxide (NO) production in a dose-dependent manner, with the highest concentration lowering NO to 26.5% of the level observed in unstimulated controls. This was accompanied by significant downregulation of inducible nitric oxide synthase (iNOS) to 67% and cyclooxygenase-2 (COX-2) mRNA expression to 75% of LPS-induced levels, as well as reductions in tumor necrosis factor alpha (TNF-α) by ≥23% and interleukin-1 beta (IL-1β) by up to 81%. The oil simultaneously upregulated heme oxygenase-1 (HO-1) mRNA by 10–25%, suppressed interferon beta (IFN-β) transcription, decreased superoxide anion generation by up to 20%, and lowered lipid peroxidation below baseline values of unstimulated macrophages, indicating an interplay between its antioxidant and anti-inflammatory effects [16]. Moreover, ethanol extracts of *C. zizanioides* were assessed using the albumin denaturation assay, a model relevant to chronic inflammatory pathologies such as rheumatoid arthritis. The extract, obtained through 48 h maceration in ethanol (1:10 g/mL) and containing sesquiterpene alcohols such as vetiverol and vetivones, inhibited protein denaturation in a concentration-dependent manner (0–240 μg/mL) with an IC_{50} of 157.63 ± 4.89 μg/mL, representing approximately 2.8-fold lower potency than diclofenac (IC_{50} = 55.76 ± 2.35 μg/mL). This inhibitory effect is consistent with prior observations that vetiver essential oil reduces fibroblast collagen III production and leukocyte migration, both of which contribute to its anti-inflammatory potential [19–21]. These findings, as outlined in Table 1, collectively indicate that *C. zizanioides* extracts effectively reduce pro-inflammatory cytokine expression, specifically TNFα and IL-1β, across various cell types, supporting its traditional medicinal use as an anti-inflammatory agent. However, it is important to note that these in vitro findings, while promising, are preliminary and require further validation using in vivo models and human studies to confirm their physiological relevance.

Animal model studies further validate the anti-inflammatory efficacy of *C. zizanioides*. Methanolic root extract administered orally at 300 or 600 mg/kg significantly inhibited carrageenan-induced paw edema in Wistar rats, with the higher dose achieving 66.17% inhibition at 6 h post-injection, closely matching indomethacin's 72.30% inhibition rate. In a cotton pellet-induced granuloma model, methanolic root extract reduced granuloma formation by 53.69%, suggesting suppression of fibroblast proliferation and extracellular matrix deposition in chronic inflammation [18]. Another study examined the root essential oil in carrageenan-induced peritonitis and paw edema models in mice. Intraperitoneal administration at 25–100 mg/kg reduced leukocyte migration by up to 62.5% and significantly inhibited all three inflammatory phases in paw edema swelling. In the formalin test, root essential oil suppressed the inflammatory (late) phase by 86% at 100 mg/kg without affecting the neurogenic (early) phase, indicating a peripheral site of action. Gas chromatography–mass spectrometry (GC-MS) analysis revealed major sesquiterpenes including khusimol, vetiselinenol, and cyclocopancamphan-12-ol. Notably, the root essential oil showed no significant activity in lipid peroxidation inhibition assays, suggesting its effects are not mediated by direct radical scavenging [22].

Chronic inflammation models further highlight *C. zizanioides*' anti-inflammatory properties. A study compared ethanolic extracts from aerial parts (CA) and roots (CR) in a complete Freund's adjuvant (CFA)-induced rheumatoid arthritis rat model. UHPLC/HRMS profiling identified CA as rich in phenolic compounds (42 identified, including flavonoid C-glycosides, lignans, and flavolignans), while CR contained fewer phenolics (13 identified) but higher levels of triterpenes and sesquiterpenes. Oral administration of both extracts significantly reduced serum anti-citrullinated protein antibodies (ACPA), IL-6, and TNF-α while restoring IL-10 levels, with CA exhibiting stronger modulation [8]. Molecular analyses revealed that both extracts downregulated Janus kinase 2 (JAK2) and signal transducer and activator of transcription 3 (STAT3) while upregulating suppressors of cytokine signaling 3 (SOCS3). Reductions in extracellular signal-regulated kinases 1/2 (ERK1/2) and TNF receptor-associated factor 6 (TRAF6) were also observed, alongside suppression of c-FOS, nuclear factor of activated T-cells cytoplasmic 1 (NFATC1), nuclear factor kappa B (NF-κB), and receptor activator of nuclear factor kappa-B ligand (RANKL). Histopathological analysis confirmed marked attenuation of synovial inflammation, pannus formation, and cartilage destruction, particularly in CA-treated animals [8].

The mechanistic basis for these effects appears to involve a convergence of anti-inflammatory, antioxidant, and immunomodulatory actions. By downregulating NF-κB activation, through inhibition of nuclear translocation and DNA binding, *C. zizanioides* extracts reduce transcription of key inflammatory mediators such as TNF-α, IL-1β, IL-6, COX-2, and iNOS [8,16,23]. This modulation of transcriptional activity is reinforced by interference with upstream signaling pathways, including suppression of JAK2/STAT3, ERK1/2, and TRAF6, thereby attenuating both cytokine production and osteoclastogenesis [8,24–26]. Limitation of leukocyte recruitment to inflammatory sites, as shown in carrageenan-induced peritonitis [22] and further supported by reductions in chemokine-driven adhesion molecule expression [9], disrupts a critical step in the amplification of inflammation. Additionally, phenolic-rich extracts enhance cellular antioxidant defenses via increased activity of superoxide dismutase (SOD), glutathione peroxidase (GPX), and catalase (CAT), mitigating reactive oxygen species (ROS)-mediated activation of redox-sensitive transcription factors [9,16]. Certain constituents, including sesquiterpene alcohols such as khusimol and isovalencenol, further influence tissue remodeling processes by reducing fibroblast collagen III synthesis and limiting T-cell proliferation [9]. Flavonoids present in aerial extracts may also engage GABA-A receptors, producing central nervous system depressant effects that alleviate pain and stress associated with chronic inflammation [9,27,28].

Taken together, these studies indicate that *C. zizanioides* exerts multi-target anti-inflammatory effects through coordinated suppression of pro-inflammatory mediators, inhibition of immune cell infiltration, modulation of key intracellular signaling cascades, attenuation of oxidative stress, and regulation of tissue remodeling pathways. The differential efficacy of aerial versus root extracts highlights the importance of phytochemical composition in determining pharmacological activity, with phenolic-rich aerial parts and sesquiterpene-rich roots offering distinct yet complementary modes of action. While preclinical studies provide strong support for the anti-inflammatory potential of *C. zizanioides*, further research is necessary to explore its clinical relevance. Expanding on human clinical data, comparing its effects with existing anti-inflammatory drugs, and conducting dose–response and toxicity analyses will provide a more comprehensive evaluation of its therapeutic potential. Additionally, mechanistic studies investigating the interaction of its bioactive compounds with molecular targets could enhance our understanding of its pharmacological properties. A clearer understanding of pharmacokinetics, bioavailability, and safety profiles in human subjects should be explored to translate *C. zizanioides* from promising preclinical findings into clinical therapies, serving as a viable natural alternative for managing inflammatory disorders. The cytokines and signaling pathways modulated by *C. zizanioides* are summarized in Figure 1.

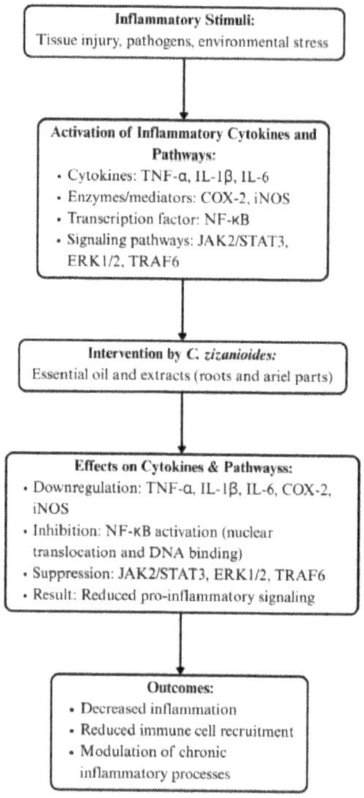

Figure 1. Cytokines and signaling pathways modulated by *C. zizanioides*.

3. Analgesic Effects

Pain relief is a fundamental therapeutic goal in medical practice [29,30]. Conventional analgesic medications, such as opioids and nonsteroidal anti-inflammatory drugs (NSAIDs), are commonly used for pain management, but they are associated with adverse effects and the potential for abuse [31,32]. These concerns have fueled interest in alternative analgesics derived from natural sources, particularly medicinal plants. *C. zizanioides* has traditionally been employed for pain relief, and scientific investigations have begun to validate its ethnomedicinal application, revealing promising analgesic properties [33,34]. Analgesic effects of various *C. zizanioides* extracts have been demonstrated in a few in vitro studies. Studies have shown that both aqueous and alcoholic root extracts exhibit analgesic actions comparable to standard drugs like aspirin or morphine in tests such as acetic acid-induced writhing, hot plate, and tail immersion assays [22]. These results indicate significant attenuation of chemically and thermally induced nociceptive pain behaviors. Additionally, essential oil compositions containing *C. zizanioides* oils have displayed dose-dependent pain inhibition by lowering pain score parameters [22]. Research has explored the antinociceptive activity of *C. zizanioides* essential oil. A study by Lima et al. characterized *C. zizanioides* essential oil using GC/MS, identifying major constituents such as khusimol (19.57%), E-isovalencenol (13.24%), α-vetivone (5.25%), and β-vetivone (4.87%) [22]. Intraperitoneal administration of *C. zizanioides* essential oil at doses of 50 and 100 mg/kg significantly reduced the number of writhes by 51.9% and 64.9%, respectively, and decreased paw licking during the second phase of the formalin test by 56.7% and 86.2%, respectively, compared to control groups. However, the *C. zizanioides* essential oil did not exhibit significant effects in hot plate and rota-rod tests, suggesting a peripheral rather than central mechanism of action [22]. Another study on *C. zizanioides* leaf extract demonstrated

66.08% analgesic activity in animal models, compared to 91.11% for diclofenac sodium [35]. Collectively, these findings support the analgesic properties of *C. zizanioides*, reinforcing its traditional use in pain management. A detailed overview of these findings is presented in Table 2, highlighting the analgesic efficacy and experimental models used in recent studies. It is important to note that while these studies provide promising results, they stem from preclinical animal studies with limited sample sizes and variability in dosage, preparation, and administration routes. Therefore, the reproducibility and consistency of the findings are limited. As a result, there is a gap in the translation to human clinical therapies. More studies are warranted to ensure consistency in the results, and human clinical trials are crucial to strengthen the evidence supporting the analgesic efficacy of *C. zizanioides*.

Table 2. Summary of the analgesic effects of *C. zizanioides*.

Extract Type	Bioactive Compound Involved	Model/Test Used	Key Findings	Comparative Efficacy	References
Root essential oil	khusimol, E-isovalencenol, α-/β-vetivone	• Acetic acid writhing test (50 and 100 mg/kg) • Formalin test: Phase 2 licking • Hot plate and rota-rod test • Carrageenan-induced leukocyte migration (peritoneal) • Paw edema	• Writhes: At 50/100 mg/kg: writhes inhibited by 51.9% and 64.9% • Formalin Phase 2 licking reduced by 56.7% and 86.2%. • No effect in hot plate or rotarod tests. • ↓ Inflammation (dose-dependent leukocyte inhibition and edema suppression)	Aspirin inhibited writhing ~75%; vetiver EO achieved ~65% inhibition. Demonstrated peripheral-only analgesic action.	[22]
Ethanolic leave extract	GC/MS profiled 63 compounds including esters, sesquiterpenes, alcohols, hydrocarbons. Major compounds: 9,19-Cyclolanostan-3-ol acetate (8.8%), 13-Docosenamide (8.35%), γ-Sitosterol (5.2%)	• Acetic acid writhing (oral, 500 mg/kg)	~66.08% analgesic activity	Compared to 91.11% for diclofenac sodium.	[35]

↓ indicates a decrease (downregulation) in the corresponding molecule or effect.

The antinociceptive mechanisms of *C. zizanioides* are believed to involve multiple pathways, including modulation of inflammatory mediators, NO pathways, and possibly central pain signaling. *C. zizanioides* essential oil has been found to inhibit carrageenan-induced leukocyte migration and paw edema in rodents, suggesting suppression of key pain mediators such as prostaglandins. A study reported that *C. zizanioides* essential oil downregulates inducible nitric oxide synthase (iNOS) expression in macrophages, thereby reducing NO production, a pathway integral to pain modulation [22]. Although direct evidence of opioid receptor involvement remains limited, some essential oils with similar phytochemical compositions exert analgesic effects via opioid receptor interactions. One study found that *C. zizanioides* essential oil exhibited antinociceptive properties comparable to morphine in an acetic acid-induced writhing test. This effect was partially reversed by naloxone, an opioid receptor antagonist, indicating the potential involvement of opioid receptors in its analgesic action [22]. Additionally, *C. zizanioides* essential oil reduced pain behaviors in the second phase of the formalin test, associated with inflammatory pain, but did not show significant effects in the Hargreaves' test, a measure of thermal nociception [22]. *C. zizanioides* shows the most promising therapeutic application in inflammatory and chemically induced pain models, with significant reductions in writhing and formalin test scores, while limited effects in thermal or central models suggest a targeted analgesic action. Moreover, neurotransmitter systems such as glutamate, GABA, and serotonin are critical in pain perception [36,37], and while studies specifically addressing *C. zizanioides* are scarce, findings from other essential oils suggest a possible role in modulating these pathways [9,38–41].

Phytochemical constituents such as monoterpenes, sesquiterpenes, and phenolic acids, which are found in *C. zizanioides* roots and essential oils, may contribute to these analgesic actions [22,35,42–44]. Despite promising activity in animal models, clinical evaluations of human efficacy and safety remain lacking. Future studies should focus on isolating bioactive compounds, understanding pharmacological interactions, conducting clinical trials on specific pain indications, and developing optimized drug delivery systems to enhance the therapeutic potential of *C. zizanioides*. Overall, *C. zizanioides* demonstrates notable analgesic potential through multiple molecular mechanisms, supporting its traditional use in pain relief. With future clinical research advancements, *C. zizanioides* holds promise as a safe, complementary analgesic targeting diverse pain pathways.

4. Antioxidant Properties

Antioxidant activity is a key therapeutic attribute of *C. zizanioides*, primarily mediated through its ability to neutralize free radicals and modulate endogenous antioxidant defense systems [35,45–48]. In vitro and in vivo studies have demonstrated that *C. zizanioides* extracts exhibit significant free radical scavenging, metal chelation, and enzymatic antioxidant modulation, all of which contribute to its protective effects against oxidative stress-related pathologies [16,49–52]. These diverse antioxidant mechanisms are summarized in Table 3, which outlines an overview of key studies and outcomes. Several in vitro assays, including DPPH, ABTS, and phosphomolybdenum, have been employed to evaluate these effects. Ethanolic, ethyl acetate, chloroform, ether, and aqueous extracts of the roots consistently display dose-dependent DPPH radical scavenging. Reported IC_{50} values span from as low as 10.73 μg/mL for ethanolic extract, comparable to ascorbic acid (4.61 μg/mL) [53], to higher values such as 157.38 μg/mL for ethanol and 112.79 μg/mL for ethyl acetate extracts [54]. Root essential oils show similar efficacy, producing 50% inhibition at concentrations around 184–185 μg/mL, with activity benchmarked against BHT and ascorbic acid [55]. The phytochemical composition of these extracts is central to their activity. Essential oils are particularly rich in sesquiterpenes such as longiverbenone, longipinocarvone, Cedr-8-en-13-ol, khusimol, α-vetivone, and bicyclo-vetivenol, whereas ethanolic extracts contain over a hundred compounds, including khusenic acid, ascorbic acid, junipen, γ-himachalene, and α-guaiene [35,53–58]. Phenolic acids, both soluble and bound forms, such as p-coumaric, p-dihydroxybenzoic, and ferulic acid, make important contributions through hydrogen donation and radical scavenging [56,57,59]. Flavonoids and other phenolic constituents further enhance activity by donating electrons and stabilizing reactive oxygen species [54,60]. Beyond radical scavenging, *C. zizanioides* extracts also exhibit ferric-reducing capacity. FRAP assays demonstrate that essential oils reduce Fe^{3+} to Fe^{2+} with EC_{50} values near 185 μg/mL, while phosphomolybdenum assays reveal total antioxidant capacities of up to 75.5% at 100 ppm, correlating closely with phenolic content [55,56]. These findings suggest that sesquiterpenes, terpenic alcohols, and phenolics act synergically, providing both radical-scavenging and electron-transfer mechanisms to suppress oxidative chain reactions.

Table 3. Summary of the mechanisms underlying the antioxidant activity of *C. zizanioides*.

Study Design	Extract Type and Bioactive Compounds Identified	Assay/Model	Key Findings	Mechanism	References
In vitro	Ethanolic leaf extract of *C. zizanoides*; 63 phytoconstituents identified including 9,19-Cyclolanostan-3-ol acetate (3β) and Phytol	DPPH radical scavenging assay	Moderate antioxidant activity (IC50 = 257.23 μg/mL) compared to ascorbic acid.	Antioxidant via radical scavenging and SOD, catalase, GPx	[35]
In vitro	Aqueous and ethanolic whole plant extracts; glycosides, carbohydrates, phenols, flavonoids, saponins, gums, mucilage	FRAP, Nitric oxide scavenging, Hydrogen peroxide scavenging, DPPH-RSA	Ethanolic extract showed higher antioxidant activity than aqueous; correlated with phenolic and flavonoid content.	Antioxidant activity via free radical scavenging and electron donation	[49]
In vitro	Crude oil; β-vetivenene, β-vetivone, α-vetivone, khusimol, bicyclovetivenol	DPPH radical scavenging, Fe^{2+} metal chelating	Strong DPPH scavenging (~93% at 10 μL/mL); weak metal chelation.	Free radical scavenging via terpenoid constituents	[45]
In vitro	Essential oil; cedr-8-en-13-ol, α-amorphene, β-vatirenene, α-gurjunene	LPS-stimulated RAW 264.7 macrophages; superoxide anion, MDA, SOD assays	↓ Superoxide anion (12–20%), ↓ MDA, ↓ NO, ↓ apoptosis.	Anti-inflammatory: ↓ HO-1, iNOS, COX-2, TNF-α, IL-1β, IFN-β; Antioxidant: ↓ oxidative stress and lipid peroxidation	[16]
In vitro	*C. zizanioides* oil; complex mixture of terpenoids	DPPH radical scavenging assay	Strong antioxidant activity (93% scavenging at 10 μL/mL), ~α-tocopherol, >BHT.	Free radical scavenging by components of *C. zizanioides* oil	[46]
In vitro	Hexane root extracts of two genotypes; phenolics and flavonoids	FRAP, DPPH, TAC, RP, TPC; oxidative stress in erythrocytes (H_2O_2, t-BHP)	KS1 genotype (spent root KSD) showed the highest antioxidant activity (FRAP, DPPH, TPC); protected GSH and ↓ MDA under H_2O_2 stress.	Free radical scavenging via phenolics; antioxidant protection of erythrocytes	[50]
In vitro	Root essential oil extracted via CXE, HD, IVD, and SFE; major components valerenol, valerenal, β-cadinene, β-vetivenene	DPPH radical scavenging	CXE oil showed moderate antioxidant activity (IC50 3.71 mg/mL).	Free radical scavenging via terpenoid constituents	[61]
In vitro	Root ethanolic extract; contains alkaloids, flavonoids, tannins, phenols, saponins, triterpenoids	Reducing power assay, superoxide anion scavenging, deoxyribose degradation, total antioxidant capacity, total phenolics and flavonoids	Dose-dependent strong antioxidant activity; superoxide IC50 130.36 μg/mL; high phenolic and flavonoid content.	Free radical scavenging; hydroxyl radical neutralization; lipid peroxidation inhibition	[62]
In vitro	Silver nanoparticles synthesized from aqueous extract of *Vetiveria zizanioides*	DPPH free radical scavenging assay	Dose-dependent antioxidant activity; max inhibition 72.4% at 50 μL; comparable to standard.	Free radical scavenging by nanoparticle-mediated electron donation	[63]
In vitro	Aqueous, methanolic, and ethanol root extracts; bioactive compounds include phenolics, flavonoids, alkaloids, saponins, tannins.	DPPH, FRAP, ABTS radical scavenging assays	Dose-dependent antioxidant activity; high phenolics correlated with strong scavenging; ABTS and FRAP confirm potent activity.	Free radical scavenging via electron donation and H-atom transfer by phenolics and flavonoids	[64]

Table 3. *Cont.*

Study Design	Extract Type and Bioactive Compounds Identified	Assay/Model	Key Findings	Mechanism	References
In vitro	Water-soluble, glycoside, and cell wall-bound phenolic acids; major: p-coumaric, p-dihydroxybenzoic, ferulic acids	ABTS assay (TEAC)	Alkaline water-soluble (cell wall-bound) fraction: highest phenolics and antioxidant activity; strong correlation (r = 0.988) with TEAC.	Antioxidant via free radical scavenging by phenolic acids; higher lignin-bound phenolics ↑ stress mitigation	[59]
In vitro	Essential oil; major compounds: khusimol, isovalencenol, 2-isopropyl-5-methyl-9-methylene-bicyclo [4.4.0]decene, α-vetivol, beta-maalene, vetiselinenol, γ-selinenes, zizanol, khusiol, β-vatirenes	Phosphomolybdenum assay	Essential oil showed 75.5% total antioxidant capacity at 0.1 mg/mL.	Antioxidant activity likely mediated by phenolic and other bioactive constituents	[56]
In vitro	Essential oil; major compounds: Khusimol (25.60%), Bicyclo-vetivenol (11.47%), α-Vetivone (7.76%)	DPPH radical scavenging assay	Dose-dependent antioxidant activity; highest activity 60.43% at 0.5 mg/mL, lowest 52.74% at 0.03125 mg/mL; activity lower than BHT.	Antioxidant activity mainly from terpenic alcohols and phenolics; both major and minor constituents contribute	[57]
In vivo	Crude water extract (roots)	Biochemical assays for serum ALT, AST, ALP; liver antioxidant enzymes: GSH, GR, GST, GPx; lipid peroxidation (MDA)	Pre/post Vetiveria extract ↓ ALT, AST, ALP; ↑ GSH, GR, GST; ↓ lipid peroxidation; pre-treatment > post-treatment.	Protection via maintaining hepatic antioxidants, ↑ GR and GST, ↓ oxidative stress and lipid peroxidation; partially mimics NAC	[58]
In vitro	Essential oil; main components: longiverbenone (27.31%), longipinocarvone (26.88%), Cedr-8-en-13-ol (26.26%)	DPPH free radical scavenging assay, FRAP assay	Dose-dependent DPPH and FRAP activity; FRAP IC50 = 184.8 ± 1.02 μg/mL; activity linked to phenolics.	Antioxidant activity from high EO phenolics; electron donation and free radical neutralization	[55]
In vitro	Ethanol and ethyl acetate extracts; rich in flavonoids and phenolic compounds	DPPH free radical scavenging assay	Both extracts: dose-dependent scavenging; 140 μg/mL inhibition: ethanol 40.7%, ethyl acetate 59.3%; IC50: ethanol 157.38, ethyl acetate 112.79 μg/mL.	Antioxidant activity via H-donation by phenolics and flavonoids, quenching free radicals	[54]
In vitro	Ethanolic root extract; major compounds include Khusenic acid, Ascorbic acid, Junipen, gamma-Himachalene, alpha-Guaiene	DPPH radical scavenging assay	Strong antioxidant activity (IC50 10.73 μg/mL; ascorbic acid 4.61 μg/mL)	Antioxidant-mediated free radical scavenging contributes to cytotoxicity in cancer cells	[53]

↑ indicates an increase (upregulation), and ↓ indicates a decrease (downregulation) in the corresponding molecule or effect.

Animal studies support the physiological relevance of these findings. In a murine model of acetaminophen-induced hepatotoxicity, crude aqueous root extracts restored hepatic glutathione (GSH) and the activities of glutathione reductase (GR) and glutathione S-transferase (GST) while normalizing GPx activity. Pre-treatment offered slightly greater protection than post-treatment. The extracts also reduced malondialdehyde (MDA) levels in liver tissue, indicating inhibition of lipid peroxidation and suggesting hepatoprotective effects mediated by phenolics and terpenic alcohols [58]. Studies on essential oils sourced from different regions, including Comoros and Egypt, show consistent dose-dependent radical scavenging [56,57]. Strong correlations between phenolic content and antioxidant activity in water-soluble fractions further highlight the pivotal role of phenolics in free radical neutralization [59].

Chemical profiling across investigations highlights the diversity of compounds responsible for these effects. Essential oils are dominated by sesquiterpenes, terpenic alcohols, and minor ketones, while ethanolic extracts are enriched with phenolics and flavonoids. This complex phytochemical composition enables synergistic interactions, combining hydrogen donation, electron transfer, and modulation of enzymatic defenses to produce strong antioxidant activity.

Despite the promising evidence, limitations exist. Most of the available evidence comes from preliminary in vitro studies or short-term in vivo models, leaving their direct applicability to human health uncertain. Differences in extraction methods, solvents, and assay conditions contribute to variability in reported IC_{50} and EC_{50} values. Moreover, the bioavailability, pharmacokinetics, and clinical efficacy of these compounds have not yet been clarified. While comparisons with reference antioxidants such as ascorbic acid and BHT are informative, they cannot fully predict performance in complex biological systems. Future work should therefore prioritize standardized extraction and testing protocols, detailed chemical profiling, and broader in vivo models. Greater attention should be given to synergistic interactions among bioactive compounds, dose–response dynamics, and long-term safety. Ultimately, clinical studies will be needed to validate the therapeutic potential of *C. zizanioides* as an antioxidant.

5. Antimicrobial Effects

The antibacterial properties of *C. zizanioides* have been investigated, demonstrating its efficacy against a broad range of pathogenic bacteria. These effects are typically assessed using methods such as minimum inhibitory concentration (MIC) assays, agar disk diffusion, and agar well diffusion to quantify antimicrobial effects, determine bacteriostatic versus bactericidal activity, and compare efficacy across microbial species [65–68]. In MIC assays, serial dilutions of *C. zizanioides* essential oils or crude extracts are incubated with bacterial cultures to identify the lowest concentration preventing visible growth, whereas diffusion-based approaches involve applying defined volumes of extract to agar plates and measuring inhibition zones after overnight incubation [65,66,69–71]. Although these assays provide a preliminary assessment, they do not account for pharmacokinetics, bioavailability, or host immune factors that affect clinical efficacy. Comparative analyses with standard antibiotics further substantiate the antimicrobial efficacy of *C. zizanioides* extracts, which have been shown to exhibit broad-spectrum activity against multidrug-resistant bacteria, viruses, and yeasts, reinforcing their traditional medicinal applications [69,70]. Recent studies further demonstrated strong inhibitory activity of *C. zizanioides* essential oil against methicillin-resistant Staphylococcus aureus (MRSA), with low MIC values, time-kill assays confirming bactericidal activity, and mechanistic evidence of protein, DNA, and RNA leakage due to membrane disruption [72,73]. Additionally, methanol and ethyl acetate extracts of *C. zizanioides* showed significant activity against Gram-negative bacteria such as *E. coli* and *P. mirabilis*, with inhibition zones ranging between 18 and 27 mm, and in polyherbal formulations, the activity extended synergistically to include *Pseudomonas aeruginosa* and *Enterobacter* sp. with inhibition zones up to 40 mm [74,75]. Commercial *C. zizanioides* oils have also been reported to be especially active against Gram-positive bacteria, including MRSA and Bacillus strains, while showing moderate antifungal activity against *Candida albicans* and strong inhibition of *Candida glabrata*, a clinically relevant azole-resistant pathogen [76,77].

Experimental studies consistently demonstrate that root-derived extracts and essential oils of *C. zizanioides* are particularly effective against Gram-positive bacteria such as *Staphylococcus aureus*, *Bacillus subtilis*, and *Bacillus cereus*, with inhibition zones ranging from 19.3 to 28.3 mm and MIC values as low as 0.0625% (v/v) for essential oils [65,69,70]. In contrast, Gram-negative bacteria, including *Escherichia coli* and *Pseudomonas aeruginosa*, generally exhibit reduced susceptibility, with inhibition zones as small as 9 mm, reflecting structural differences in cell envelopes [65,66]. Dose-dependent inhibition is observed in ethanolic and methanolic extracts, which display increasing activity at higher concentrations against *S. aureus*, *E. coli*, and *P. aeruginosa* [66,78]. Interestingly, *B. subtilis* shows minimal sensitivity to ethanolic extracts, illustrating species-specific variability in response to phytochemical constituents [66].

The antibacterial effects are largely attributed to the presence of diverse bioactive phytochemicals, including alkaloids, flavonoids, saponins, terpenoids, tannins, phenolics, and sesquiterpenes [65,66,79,80]. Essential oils, rich in cycloisolongifolene derivatives and other volatile compounds, act primarily by compromising bacterial cell membrane integrity, disrupting enzymatic activity, and interfering with metabolic pathways. The bactericidal potential is confirmed by MBC/MIC ratios ≤ 4 across tested strains, with Gram-positive bacteria consistently displaying the highest sensitivity [65]. A study reported that *C. zizanioides* exerts its antibacterial effect by disrupting bacterial membrane integrity, leading to the leakage of intracellular proteins, DNA, and RNA, with the effect being more pronounced at higher concentrations [73]. Comparisons with standard antibiotics reveal that *C. zizanioides* essential oils can achieve statistically significant inhibition relative to tetracycline and erythromycin, emphasizing their potency and wide-ranging antimicrobial capacity [65].

The antibacterial potential of *C. zizanioides* is particularly important when it comes to the efficacy against antibiotic-resistant bacteria. Hexane and ethanol extracts of *C. zizanioides* roots inhibit 55–70% of *methicillin-resistant S. aureus* (MRSA) isolates at remarkably low MIC values, with hexane extracts effective at 0.012 mg/mL. Such results underscore the potential of lipophilic constituents to overcome resistance mechanisms [72,80,81]. Furthermore, combinations of *C. zizanioides* essential oils with other plant oils, such as Pogostemon patchouli, demonstrate synergistic activity against MRSA, with ΣFIC indices ranging from 0.35 to 0.38, supporting the potential for combination therapies to enhance efficacy [82]. Innovative strategies leveraging nanotechnology have also been explored [83,84]. Silver nanoparticles synthesized using aqueous root extracts of *C. zizanioides* exhibit potent antibacterial effects against both *S. aureus* and *P. aeruginosa* bacteria. Nanoparticles, averaging 10–20 nm in diameter, are thought to act via enhanced surface interactions that compromise membrane integrity and enzymatic functions, illustrating a contemporary approach to improving pharmacodynamic properties of plant-derived antimicrobials [79,85].

While these are promising findings, it is important to note that certain methodological limitations are presented, such as variability in extract composition, concentration, and microbial strains tested. Moreover, few in vivo studies or clinical trials exist to confirm the antimicrobial efficacy and safety of *C. zizanioides* extracts in humans. Therefore, translation from in vitro results to clinical applications remains a critical gap. A detailed summary of the experimental assays and proposed mechanisms of action is provided in Table 4. Notwithstanding these limitations, variation in antimicrobial efficacy is influenced by extract type, solvent polarity, seasonal fluctuations in phytochemical content, and bacterial strain-specific susceptibility [65,66,78]. While in vitro studies demonstrate reproducible inhibitory effects, further research is required to assess pharmacokinetics, safety, and clinical applicability, alongside isolation and structural characterization of the active compounds responsible [65,66,79]. A detailed summary of the experimental assays and proposed mechanisms of action is provided in Table 4.

Table 4. Summary of the mechanisms underlying the antimicrobial activity of *C. zizanioides*.

Study Design/Assay	Extract Type and Bioactive Compounds Involved	Target Microorganism	Key Findings	Mechanism of Action	References
MIC assay	Ethanol, hexane, methanol extracts; Essential oil; Phenolic acids; Flavonoids; Terpenoids	*S. aureus*, *B. subtilis*, *MRSA*, *E. coli*, *P. aeruginosa*, *Candida* spp.	• Low MIC values indicate strong potency • Effective against drug-resistant pathogens • Comparable to standard antibiotics	• Disruption of bacterial cell walls and membranes; interference with metabolic enzymes • Inhibition of efflux pumps and virulence factors • Membrane permeability increase leading to leakage of DNA, RNA, proteins	[36–38,44–46,48,49,51]
Mechanistic Observations (descriptive studies)	Essential oils, methanol/ethanol extracts; Terpenoids; Flavonoids; Phenolic acids	Broad spectrum (Gram-positive, Gram-negative, fungi)	• Disrupts biofilm formation • Impairs protein synthesis, nucleic acid metabolism • Inhibits pili/fimbriae • Reduces toxin and enzyme production	• Disrupts nutrient uptake • Disrupts nutrient uptake • Induces ROS accumulation • Damages proteins, lipids, DNA • Increases membrane permeability leading to cell death	[65,73–75,79]
Disk diffusion/Well diffusion	Essential oils; Root, leaf, and methanol extracts; Terpenoids; Flavonoids; Phenolic acids	*S. aureus*, *MRSA strains*, *E. coli*, *K. pneumoniae*, *P. aeruginosa*, *Candida* spp.	• Inhibition zones 12–35 mm • Strong antibacterial and antifungal activity • Effective against multidrug-resistant strains	• Diffusion of active compounds into agar • Direct interaction with microbial membranes • Membrane disruption • Cell content leakage • Inhibition of microbial growth at the zone of contact	[73–77,81,82]
Nanoparticle-mediated assay	Silver nanoparticles synthesized using root aqueous extract; Phytochemicals	*S. aureus*, *P. aeruginosa*	Potent antibacterial effect; Effective at low concentration (25 μg/mL)	• Alteration of bacterial cell membrane permeability • Enzyme inhibition • Increased ROS generation	[85]

6. Wound Healing Potential

Skin health relies on a delicate balance of lipid metabolism, extracellular matrix regulation, and coordinated cellular signaling [86–88]. Natural bioactive compounds, particularly those derived from *C. zizanioides*, have recently attracted attention for their multifaceted roles in dermatology, ranging from anti-aging effects to tissue repair and remodeling. While current research is limited, preliminary results are promising. Tollenaere et al. (2021) demonstrated that *C. zizanioides* root extracts can stimulate lipid synthesis across multiple skin cell types, including keratinocytes, sebocytes, and adipocytes [89]. The extract not only increased sebum production in sebocytes by 31 percent but also improved lipid composition, elevating levels of lauric acid and sapienic acid by 42 and 43 percent, respectively [89]. Moreover, it enhanced epidermal barrier integrity through the upregulation of sphingomyelin, phosphoglycerides, ceramides, and cerebrosides, alongside key proteins involved in lipid transport and cornification, including CERT, involucrin, and dermokine. Importantly, *C. zizanioides* extract promoted both adipogenesis and lipogenesis in pre-adipocytes and full-thickness skin explants, resulting in a greater number and size of adipocytes [89]. Clinically, these effects translated into improved skin plumpness, reduced signs of fatigue, and a measurable decrease in wrinkle depth. Chemical analysis revealed a complex mixture of sesquiterpenes, including zizanoic acid, oplopanone, and teuhetenone, as well as several previously unreported compounds, underscoring the potential of these constituents as active agents in lipid regulation and skin rejuvenation [89]. In parallel, another study explored the effects of *C. zizanioides* essential oil in pre-inflamed human dermal fibroblasts, designed to model chronic inflammation and fibrosis. The study revealed that the essential oil exerted pronounced antiproliferative effects and significantly reduced collagen III production, a critical component of the extracellular matrix involved in tissue remodeling [90]. Genome-wide analyses showed that 163 genes were significantly regulated, most of which were associated with metabolic processes and tissue remodeling pathways, including cholesterol biosynthesis and inositol pyrophosphate metabolism [90]. These findings suggest that *C. zizanioides* essential oil may influence both structural and metabolic aspects of skin physiology, complementing the lipid-enhancing properties observed with *C. zizanioides* root extracts [90]. While no significant modulation of inflammatory biomarkers was observed in fibroblasts, previous studies in macrophages indicate that *C. zizanioides* essential oil possesses anti-inflammatory and antioxidant activities, highlighting potential cell-type-specific effects [16,22].

Beyond lipid metabolism and fibroblast activity, the broader molecular framework for skin repair involves a coordinated interplay of signaling pathways that govern inflammation, angiogenesis, re-epithelialization, and oxidative stress. Flavonoids promote wound healing through multiple pathways, including Wnt/β-catenin, Hippo, TGF-β, JNK, NF-κB, MAPK/ERK, PI3K/Akt, and Nrf2/ARE. These pathways collectively regulate cellular proliferation, differentiation, and antioxidant defenses [91]. Since, *C. zizanioides* extract have shown to be rich in flavonoids, it is possible that the mechanisms influenced by *C. zizanioides* extracts appear to converge with these pathways, as seen in Figure 2, particularly through modulation of lipid and cholesterol metabolism, suggesting a potential role in enhancing barrier function, tissue remodeling, and oxidative stress protection. However, these mechanisms should be studied further with *C. zizanioides* flavonoids rich extracts. Therefore, these studies highlight the potential of *C. zizanioides* derived compounds as multifunctional agents in skin health. Root extracts primarily support lipid synthesis, sebum quality, and adipocyte function, contributing to hydration, plumpness, and visible anti-aging effects. Essential oil influences fibroblast activity and gene expression, affecting tissue remodeling and metabolic regulation. When considered alongside the mechanistic insights from flavonoid-mediated wound healing, it becomes evident that *C. zizanioides* bioactive may act across multiple levels, including lipid metabolism, extracellular matrix remodeling, and oxidative stress regulation, to strengthen barrier integrity, enhance resilience, and support tissue repair. As highlighted in Table 5, these findings provide a strong foundation for further research into *C. zizanioides* extracts and essential oil as therapeutic agents in dermatology, with potential applications in anti-aging, metabolic skin support, and wound healing.

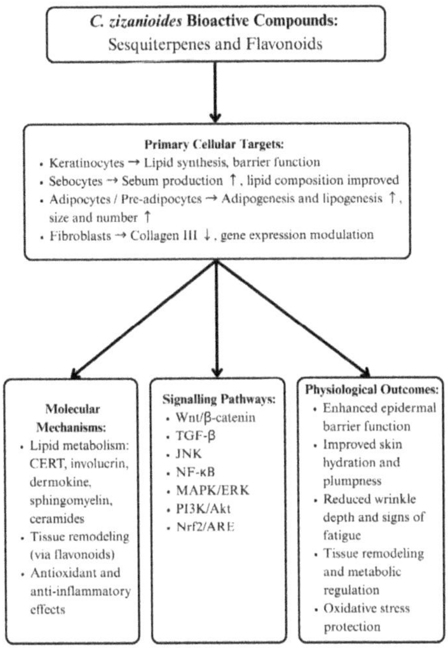

Figure 2. Wound healing potential of *C. zizanioides*. ↑ indicates an increase (upregulation), and ↓ indicates a decrease (downregulation) in the corresponding molecule or effect.

Table 5. Summary of the mechanisms underlying the wound healing activity of *C. zizanioides*.

Extract Type	Target Cells and Pathways	Mechanisms of Action	Observed Effects	Clinical Implications	References
Root Extract	Keratinocytes, sebocytes, adipocytes, skin explants	- ↑ Lipid synthesis (lauric acid ↑ 42%, sapienic acid ↑ 43%) - ↑ Sebum production (+31%) - ↑ Barrier lipids (sphingomyelin, ceramides, cerebrosides) - ↑ Proteins (CERT, involucrin, dermokine) - ↑ Adipogenesis and lipogenesis	- Improved epidermal barrier - Greater number/size of adipocytes - Enhanced plumpness - ↓ Wrinkle depth, ↓ fatigue signs	Hydration, anti-aging, barrier strengthening	[89]
Essential Oil	Pre-inflamed fibroblasts	- ↓ Collagen III (extracellular matrix remodeling) - Antiproliferative effect - Regulation of 163 genes (cholesterol biosynthesis, inositol pyrophosphate metabolism)	- Modulation of fibroblast activity - Regulation of tissue remodeling and metabolic pathways	Anti-fibrotic potential, metabolic skin support, wound repair	[90]
Flavonoids	Multiple signaling pathways (Wnt/β-catenin, Hippo, TGF-β, Hedgehog, JNK, NF-κB, MAPK/ERK, PI3K/Akt, Nrf2/ARE)	- Regulate inflammation - Promote angiogenesis - Stimulate reepithelialization - Enhance antioxidant defenses	- Coordinated wound healing responses	Basis for linking *C. zizanioides* activity to broader skin repair mechanisms	[91]

↑ indicates an increase (upregulation), and ↓ indicates a decrease (downregulation) in the corresponding molecule or effect.

7. Anticancer Activity

C. zizanioides has demonstrated notable anticancer potential through a combination of cytotoxic, apoptotic, and ROS-mediated mechanisms, both in vitro and in vivo. In vitro studies on various cancer cell lines have consistently shown that *C. zizanioides* extracts and derived compounds exert selective cytotoxicity. For instance, hydrodistilled *C. zizanioides* oil exhibited moderate cytotoxic activity against human lung (A549) and hepatocellular carcinoma (HepG2) cell lines, with IC50 values of 32.2 μg/mL and 37.6 μg/mL, respectively [92]. Similarly, *C. zizanioides* oil was evaluated against colon (WiDr), triple-negative breast cancer (4T1), and estrogen receptor-positive breast cancer (T47D) cells. *C. zizanioides* oil induced cell cycle arrest, primarily in G2/M phase for WiDr and 4T1 cells, with IC50 values of 302 μg/mL, 60 μg/mL, and 112 μg/mL for WiDr, 4T1, and T47D cells, respectively, highlighting selective cytotoxicity toward TNBC cells [93]. Mitotic index analysis further confirmed that crude *C. zizanioides* oil from agricultural sources suppressed proliferation of HeLa cervical cancer cells more effectively than commercial essential oil, with MI reductions from 5.57% (control) to 1.70% [94].

Mechanistically, the anticancer activity of *C. zizanioides* is frequently linked to the induction of oxidative stress through increased reactive oxygen species (ROS) levels. *C. zizanioides* oil treatment elevated intracellular ROS in WiDr and T47D cells, correlating with apoptosis induction, while 4T1 cells underwent G2/M arrest without significant ROS increase, suggesting alternative mechanisms of cytotoxicity [93]. Sesquiterpene components such as β-caryophyllene, khusimol, and β-vetispirene have been identified as major contributors to these effects. Notably, β-vetispirene selectively inhibited AKR1C1, an enzyme overexpressed in lung cancer that maintains intracellular ROS below apoptotic thresholds, without affecting the highly homologous AKR1C2, thereby promoting ROS-mediated apoptosis [95]. The selective targeting of AKR1C1 highlights a genetic and molecular basis for the anticancer efficacy of specific *C. zizanioides* compounds. Furthermore, *C. zizanioides* oil's cytotoxicity in 4T1 TNBC cells has been attributed to its agonistic interaction with the Cannabinoid CB2 receptor (CNR2), a gene highly expressed in these cells and associated with poor patient survival, while estrogen receptor-expressing T47D cells displayed less sensitivity due to lower receptor engagement [93].

Several studies have confirmed the low cytotoxicity of *C. zizanioides* on normal cells, including HEK 293 kidney cells and L929 fibroblasts, indicating a favorable therapeutic index for potential anticancer applications [96,97]. Cytotoxicity assays consistently showed cell viability above 75% at concentrations where anticancer activity was observed in tumor cells. These findings support the selective activity of bioactive sesquiterpenes and essential oil components against malignant cells. Although in vivo studies directly evaluating tumor suppression are limited, the immunomodulatory properties of valencene, a major fraction of *C. zizanioides* root extract, suggest potential indirect anticancer benefits through immune system enhancement. Valencene, a compound found in *C. zizanioides* extract, significantly increased phagocytic index, nitric oxide scavenging, and antibody production in Swiss albino mice, while reducing pro-inflammatory TNF-α production, suggesting a dual role in enhancing antitumor immunity while minimizing systemic inflammation [96]. These immune effects may complement direct cytotoxic mechanisms in vivo, although formal tumor xenograft studies remain to be conducted.

Collectively, the current literature, as outlined in Table 6, indicates that *C. zizanioides* exerts anticancer activity through multiple convergent mechanisms, including selective induction of apoptosis, ROS-mediated oxidative stress, cell cycle arrest, and immune modulation. The identification of genetic targets such as AKR1C1 and CNR2 highlights molecular pathways that may be exploited for targeted therapy. The selective cytotoxicity, low toxicity to normal cells, and molecular specificity highlight the therapeutic promise of *C. zizanioides* and its sesquiterpene constituents for future anticancer drug development. While these in vitro studies provide important mechanistic insights, their methodological limitation lies in the simplified cellular models, and they may not fully translate to the complex human physiological system. Therefore, clinical studies are essential when evaluating therapeutic potential.

Table 6. Summary of the mechanisms underlying the anticancer activity of *C. zizanioides*.

Extract Type	Bioactive Compound Involved	Cell Line Model	Effect	References
Essential oil	Cedr-8-en-13-ol, α-/β-pinene, α-/γ-terpinene	B16 melanoma cells	Non-cytotoxic; ↓ melanin, tyrosinase activity, ↑ antioxidant enzymes, ↓ oxidative stress	[98]
Ethanolic root extract	Ellagic acid, Ascorbic acid, Linoleic acid, α-/β-Sitosterol	SCC-29B (oral), DU-145 (prostate), Vero	Cytotoxic to SCC-29B, DU-145; induces DNA damage, apoptosis; minimal Vero toxicity	[53]
C. zizanioides acetate oil	Not specified	In vivo: Swiss albino mice treated with cisplatin	Protects from nephrotoxicity, DNA/chromosomal damage; restores GSH/enzymes; ↑ antioxidant defenses	[99]
Essential oil	Sesquiterpene lactones	HeLa (human cervical cancer)	Cytotoxic (IC50 0.05%); induces apoptosis, ROS, mitochondrial depolarization	[100]
Methanol extract (polyherbal including *C. zizanioides*)	Carbohydrates, alkaloids, steroids, saponins, flavonoids, tannins	HeLa, MCF-7	Cytotoxic to HeLa; low toxicity to MCF-7; likely antioxidant mediated	[75]
Ethanolic root extract	Longifolene	DU-145 (prostate cancer), SCC-29B (oral cancer), Vero (healthy kidney)	Cytotoxic to prostate/oral cancer; minimal Vero toxicity	[101]
Aqueous root extract	Valencene	L929 fibroblasts (cytotoxicity)	Cytotoxic; ↑ TNF-α, IL-6; immunomodulatory	[96]
Essential Oil	β-caryophyllene, α-humulene, caryophyllene oxide	In vitro: WiDr (colon), 4T1 (TNBC), T47D (luminal breast) cancer cells; MTT assay	Cytotoxic; ↑ ROS, apoptosis; docking confirms binding	[93]
C. zizanioides crude oil and commercial essential oil	Not specified	In vitro: HeLa cervical cancer cells; Mitotic index (MI) assay	Stronger anticancer (MI 1.70%) vs. commercial EO (3.26%) and control (5.57%); ↓ MI → ↑ antimitotic activity; active components not identified.	[94]
C. zizanioides oil	Khusimol, aristol-1(10)en-9-ol, cyclocopacamphenol, bicyclo[5.2.0]nonane-2-methylene-4,8,8-trimethyl-4-vinyl	In vitro: Human lung (A549) and hepatocellular (HepG2) cancer cell lines; MTT assay	Moderate cytotoxicity; inhibits proliferation	[92]
C. zizanioides oil	Beta vetispirene	In silico: Molecular docking; bioinformatics	Selectively inhibits AKR1C1/2; ↑ ROS, apoptosis (lung cancer)	[95]
Methanolic root extract	Not specified	In vitro: MTT assay on HEK 293 cells	Non-cytotoxic; no viability reduction	[97]

↑ indicates an increase (upregulation), and ↓ indicates a decrease (downregulation) in the corresponding molecule or effect.

8. Conclusions

The pharmacological potential of *C. zizanioides* has been extensively explored, validating its traditional ethnomedical applications through modern scientific investigations. Its anti-inflammatory activity involves modulation of cytokines and inflammatory mediators, while its strong antioxidant capacity stems from free radical scavenging, metal chelation, and enhancement of endogenous antioxidant enzymes. *C. zizanioides* also exhibits broad-spectrum antimicrobial effects, alongside selective anticancer properties characterized by apoptosis induction and tumor growth inhibition. Additionally, *C. zizanioides* supports effective wound healing through the stimulation of fibroblast activity and extracellular matrix. Therefore, these pharmacological activities position *C. zizanioides* as a promising candidate for therapeutic development across diverse areas such as managing inflammatory disorders, infections, cancer, and tissue repair. These effects are attributed to its diverse bioactive constituents, including sesquiterpenes, flavonoids, and steroids, which exert their therapeutic influence by modulating molecular pathways and immune responses. However, most of the current evidence is derived from in vitro and animal studies with limited clinical data to confirm efficacy and safety in humans. Moreover, the variability in extract preparation, dosage, and bioavailability presents challenges for standardization and reproducibility. To bridge these gaps, future research must focus on clinical trials, pharmacokinetics, and toxicological studies. Only then can the full therapeutic potential of *C. zizanioides* be realized and translated into clinical applications.

9. Future Directions

To advance the use and integration of *C. zizanioides* into modern pharmacotherapy, several key research areas must be addressed. Comprehensive phytochemical profiling is necessary to characterize the full spectrum of bioactive compounds and elucidate their synergistic interactions within biological pathways. Pharmacokinetic studies should be conducted to determine the bioavailability, metabolic fate, and systemic distribution of active molecules, thereby providing insights into their therapeutic efficacy and safety. Additionally, specificity and interaction potential with other therapeutic agents should be assessed to facilitate rational drug design. Improving delivery systems, particularly for hydrophobic bioactive constituents, remains a critical challenge, and novel formulation strategies should be explored to enhance solubility and bioavailability. Most importantly, rigorous, well-controlled clinical trials must be initiated to establish the safety, efficacy, and therapeutic applicability of *C. zizanioides* across various medical indications. As research continues to elucidate the pharmacological mechanisms and clinical benefits of *C. zizanioides*, its traditional medicinal use can be effectively translated into evidence-based modern medicine, expanding its role in contemporary healthcare.

Author Contributions: Writing—original draft, conceptualization, investigation, literature analysis: C.J.G., A.F.M. and I.A.A.-Y. Tables and figures generation: S.A.A.R. and A.F.M. Critical revision: C.J.G., A.F.M., I.A.A.-Y. and S.A.A.R. All authors have read and agreed to the published version of the manuscript.

Funding: The work in this paper was supported, in part, by the Open Access Program from the American University of Sharjah. This paper represents the opinions of the authors and does not mean to represent the position or opinions of the American University of Sharjah.

Institutional Review Board Statement: Not applicable.

Informed Consent Statement: Not applicable.

Data Availability Statement: No new data were created or analyzed in this study. Data sharing is not applicable to this article.

Conflicts of Interest: The authors declare that they have no competing interests.

References

1. Rao, R.R.; Suseela, M.R. *Vetiveria zizanioides* (Linn.) Nash—A multipurpose eco–friendly grass of India. *ICV–2 Held Cha–Am Phetchaburi Thail.* **2000**, 18–22.
2. Anjum, D.S.; Varshney, D.S. Usheera (*Vetiveria zizanioides* Linn.): Ayurvedic Insights and Modern Pharmacological Evidence. *TIJER—Int. Res. J.* **2025**, *12*, a280–a292. Available online: https://tijer.org/tijer/viewpaperforall.php?paper=TIJER2507029 (accessed on 12 June 2025).
3. Pareek, A.; Kumar, A. Ethnobotanical and pharmaceutical uses of *Vetiveria zizanioides* (linn) nash: A medicinal plant of Rajasthan. *Int. J. Life Sci. Pharma Res.* **2013**, 3, L-12–L-18. Available online: https://www.semanticscholar.org/paper/ETHNOBOTANICAL-AND-PHARMACEUTICAL-USES-OF-VETIVERIA-Pareek-Kumar/801c3de40786552ef952fca56eede9e3eb92732d (accessed on 12 June 2025).
4. Barcellos-Silva, I.G.C.; Dos Santos, F.K.F.; Kharkwal, H.; Chander, S.; Kharkwal, A.C.; Awasthi, R.; Dhiman, N.; Sharma, B.; Kulkarni, G.T.; Larssen, H.; et al. Vetiver, *Vetiveria zizanioides* (L.) Nash: Biotechnology, Biorefineries, and the Production of Volatile Phytochemicals. *Plants* **2025**, *14*, 1435. [CrossRef] [PubMed] [PubMed Central]
5. Belhassen, E.; Filippi, J.-J.; Brévard, H.; Joulain, D.; Baldovini, N. Volatile constituents of vetiver: A review. *Flavour. Fragr. J.* **2014**, *30*, 26–82. [CrossRef]
6. Ramírez-Rueda, R.; Marinho, J.; Salvador, M. Bioguided identification of antimicrobial compounds from *Chrysopogon zizaniodes* (L.) roberty root essential oil. *Future Microbiol.* **2019**, *14*, 1179–1189. [CrossRef]
7. Devprakash, D.; Prashant Singh, P.S.; Srinivasan, K.K.; Subburaju, T.; Singh, S.K. Antifungal activity of alcoholic and aqueous extracts of *Vetiveria zizanioides*. *J. Pharm. Res. Opin.* **2011**, *1*, 85–88.
8. Arafat, M.A.M.; Khalil, M.N.A.; Mohamed, O.G.; Abd El-Ghafar, O.A.M.; Tripathi, A.; Mahrous, E.A.; Abd El-kader, E.M.; El-Hawary, S. Vetiver aerial parts and roots ameliorate rheumatoid arthritis in complete Freund's adjuvant rat model, a phytochemical profiling and mechanistic study. *J. Ethnopharmacol.* **2023**, *317*, 116764. [CrossRef]
9. Grover, M.; Behl, T.; Bungau, S.; Aleya, L. Potential therapeutic effect of *Chrysopogon zizanioides* (Vetiver) as an anti-inflammatory agent. *Env. Sci. Pollut. Res. Int.* **2021**, *28*, 15597–15606. [CrossRef]
10. Ricciotti, E.; FitzGerald, G.A. Prostaglandins and inflammation. *Arterioscler. Thromb. Vasc. Biol.* **2011**, *31*, 986–1000. [CrossRef]
11. Chen, L.; Deng, H.; Cui, H.; Fang, J.; Zuo, Z.; Deng, J.; Li, Y.; Wang, X.; Zhao, L. Inflammatory responses and inflammation-associated diseases in organs. *Oncotarget* **2017**, *9*, 7204–7218. [CrossRef] [PubMed] [PubMed Central]
12. Nunes, C.D.R.; Barreto Arantes, M.; Menezes de Faria Pereira, S.; Leandro da Cruz, L.; de Souza Passos, M.; Pereira de Moraes, L.; Vieira, I.J.C.; Barros de Oliveira, D. Plants as Sources of Anti-Inflammatory Agents. *Molecules* **2020**, *25*, 3726. [CrossRef] [PubMed] [PubMed Central]
13. Beg, S.; Swain, S.; Hasan, H.; Barkat, M.A.; Hussain, M.S. Systematic review of herbals as potential anti-inflammatory agents: Recent advances, current clinical status and future perspectives. *Pharmacogn. Rev.* **2011**, *5*, 120–137. [CrossRef] [PubMed] [PubMed Central]
14. Gonfa, Y.H.; Tessema, F.B.; Bachheti, A.; Rai, N.; Tadesse, M.G.; Singab, A.N.; Chaubey, K.K.; Bachheti, R.K. Anti-inflammatory activity of phytochemicals from medicinal plants and their nanoparticles: A review. *Curr. Res. Biotechnol.* **2023**, *6*, 100152. [CrossRef]
15. Lunz, K.; Stappen, I. Back to the Roots-An Overview of the Chemical Composition and Bioactivity of Selected Root-Essential Oils. *Molecules* **2021**, *26*, 3155. [CrossRef]
16. Chou, S.T.; Lai, C.P.; Lin, C.C.; Shih, Y. Study of the chemical composition, antioxidant activity and anti-inflammatory activity of essential oil from *Vetiveria zizanioides*. *Food Chem.* **2012**, *134*, 262–268. [CrossRef]
17. Zhang, J.M.; An, J. Cytokines, inflammation, and pain. *Int. Anesthesiol. Clin.* **2007**, *45*, 27–37. [CrossRef] [PubMed] [PubMed Central]
18. Narkhede, M.B.; Wagh, A.E.; Rathi, A.M. Anti-inflammatory activity of *Vetiveria zizanioides* (linn.) Root. *J. Pharm. Res.* **2012**, *5*, 2016–2017.
19. Mai, H.P.; Lê, K.K.L. Evaluation of the anti-inflammatory activity of Vetiver root extract in Dak Lak, Viet Nam. *HCMUE J. Sci.* **2024**, *21*, 407. [CrossRef]
20. Frow, E.K.; Reckless, J.; Grainger, D.J. Tools for anti-inflammatory drug design: In vitro models of leukocyte migration. *Med. Res. Rev.* **2004**, *24*, 276–298. [CrossRef]

21. Singh, D.; Rai, V.; Agrawal, D.K. Regulation of Collagen I and Collagen III in Tissue Injury and Regeneration. *Cardiol. Cardiovasc. Med.* **2023**, *7*, 5–16. [CrossRef] [PubMed] [PubMed Central]
22. Lima, J.T.; Rocha, R.F.; Moreira, J.C.F.; Araújo, A.A.S. Phytochemical screening, antinociceptive and anti-inflammatory activities of Chrysopogon zizanioides essential oil. *Rev. Bras. Farmacogn.* **2012**, *22*, 443–450. [CrossRef]
23. Tak, P.P.; Firestein, G.S. NF-kappaB: A key role in inflammatory diseases. *J. Clin. Investig.* **2001**, *107*, 7–11. [CrossRef] [PubMed] [PubMed Central]
24. Boyce, B.F.; Xiu, Y.; Li, J.; Xing, L.; Yao, Z. NF-κB-Mediated Regulation of Osteoclastogenesis. *Endocrinol. Metab.* **2015**, *30*, 35–44. [CrossRef]
25. Ageeva, T.; Rizvanov, A.; Mukhamedshina, Y. NF-κB and JAK/STAT Signaling Pathways as Crucial Regulators of Neuroinflammation and Astrocyte Modulation in Spinal Cord Injury. *Cells* **2024**, *13*, 581. [CrossRef]
26. Nakano, R.; Kitanaka, T.; Namba, S.; Kitanaka, N.; Suwabe, Y.; Konno, T.; Yamazaki, J.; Nakayama, T.; Sugiya, H. Non-Transcriptional and Translational Function of Canonical NF-κ B Signaling in Activating ERK1/2 in IL-1 β-Induced COX-2 Expression in Synovial Fibroblasts. *Front. Immunol.* **2020**, *11*, 579266. [CrossRef] [PubMed]
27. Hanrahan, J.R.; Chebib, M.; Johnston, G.A. Flavonoid modulation of GABA(A) receptors. *Br. J. Pharmacol.* **2011**, *163*, 234–245. [CrossRef] [PubMed] [PubMed Central]
28. Wasowski, C.; Marder, M. Flavonoids as GABAA receptor ligands: The whole story? *J. Exp. Pharmacol.* **2012**, *4*, 9–24. [CrossRef] [PubMed]
29. Ballantyne, J.C.; Shin, N.S. Efficacy of opioids for chronic pain: A review of the evidence. *Clin. J. Pain* **2008**, *24*, 469–478. [CrossRef]
30. Milani, D.A.Q.; Davis, D.D. Pain management medications. In *StatPearls [Internet]*; StatPearls Publishing: St. Petersburg, FL, USA, 2023. Available online: https://www.ncbi.nlm.nih.gov/books/NBK560692/ (accessed on 12 June 2025).
31. O'Brien, T.; Chrirstrup, L.L.; Drewes, A.M.; Fallon, M.T.; Kress, H.G.; McQuay, H.J.; Mikus, G.; Morlion, B.J.; Perez-Cajaraville, J.; Pogatzki-Zahn, E.; et al. European Pain Federation position paper on appropriate opioid use in chronic pain management. *Eur. J. Pain* **2017**, *21*, 3–19. [CrossRef] [PubMed]
32. Harirforoosh, S.; Asghar, W.; Jamali, F. Adverse effects of nonsteroidal antiinflammatory drugs: An update of gastrointestinal, cardiovascular and renal complications. *J. Pharm. Pharm. Sci.* **2013**, *16*, 821–847. [CrossRef] [PubMed]
33. Borde, B.; Dubey, S.; Agrawal, S.; Bhalerao, A.; Kadam, K.; Borde, G. Vetiver zizanioides and their Pharmacological Activity. *Introduction* **2023**, *28*, 240–247. Available online: https://ijppr.humanjournals.com/wp-content/uploads/2023/12/18.Bhagyashri-Borde-Suchita-Dubey-Shivam-Agrawal-Ankush-Bhalerao-Kamlesh-Kadam-Dr.-Geeta-Borde.pdf (accessed on 12 June 2025).
34. McCurdy, C.R.; Scully, S.S. Analgesic substances derived from natural products (natureceuticals). *Life Sci.* **2005**, *78*, 476–484. [CrossRef] [PubMed]
35. Akash, S.; Akter, M.A.; Talukder, C.; Mim, S.A.; Rahman, A.; Obaid, A.A.; Islam, M.; Himu, J.I.; Aziz, T.; Rashid, M.A.; et al. Antioxidant, Antidiabetic, Analgesic, and Antibacterial Properties of Chrysopogon zizanioides Leaf Extract: An In Vivo, In Vitro, and In Silico Evaluation. preprint (Version 1). 2024. Available online: https://www.researchsquare.com/article/rs-5219178/v1 (accessed on 12 June 2025).
36. Paredes, S.; Cantillo, S.; Candido, K.D.; Knezevic, N.N. An Association of Serotonin with Pain Disorders and Its Modulation by Estrogens. *Int. J. Mol. Sci.* **2019**, *20*, 5729. [CrossRef] [PubMed] [PubMed Central]
37. Yang, S.; Zhang, B.; Wang, D.; Hu, S.; Wang, W.; Liu, C.; Wu, Z.; Yang, C. Role of GABAergic system in the comorbidity of pain and depression. *Brain Res. Bull.* **2023**, *200*, 110691. [CrossRef]
38. Wang, Z.J.; Heinbockel, T. Essential Oils and Their Constituents Targeting the GABAergic System and Sodium Channels as Treatment of Neurological Diseases. *Molecules* **2018**, *23*, 1061. [CrossRef] [PubMed] [PubMed Central]
39. Böhm, C.; Wiessler, A.L.; Janzen, D.; Nausester, J.; Slavik, B.; Loos, H.M.; Villmann, C.; Buettner, A. Modulatory effect of various essential oils on different GABAA receptor subtypes present in the central nervous system. *Phytomedicine Plus* **2025**, *5*, 100852. [CrossRef]
40. Fung, T.K.H.; Lau, B.W.M.; Ngai, S.P.C.; Tsang, H.W.H. Therapeutic Effect and Mechanisms of Essential Oils in Mood Disorders: Interaction between the Nervous and Respiratory Systems. *Int. J. Mol. Sci.* **2021**, *22*, 4844. [CrossRef] [PubMed] [PubMed Central]
41. Hao, C.W.; Lai, W.S.; Ho, C.T.; Sheen, L.Y. Antidepressant-like effect of lemon essential oil is through a modulation in the levels of norepinephrine, dopamine, and serotonin in mice: Use of the tail suspension test. *J. Funct. Foods* **2013**, *5*, 370–379. [CrossRef]
42. Guimarães, A.G.; Quintans, J.S.; Quintans, L.J., Jr. Monoterpenes with analgesic activity—A systematic review. *Phytother. Res.* **2013**, *27*, 1–15. [CrossRef] [PubMed]
43. de Cássia Da Silveira e Sá, R.; Andrade, L.N.; De Sousa, D.P. Sesquiterpenes from essential oils and anti-inflammatory activity. *Nat. Prod. Commun.* **2015**, *10*, 1934578X1501001033. [CrossRef]
44. Kiokias, S.; Oreopoulou, V. A Review of the Health Protective Effects of Phenolic Acids against a Range of Severe Pathologic Conditions (Including Coronavirus-Based Infections). *Molecules* **2021**, *26*, 5405. [CrossRef] [PubMed] [PubMed Central]

45. Kim, H.J.; Chen, F.; Wang, X.; Chung, H.Y.; Jin, Z. Evaluation of antioxidant activity of vetiver (*Vetiveria zizanioides* L.) oil and identification of its antioxidant constituents. *J. Agric. Food Chem.* **2005**, *53*, 7691–7695. [CrossRef] [PubMed]
46. Chen, F.; Wang, X.; Kim, H. Antioxidant, Anticarcinogenic and Termiticidal Activities of Vetiver Oil. 2003. Available online: https://www.semanticscholar.org/paper/Antioxidant-%2C-Anticarcinogenic-and-Termiticidal-of-Chen-Wang/6cdfdbaf5bd8a1924ed96c5a97e822871bcca064 (accessed on 12 June 2025).
47. Lü, J.M.; Lin, P.H.; Yao, Q.; Chen, C. Chemical and molecular mechanisms of antioxidants: Experimental approaches and model systems. *J. Cell Mol. Med.* **2010**, *14*, 840–860. [CrossRef] [PubMed] [PubMed Central]
48. Chaudhary, P.; Janmeda, P.; Docea, A.O.; Yeskaliyeva, B.; Abdull Razis, A.F.; Modu, B.; Calina, D.; Sharifi-Rad, J. Oxidative stress, free radicals and antioxidants: Potential crosstalk in the pathophysiology of human diseases. *Front. Chem.* **2023**, *11*, 1158198. [CrossRef] [PubMed]
49. Dahiya, D.P.; Sreenivasan, K.K.; Subburaju, T. Comparative antioxidant studies of ethanol extract and fresh aqueous extract of *Vetiveria zizanioides*. *Int. J. Pharm. Pharm. Sci.* **2011**, *3*, 325–331.
50. Luqman, S.; Kumar, R.; Kaushik, S.; Srivastava, S.; Darokar, M.P.; Khanuja, S.P. Antioxidant potential of the root of *Vetiveria zizanioides* (L.) Nash. *Indian J. Biochem. Biophys.* **2009**, *46*, 122–125.
51. El Ayadi, A.; Salsbury, J.R.; Enkhbaatar, P.; Herndon, D.N.; Ansari, N.H. Metal chelation attenuates oxidative stress, inflammation, and vertical burn progression in a porcine brass comb burn model. *Redox Biol.* **2021**, *45*, 102034. [CrossRef] [PubMed] [PubMed Central]
52. Wassmann, S.; Wassmann, K.; Nickenig, G. Modulation of oxidant and antioxidant enzyme expression and function in vascular cells. *Hypertension* **2004**, *44*, 381–386. [CrossRef]
53. Grover, M.; Behl, T.; Virmani, T. Phytochemical Screening, Antioxidant Assay and Cytotoxic Profile for Different Extracts of Chrysopogon zizanioides Roots. *Chem. Biodivers.* **2021**, *18*, e2100012. [CrossRef]
54. Tarai, D.K.; Nayak, S.; Karan, S. In vitro free radical scavenging activity of *Vetiveria zizanioides*. *J. Pharm. Res.* **2010**, *3*, 681–683.
55. Jeddi, M.; Fikri-Benbrahim, K.; El Hachlafi, N.; Benkhaira, N.; Aboussemdai, A.; Ouaritini, Z.B. Chemical Composition of Thymus vulgaris, Origanum compactumand *Vetiveria zizanoides* Essential oils and their Antibacterial and Antioxidant Activities. *Trop. J. Nat. Prod. Res.* **2023**, *7*, 2244–2250. [CrossRef]
56. Samaan, M.; Ebid, M.; Thabet, M. GC-MS analysis, antioxidant capacity and antimicrobic action of *Vetiveria zizanioides*. Essential oil cultivated in north Egypt. *J. Plant Prod.* **2022**, *13*, 699–704. [CrossRef]
57. Soidrou, S.H.; Farah, A.; Satrani, B.; Ghanmi, M.; Lachkar, M.; Mohamed, A.S. Chemical composition, antioxidant and antimicrobial activity of *Vetiveria zizanioides* roots essential oil harvested in Ndzuwani, Comoros. *World J. Pharm. Pharm. Sci.* **2020**, *9*, 82–95.
58. Hewawasam, R.P.; Jayatilaka, K.A.P.W. Antioxidant effect of crude water extract of *Vetiveria zizanioides* (Gramineae) in mice with acetaminophen induced hepatotoxicity. *Int. J. Pharmacogn.* **2015**, *2*, 11–20. [CrossRef]
59. Prajna, J.; Richa, J.; Dipjyoti, C. HPLC quantification of phenolic acids from *Vetiveria zizanioides* (L.) Nash and its antioxidant and antimicrobial activity. *J. Pharm.* **2013**, *2013*, 270472. [CrossRef]
60. Hassanpour, S.H.; Doroudi, A. Review of the antioxidant potential of flavonoids as a subgroup of polyphenols and partial substitute for synthetic antioxidants. *Avicenna J. Phytomed.* **2023**, *13*, 354–376. [CrossRef] [PubMed] [PubMed Central]
61. David, A.; Wang, F.; Sun, X.; Li, H.; Lin, J.; Li, P.; Deng, G. Chemical composition, antioxidant, and antimicrobial activities of *Vetiveria zizanioides* (L.) nash essential oil extracted by carbon dioxide expanded ethanol. *Molecules* **2019**, *24*, 1897. [CrossRef]
62. Subhadradevi, V.; Asokkumar, K.; Umamaheswari, M.; Sivashanmugam, A.T.; Sankaranand, R. In vitro antioxidant activity of *Vetiveria zizanioides* root extract. *Tanzania J. Health Res.* **2010**, *12*, 138–143. [CrossRef]
63. Ali, S.; Arthanari, A.; Shanmugam, R. Antioxidant activity of silver nanoparticles synthesized using *Vetiveria zizanioides*-in vitro study. *J. Res. Med. Dent. Sci.* **2021**, *9*, 199–203. Available online: https://www.jrmds.in/articles/antioxidant-activity-of-silver-nanoparticles-synthesized-using-vetiveria-zizanioidesin-vitro-study.pdf (accessed on 12 June 2025).
64. Muthukrishnan, S.; Manogaran, P. Phytochemical analysis and free radical scavenging potential activity of *Vetiveria zizanioides* Linn. *J. Pharmacogn. Phytochem.* **2018**, *7*, 1955–1960. Available online: https://www.phytojournal.com/archives/2018/vol7issue2/PartAB/7-2-24-402.pdf (accessed on 12 June 2025).
65. Kandsi, F.; Abdnim, R.; Benkhaira, N.; Zahra Lafdil, F.; Bnouham, M.; Yamani, B.; Naceiri Mrabti, H.; Wondmie, G.F.; Bin Jardan, Y.A.; Ibenmoussa, S.; et al. Integrated assessment of phytochemicals, Antilipase, hemoglobin antiglycation, antihyperglycemic, antifungal and antibacterial properties of *Vetiveria zizanioides* (L.) Nash. *Int. J. Food Prop.* **2024**, *27*, 1150–1166. [CrossRef]
66. Devi, S.V.; Kumar, K.A.; Maheswari, M.U.; Shanmugam, A.T.S.; Anand, R.S. In vitro antibacterial activity of ethanolic extract of *Vetiveria zizaniodes* root. *Int. J. Pharm. Sci. Res.* **2010**, *1*, 120–124. [CrossRef]
67. Gajic, I.; Kabic, J.; Kekic, D.; Jovicevic, M.; Milenkovic, M.; Mitic Culafic, D.; Trudic, A.; Ranin, L.; Opavski, N. Antimicrobial Susceptibility Testing: A Comprehensive Review of Currently Used Methods. *Antibiotics* **2022**, *11*, 427. [CrossRef]

68. Hulankova, R. Methods for Determination of Antimicrobial Activity of Essential Oils In Vitro-A Review. *Plants* **2024**, *13*, 2784. [CrossRef] [PubMed] [PubMed Central]
69. Oliveira, T.A.S.; Vieira, T.M.; Esperandim, V.R.; Martins, C.H.G.; Magalhães, L.G.; Miranda, M.L.D.; Crotti, A.E.M. Antibacterial, Antiparasitic, and Cytotoxic Activities of Chemical Characterized Essential Oil of *Chrysopogon zizanioides* Roots. *Pharmaceuticals* **2022**, *15*, 967. [CrossRef]
70. Dos Santos, D.S.; Oberger, J.V.; Niero, R.; Wagner, T.; Delle Monache, F.; Cruz, A.B.; Martin-Quintal, Z.; Cechinel Filho, V. Seasonal phytochemical study and antimicrobial potential of *Vetiveria zizanioides* roots. *Acta Pharm.* **2014**, *64*, 495–501. [CrossRef]
71. Balouiri, M.; Sadiki, M.; Ibnsouda, S.K. Methods for in vitro evaluating antimicrobial activity: A review. *J. Pharm. Anal.* **2016**, *6*, 71–79. [CrossRef]
72. Shinjyo, Y.; Midorikawa, N.; Matsumoto, T.; Sugaya, Y.; Ozawa, Y.; Oana, A.; Horie, C.; Yoshikawa, H.; Takahashi, Y.; Hasegawa, T.; et al. Analysis of cell death in *Bacillus subtilis* caused by sesquiterpenes from *Chrysopogon zizanioides* (L.) Roberty. *J. Gen. Appl. Microbiol.* **2022**, *68*, 62–70. [CrossRef] [PubMed]
73. Jha, V.; Parab, M.; Jain, T.; Gharat, Y.; Koli, J. Essential Oil of Chrysopogon Zizanioides Increases Membrane Permeability, Disturbs Cell Membrane Integrity, and Suppresses the Growth of Methicillin-Resistant Staphylococcus Aureus (MRSA). *J. Pharm. Drug Res.* **2023**, *6*, 638–649.
74. Nantachit, K.; Bunchoo, M.; Khantava, B.; Khamvan, C. Antimicrobial activity of alkaloid from roots of *Vetiveria zizanoides* (L.) Nash ex Small. *Thai Pharm. Health Sci. J.* **2010**, *5*, 99–102.
75. Devanathadesikan Seshadri, V.; Vijayaraghavan, P.; Kim, Y.O.; Kim, H.J.; Ahmed Al-Ghamdi, A.; Elshikh, M.S.; Al-Dosary, M.A.; Alsubaie, Q.D. In vitro antioxidant and cytotoxic activities of polyherbal extracts from *Vetiveria zizanioides*, *Trichosanthes cucumerina*, and *Mollugo cerviana* on HeLa and MCF-7 cell lines. *Saudi J. Biol. Sci.* **2020**, *27*, 1475–1481. [CrossRef] [PubMed]
76. Luqman, S.; Srivastava, S.; Darokar, M.P.; Khanuja, S.P. Detection of Antibacterial Activity in Spent Roots of Two Genotypes of Aromatic Grass *Vetiveria zizanioides*. *Pharm. Biol.* **2005**, *43*, 732–736. [CrossRef]
77. Burger, P.; Landreau, A.; Watson, M.; Janci, L.; Cassisa, V.; Kempf, M.; Azoulay, S.; Fernandez, X. Vetiver Essential Oil in Cosmetics: What Is New? *Medicines* **2017**, *4*, 41. [CrossRef] [PubMed] [PubMed Central]
78. Premjanu, N.; MR, R.K.; Srividya, S.S.; Malathy, B.R.; Revathy, R. Antibiofilm activity of ethanolic root extract of *Vetiveria zizanioides* against dental pathogens. *Curr. Trends Biotechnol. Pharm.* **2024**, *18*, 88–105. [CrossRef]
79. Ghosh, T.; Biswas, M.K.; Maity, D.; Dutta, P. Study of Antibacterial Activity of *Chrysopogon zizanioids* (vetiver) and its Anti-venom Potential Aspect. *Res. J. Agric. Sci.* **2020**, *11*, 889–894.
80. De Zoysa, M.H.N.; Rathnayake, H.; Hewawasam, R.P.; Wijayaratne, W.M.D.G.B. Determination of In Vitro Antimicrobial Activity of Five Sri Lankan Medicinal Plants against Selected Human Pathogenic Bacteria. *Int. J. Microbiol.* **2019**, *2019*, 7431439. [CrossRef]
81. Rathnayake, H.; De Zoysa, M.H.N.; Hewawasam, R.P.; Wijayaratne, W.M.D.G.B. Comparison of In Vitro Antibacterial Activity of *Epaltes divaricata* and *Vetiveria zizanioides* against Methicillin-Resistant Staphylococcus aureus. *Scientifica* **2020**, *2020*, 8239053. [CrossRef]
82. Orchard, A.; Viljoen, A.; van Vuuren, S. Wound Pathogens: Investigating Antimicrobial Activity of Commercial Essential Oil Combinations against Reference Strains. *Chem. Biodivers.* **2018**, *15*, e1800405. [CrossRef]
83. Seil, J.T.; Webster, T.J. Antimicrobial applications of nanotechnology: Methods and literature. *Int. J. Nanomed.* **2012**, *7*, 2767–2781. [CrossRef]
84. Gupta, N.; Rai, D.B.; Jangid, A.K.; Kulhari, H. Use of nanotechnology in antimicrobial therapy. In *Methods in Microbiology*; Academic Press: Cambridge, MA, USA, 2019; Volume 46, pp. 143–172. [CrossRef]
85. Ram, G.D.; Kumar, S.P.; Srinivasan, T.K.; Aravind, T.; Ramya, S.; Lingaraja, D.; Bhuvaneshwari, G. Green synthesis of silver nanoparticles using Chrysopogon zizanioides root extract and their antibacterial activities. *Mater. Today Proc.* **2023**, *in press*. [CrossRef]
86. He, X.; Gao, X.; Xie, W. Research Progress in Skin Aging, Metabolism, and Related Products. *Int. J. Mol. Sci.* **2023**, *24*, 15930. [CrossRef] [PubMed] [PubMed Central]
87. Polito, M.P.; Romaldini, A.; Rinaldo, S.; Enzo, E. Coordinating energy metabolism and signaling pathways in epithelial self-renewal and differentiation. *Biol. Direct.* **2024**, *19*, 63. [CrossRef]
88. Pfisterer, K.; Shaw, L.E.; Symmank, D.; Weninger, W. The Extracellular Matrix in Skin Inflammation and Infection. *Front. Cell Dev. Biol.* **2021**, *9*, 682414. [CrossRef] [PubMed] [PubMed Central]
89. De Tollenaere, M.; Chapuis, E.; Lapierre, L.; Bracq, M.; Hubert, J.; Lambert, C.; Sandré, J.; Auriol, D.; Scandolera, A.; Reynaud, R. Overall renewal of skin lipids with Vetiver extract for a complete anti-ageing strategy. *Int. J. Cosmet. Sci.* **2021**, *43*, 165–180. [CrossRef] [PubMed] [PubMed Central]
90. Han, X.; Parker, T.L. Biological activity of vetiver (*Vetiveria zizanioides*) essential oil in human dermal fibroblasts. *Cogent Med.* **2017**, *4*, 1298176. [CrossRef]

91. Zulkefli, N.; Che Zahari, C.N.M.; Sayuti, N.H.; Kamarudin, A.A.; Saad, N.; Hamezah, H.S.; Bunawan, H.; Baharum, S.N.; Mediani, A.; Ahmed, Q.U.; et al. Flavonoids as Potential Wound-Healing Molecules: Emphasis on Pathways Perspective. *Int. J. Mol. Sci.* **2023**, *24*, 4607. [CrossRef] [PubMed]
92. Arafat, M.; Khalil, M.; Mahrous, E.A.; El Hawary, S. Volatiles composition of fresh aroma and hydrodistilled volatile oil of *Chrysopogon zizanioides* roots growing in Egypt along with the cytotoxic activities of the hydrodistilled oil. *Egypt. J. Chem.* **2024**, *67*, 299–305. [CrossRef]
93. Hanifa, M.; Wulandari, R.; Zulfin, U.M.; Nugroho, E.P.; Haryanti, S.; Meiyanto, E. Different Cytotoxic Effects of Vetiver Oil on Three Types of Cancer Cells, Mainly Targeting CNR2 on TNBC. *Asian Pac. J. Cancer Prev.* **2022**, *23*, 241–251. [CrossRef]
94. Intan, C.J.; Mualifah, M.; Dwiranti, A.; Saifudin, S. Comparative study on sources of vetiver extracts (*Vetiveria zizanioides* L.) against HeLa cells growth using mitotic index (MI) analysis. *AIP Conf. Proc.* **2024**, *3163*, 070004. [CrossRef]
95. Saputra, L.O.; Fawzy, S.Y.; Wardani, R.K.; Lukitaningsih, E. The Exploration of Vetiver (*Vetiveria zizanioides*) as Co-Chemotherapy of Lung Cancer Selectively Targets AKR1C1: Bioinformatics Approach. *Indones. J. Cancer Chemoprevention* **2022**, *13*, 114–127. [CrossRef]
96. Sunitha, V.S.; Midhun, S.J.; Sunil, M.A.; Radhakrishnan, E.K.; Mathew, J. Valencene-rich fraction from *Vetiveria zizanioides* exerts immunostimulatory effects in vitro and in mice. *Asian Pac. J. Trop. Biomed.* **2021**, *11*, 335–343. [CrossRef]
97. Divya, P.V.; Sukesh, K. Cytotoxicity of *Amaranthus viridis* and *Vetiveria zizanioides* against human embryonic kidney cells. *J. Phytopharm.* **2025**, *14*, 10–13. [CrossRef]
98. Peng, H.Y.; Lai, C.C.; Lin, C.C.; Chou, S.T. Effect of *Vetiveria zizanioides* essential oil on melanogenesis in melanoma cells: Downregulation of tyrosinase expression and suppression of oxidative stress. *Sci. World J.* **2014**, *2014*, 213013. [CrossRef]
99. Sinha, S.; Jothiramajayam, M.; Ghosh, M.; Jana, A.; Chatterji, U.; Mukherjee, A. Vetiver oil (Java) attenuates cisplatin-induced oxidative stress, nephrotoxicity and myelosuppression in Swiss albino mice. *Food Chem. Toxicol. Int. J. Publ. Br. Ind. Biol. Res. Assoc.* **2015**, *81*, 120–128. [CrossRef] [PubMed]
100. Fadhillah, S.S.; Aini, A.Q.; Mualifah, S.; Bowolaksono, A.; Dwiranti, A. Effects of *Vetiveria zizanioides* on Cytotoxicity of Human Epithelial Cervical Cancer Cells. *J. Hunan Univ. Nat. Sci.* **2022**, *49*, 329–334.
101. Grover, M.; Behl, T.; Virmani, T.; Sanduja, M.; Makeen, H.A.; Albratty, M.; Alhazmi, H.A.; Meraya, A.M.; Bungau, S.G. Exploration of Cytotoxic Potential of Longifolene/Junipene Isolated from *Chrysopogon zizanioides*. *Molecules* **2022**, *27*, 5764. [CrossRef]

biomolecules

MDPI

Article

Metabolomic Profiling and In Vitro Evaluation of Cytotoxic, Genotoxic, and Antigenotoxic Effects of *Staphylea pinnata* L. Extract from Italian Flora

Ghanya Al-Naqeb [1,2,*], Fabio Pietrolucci [3], Mauro Commisso [3], Aliki Kalmpourtzidou [1], Amanda Oldani [4], Sara Boussetta [1], Beatrice Maccarini [1], Rachele De Giuseppe [1] and Hellas Cena [1,5]

1 Laboratory of Dietetics and Clinical Nutrition, Department of Public Health, Experimental and Forensic Medicine, University of Pavia, 27100 Pavia, Italy; aliki.kalmpourtzidou01@universitadipavia.it (A.K.); sara.boussetta01@universitadipavia.it (S.B.); beatrice.maccarini01@universitadipavia.it (B.M.); rachele.degiuseppe@unipv.it (R.D.G.); hellas.cena@unipv.it (H.C.)
2 Department of Food Sciences and Nutrition, Faculty of Agriculture Food and Environment, University of Sana'a, Sana'a P.O. Box 1247, Yemen
3 Department of Biotechnology, University of Verona, 37134 Verona, Italy; fabio.pietrolucci@univr.it (F.P.); mauro.commisso@univr.it (M.C.)
4 PASS-Bio Med, Centro Grandi Strumenti, University of Pavia, 27100 Pavia, Italy; amanda.oldani@unipv.it
5 Clinical Nutrition Unit, ICS Maugeri IRCCS, 27100 Pavia, Italy
* Correspondence: ghanya.al-naqeb@unipv.it

Abstract: *Staphylea pinnata* L., (*S. pinnata*), has long been recognized in Europe as both a wild food source and a traditional medicinal. This study aimed to characterize the metabolomic profile of the leaf extract of *S. pinnata* and assess its cytotoxic, genotoxic, and antigenotoxic effects in vitro for the first time. The methanolic extract of the leaves was analyzed using Ultra-Performance Liquid Chromatography–High-Resolution Mass Spectrometry (UPLC-HRMS). To evaluate its cytotoxic, genotoxic, and antigenotoxic properties, the cytokinesis block micronucleus assay was performed on Chinese hamster ovarian K1 cells. The analysis revealed a wide variety of metabolites in the extract, with B-type procyanidins and prodelphinidins being the most abundant. The genotoxicity of the extract varied depending on its concentration; at the lowest concentration (75 μg/mL), it showed no genotoxic effects and exhibited antigenotoxic properties by reducing the frequency of micronuclei induced by mitomycin C. However, at the highest concentration (300 μg/mL), the extract demonstrated genotoxic effects. In conclusion, the *S. pinnata* extract displayed both genotoxic and antigenotoxic properties, which may be attributed to its phytochemical composition. These findings highlight the complex nature of the plant's bioactive compounds, suggesting potential therapeutic applications with careful consideration of dosage. Additional research is necessary to understand the mechanisms underlying these properties.

Keywords: *Staphylea pinnata* L.; metabolomic profile; Italian flora; cytotoxicity; genotoxicity

1. Introduction

The only surviving species of the *Staphylea* genus that is indigenous to Central and Eastern Europe is *Staphylea pinnata* L., (*S. pinnata*), also referred to as the European bladdernut. It is a member of the family Staphyleaceae [1]. This species of plant is found throughout Italy [2]. Historically, *S. pinnata* has served as a wild food source in Europe, with its seeds and flowers being used in teas and pickling [2,3]. Additionally, its spring buds have been pickled or consumed as a side dish [3]. The roasted seeds of *S. pinnata* are also used in the production of liquor and schnapps [4]. The seed kernels of *S. pinnata* have shown to be a valuable source of many bioactive components including α-tocopherol, chlorophyll, lutein, and n-6 to n-3 fatty acids [5]. Research also suggests that *S. pinnata* seeds can be used as a source of edible oil [3]. Some studies reported the use of *S. pinnata* as a traditional medicine [6]. Extract from flowers and leaves of *S. pinnata* has shown different biological activities including, antioxidants [7], cytotoxic, and antibacterial properties [8].

Plant extracts and their constituents are widely regarded as valuable sources of bioactive compounds, in high demand across various industries, including food and pharmaceuticals [9]. However, their potential to induce genotoxic effects, combined with a lack of sufficient research into their beneficial properties, significantly limits their use [10]. While herbal medicines are often considered safe due to their long history of use, it is important to recognize that some plants utilized in traditional medicine have demonstrated genotoxicity [11,12]. The genotoxic or antigenotoxic effects of most plant extracts used in traditional medicine remain uncertain.

Some *S. pinnata* research focused on specialized metabolites such as polyphenols, flavonoids, and hydroxycinnamic derivatives, which may have antibacterial, antiproliferative, and antioxidant properties [13]. A previous study identified phytochemicals like quercetin and isorhamnetin glycosides in the hydroalcoholic extract of the aerial parts, flowers, and leaves of the *S. pinnata* plant [1]. Isoquercetin and quercetin malonyl glucoside exhibited potent antioxidant, antimicrobial, and wound healing properties, making them valuable as antiaging ingredients for cosmeceutical and therapeutic applications to heal skin wounds [1].

Quercetin and isorhamnetin were reported to reduce benzo[a]pyrene-induced genotoxicity [14]. Additionally, it has been reported that low doses of quercetin can reduce cytogenetic damage caused by mitomycin C (MMC) and hydrogen peroxide [15,16]. In contrast, it was observed that quercetin and its natural glycoside isoquercetin exhibited cytotoxicity and genotoxicity by inducing genetic instability and increasing micronuclei formation in the Chinese hamster ovary K1 (CHO-K1) rodent mammalian cell line [17]. The presence of these active metabolomics in *S. pinnata* extracts even as a minor constituent may contribute to its genotoxic or antigenotoxic potential. Information regarding the potential genotoxicity of *S. pinnata* is limited in the literature as the plant is mainly recognized for its decorative and intermittent culinary purposes rather than its medicinal or toxicological characteristics. Nevertheless, it is crucial to explore the genotoxicity and antigenotoxicity of this plant extract to guarantee its safety, since it is used in traditional medicine. The primary aim of this study was to characterize the metabolomic profile of the methanolic extract from *S. pinnata* leaves using Ultra-Performance Liquid Chromatography–High-Resolution Mass Spectrometry (UPLC-HRMS). Additionally, we assessed its cytotoxic, genotoxic, and antigenotoxic effects in CHO-K1 cells through the cytokinesis block micronucleus (CBMN) assay. The automated in vitro CBMN assay was performed using an image analysis approach, with analysis carried out via a widefield fluorescence microscope.

This study is conducted under the auspices of the Italian National Biodiversity Center. Our focus is on assessing the genotoxic and antigenotoxic properties of medicinal plants native to Italy, with *S. pinnata* being one of the species examined. A significant aspect of this research is the inaugural evaluation of the genotoxicity and antigenotoxicity of *S. pinnata*. The findings from this study will yield important preliminary information regarding the safety profile of this plant.

2. Materials and Methods

2.1. Chemicals

Cell culture media, supplements, consumables, dimethyl sulfoxide (DMSO; 99.9%), and formaldehyde solution were all acquired from Euroclone S.p.A. (Milan, Italy). Methanol (HPLC grade) was supplied by Merck KGaA (Darmstadt, Germany). Mitomycin C and cytochalasin B were obtained from D.B.A. Italia s.r.l. (Milan, Italy), while 3-(4,5-dimethylthiazol-2-yl) 2,5-diphenyltetrazolium bromide (MTT) was sourced from Biosigma S.p.A. (Cona, Italy). Hoechst 33342 Staining Dye Solution from Abcam, supplied by Prodotti Gianni s.r.l. (Milano, Italy).

2.2. Sample Collections and Extraction

Three different pools of leaves of *S. pinnata* were collected from plants cultivated at the botanical garden of Padua (Italy) in the full vegetative phase on 28 September 2022. Each pool included leaves picked up by two individuals. The samples were promptly frozen in liquid nitrogen and then ground into a fine powder using an A11 basic analytical mill (IKA-Werke, Staufen, Germany). They were subsequently stored at −80 °C in plastic centrifuge tubes wrapped in foil.

2.3. Preparation of S. pinnata Leaves Methanolic Extract

Prior to the chemical and biological activities, plant extracts were prepared as follows. About 1 g of frozen powder of *S. pinnata* leaves was extracted with 10 volumes (w/v) of 100% LC-MS grade methanol (Honeywell, Seelze, Germany). The samples underwent vortexing for 30 s, followed by sonication on ice for 10 min with a 40 kHz ultrasonic bath (SOL-TEC, Milano, Italy). Subsequently, samples were centrifuged at 14,000× g for 10 min at a temperature of 4 °C. The supernatants were then collected and preserved at −20 °C. For the assessment of genotoxicity, 20 g of frozen *S. pinnata* leaf powder was subjected to extraction with methanol at a 1:10 (w/v) ratio in sealed glass containers. The mixture was agitated using a magnetic stirrer for 48 h at an ambient temperature. After this period, the methanolic extracts were filtered through quantitative filter paper (ArtiGloss, with a particle retention range of 12–15 µm) and concentrated using rotary evaporation (BUCHI R-210) at 40 °C for 30 to 50 min. To ensure complete removal of methanol, the extract was placed in a fume hood for 24 h and subsequently freeze-dried with liquid nitrogen. The final extract was then transferred to amber glass bottles and stored at 4 °C for future analysis. The yield of the extract was calculated using the appropriate formula as outlined in Equation (1):

$$\text{Yield }\% = \frac{\text{weight of obtained extract(g)}}{\text{weight of dried plant sample used(g)}} \times 100. \tag{1}$$

This method resulted in a final yield of 2% for the methanolic extract of *S. pinnata*. The concentrated extracts were then placed into small glass tubes and stored at a temperature of 4 °C for future analyses.

2.4. Metabolomic Profile of S. pinnata Methanolic Extract

The hydro-alcoholic extracts were first diluted 1:10 with LC-MS grade water (Honeywell) and subsequently filtered through 0.22 µm Minisart filters (Sartorius-Stedim Biotech, Gottingen, Germany). The metabolomic profile was then obtained using Ultra-Performance Liquid Chromatography coupled with High-Resolution Mass Spectrometry (UPLC-HRMS) (Waters, Manchester, UK). The instrumental configuration and analytical procedures were adapted from a previously published protocol, with modifications [18]. Briefly, an AC-

QUITY I CLASS Ultra Performance Liquid Chromatography (UPLC) system, coupled with an ACQUITY Photodiode Array (PDA) detector, was interfaced with an electrospray ionization (ESI) source and a Xevo G2-XS quadrupole time-of-flight (qTOF) mass spectrometer (Waters, Manchester, UK). The PDA detector utilized an eλ detector with a spectral range of 190–800 nm and a high-sensitivity flow cell. Chromatographic separation was achieved using a Waters BEH C18 reversed-phase column (2.1 mm × 100 mm, 1.7 µm), coupled to a Vanguard pre-column (2.1 mm × 5 mm, 1.7 µm), with the column compartment maintained at 30 °C. The mobile phase consisted of solvent A (0.1% formic acid in LC-MS grade water) and solvent B (100% acetonitrile), with a flow rate set at 0.350 mL/min. The gradient elution started at 1% B, held for 1 min, and then increased linearly to 40% B at 10 min, 70% B at 13.5 min, 90% B at 15 min, and 99% B at 16.5 min. The gradient was then maintained at 99% B for 3.5 min before decreasing to 1% B at 20.1 min. The system remained in an isocratic mode (1% B) for the final 4.9 min, concluding the method at 25 min. To evaluate the stability of the machine, a mixture containing six distinct authentic standards at varying concentrations (chlorogenic acid, daidzin, and naringenin at 1 ng/µL; phenylalanine at 2 ng/µL; and gallic acid, quercetin, and dihydroartemisinic acid at 3 ng/µL) was analyzed at the beginning, in the middle, and at the end of the experiment. For metabolite identification, a further data-dependent analysis was performed by performing a FAST-DDA in negative ionization mode.

The metabolites were putatively identified using the accurate mass (derived from the m/z ratio), retention time, and fragmentation pattern (MS/MS data from FAST-DDA analysis) that were compared with a proprietary database of authentic reference compounds, an in silico library of plant metabolites, and the relevant literature. Public databases, such as MoNA (https://mona.fiehnlab.ucdavis.edu/, accessed on 18 October 2023) and MassBank (https://massbank.eu/MassBank/Search, accessed on 18 October 2023), were also consulted. Positive ionization mode analysis was conducted exclusively to confirm the molecular ions observed in negative ionization mode. The metabolite abundance was obtained using Masslynx 4.1 (Waters).

2.5. Cytotoxicity Determination and Dose Selection

In this study, CHO-K1 cells were utilized to investigate cytotoxicity and genotoxicity. These cells are commonly used in such research due to their rapid growth and stable karyotype [19,20]. Previous studies have demonstrated that CHO-K1 cells exhibit a 79% sensitivity to known carcinogenic substances [21]. The CHO-K1 cells (603480) were obtained from CLS Cell Lines Service GmbH (Eppelheim, Germany). The methanolic extract of *S. pinnata* leaves was prepared by dissolving it in DMSO to create a stock solution of 50 mg/mL. To improve solubility, the extract was sonicated and stored at 4 °C until further use. CHO-K1 cells were cultured in Ham's F12 medium, supplemented with 10% fetal bovine serum (FBS), 1% penicillin/streptomycin, and 1% glutamine. The cytotoxicity of *S. pinnata* extract on CHO-K1 cells was assessed using the MTT assay. Briefly, cells were seeded into 96-well plates (Primo® Multiwall plates 96 flat bottoms, ET3096, Euroclone S.p.A., Pero, Italy) at a density of 3000 cells per well and incubated for 24 h at 37 °C in a 5% CO_2 atmosphere. Following this, the cells were treated with various concentrations of *S. pinnata* extract, ranging from 0 to 1000 µg/mL in 2-fold increments, alongside a DMSO control (0.5%) and negative controls (cell medium) for an additional 24 h. After incubation, a 5 mg/mL MTT solution was introduced, and the cells were incubated for another 4 h at 37 °C in a 5% CO_2 environment. Subsequently, the reaction solution was discarded, and 150 µL of DMSO (100%) was added to dissolve the formazan crystals. Absorbance was measured at 570 and 690 nm using a microplate reader (Synergy), and the cell viability percentage was calculated using the following formula as outlined in Equation (2):

$$\% \text{ Viability of CHO} - \text{K1 cells} = \frac{\text{Mean absorbance of the sample}}{\text{Mean absorbance of the control}} \times 100 \quad (2)$$

2.6. Genotoxicity Study Using CBMN Assay

2.6.1. Genotoxic and Antigenotoxic Study

The in vitro micronucleus (MN) test was performed using the CBMN assay as reported [22] and following OECD guidelines [23,24]. The extracts of *S. pinnata* and MMC were both dissolved in DMSO, maintaining a final DMSO concentration of no more than 0.5%. CHO-K1 cells were seeded at a density of 3000 cells per well in a 96-well plate and were incubated in a humidified environment at 37 °C with 5% CO_2 for 24 h before the introduction of the test samples. Following this incubation period, CHO-KI cells were treated with a negative control (DMOS at 0.5%) MMC at two concentrations of 0.025 and 0.50 µg/mL and *S. pinnata* extract at three concentrations of 75, 150, and 300 µg/mL, both in the presence and absence of MMC at 0.025 µg/mL, to evaluate the extract's potential to mitigate the cytotoxic and genotoxic effects induced by MMC on CHO-K1 cells. After 24 h, the media was replaced, and the cells were treated with media containing 3 µg/mL of the cytokinesis-blocking agent Cytochalasin B (Cyto B) for another 24 h. The medium was then removed, and the cells were fixed with formaldehyde at a final concentration of 4% for 15 min. After discarding the fixation solution, the cells were washed twice with phosphate-buffered saline (PBS) for 3 min each. Following the washes, the PBS was removed, and the cells were stained with 100 µL per well of diluted bisbenzimide dye (Hoechst 33342 Staining Dye Solution) for 30 min in the dark at room temperature. The concentration range of *S. pinnata* extract (75, 150, and 300 µg/mL) was determined based on MTT assay results, which showed that the highest concentration caused cytotoxicity of less than 55 ± 5%, in accordance with OECD Guidelines. Concentrations above 300 µg/mL were found to be toxic to CHO-K1 cells, with cytotoxicity exceeding 50%. Therefore, 300 µg/mL was selected as the maximum concentration for assessing genotoxicity and antigenotoxicity. A concentration of 75 µg/mL was identified as non-toxic to CHO-K1 cells and was chosen as the minimum concentration for the study. Each concentration was tested in triplicate, with over 1000 binucleated cells scored for each concentration in every run. The experiment was repeated three times.

2.6.2. Cytotoxicity Evaluation Under CBMN Assay

In vitro tests frequently fail to identify genotoxic substances unless the tested amounts cause some level of cytotoxicity. Excessive cytotoxicity, on the other hand, may produce false-positive results and cause incorrect data interpretation. Therefore, at the maximum concentration tested, it is advised that the cytotoxicity level not exceed 55 ± 5% [23,24]. Considering these factors, it is recommended to assess a minimum of three different concentrations of the test substance in the CBMN assay [24]. To address this issue in our experiments, we evaluated cytotoxicity as an integral part of the CBMN procedure, using the same cells that were utilized for counting micronuclei for the tested three concentrations used for the CBMN assay and applying two methods to assess the cytotoxicity under the CBNM assay as reported recently [22]. The first method involved calculating the percentage of cytotoxicity, which was derived from estimating the decrease in cell count following treatment, as outlined in Equation (3):

$$\% \text{ Cytotoxicity} = \frac{100 \cdot [(\text{number of cells before treatment}) - (\text{number of cells after treatment})}{\text{number of untreated cells}} \quad (3)$$

Another method utilizes CBPI, which reflects the average number of cell cycles that take place during exposure to cytochalasin B. This index is associated with a decreased ratio of binucleated to mononucleated cells. The percentage of cytotoxicity was calculated according to the formulas provided in Equations (4) and (5).

$$\text{CBPI} = 100 \cdot \frac{[N_1 + 2 \times N_2 + 3 \times N_3]}{\text{total number of cells}} \quad (4)$$

$$\% \text{ Cytotoxicity(CBPI)} = 100 - \left[100 \times \frac{\text{CBPI of treated cells} - 1}{\text{CBPI of untreated cells} - 1}\right] \quad (5)$$

N1 = number of mononucleated cells, N2 = number of binucleated cells, and N3 = number of multinucleated cells.

2.6.3. Fluorescence Microscopic Image for MN Detection

Images were acquired using an inverted Leica DMi8 widefield microscope equipped with a Leica DFC9000GT CMOS camera and a 40× dry objective (Leica HC PL Fluotar L 40×/0.60), driven by LasX software (https://www.leica-microsystems.com/products/microscope-software/p/leica-las-x-ls/downloads/, accessed on 20 January 2025). The DAPI signal was recorded in 64 fields, uniformly distributed over the well area, for each condition, and each *S. pinnata* concentration was repeated in three wells so that at least 1000 binucleated cells per condition were analyzed. Each image was serially numbered and saved in the LAS X Office.

2.6.4. Scoring of Binucleated Cells and Micronuclei

An accelerated and more accurate assessment of the micronucleus assay can be accomplished by automating the analysis of micronuclei using image processing software. In our study, binucleated cells and micronuclei were automatically identified using CellProfiler software (version 4.2.6). CellProfiler is an open-source, modular application for image analysis, specifically designed to manage extensive volumes of images. It provides a flexible platform for image analysis experts to collaborate, experiment, and develop innovative methodologies [25]. The software operates on a pipeline framework, comprising individual modules that process images in diverse manners, organized sequentially to create a thorough workflow. Initially, images captured using the widefield microscope were enhanced with the Leica Lightning deconvolution tool to improve the signal-to-noise ratio, facilitating the identification of nuclei and micronuclei by CellProfiler. Following this, images from the same experimental conditions were analyzed simultaneously by CellProfiler, as recently reported [22].

2.7. Statistical Analysis

Statistical analysis was performed using GraphPad Prism v.7. The number of independent experiments and details on statistical comparisons and levels of significance can be found in the captions of the figures.

3. Results

3.1. Metabolomic Profile of Specialized Metabolites of S. pinnata Leaf Extract

An untargeted metabolomics approach was applied to analyze the leaf extract of *S. pinnata* using UPLC-HRMS. The chromatographic profiles of the samples, acquired in both negative and positive ionization modes, are presented in Figure 1. The numbers corresponding to the metabolites tentatively identified are shown in Table 1 and further details are reported in Supplementary Table S1.

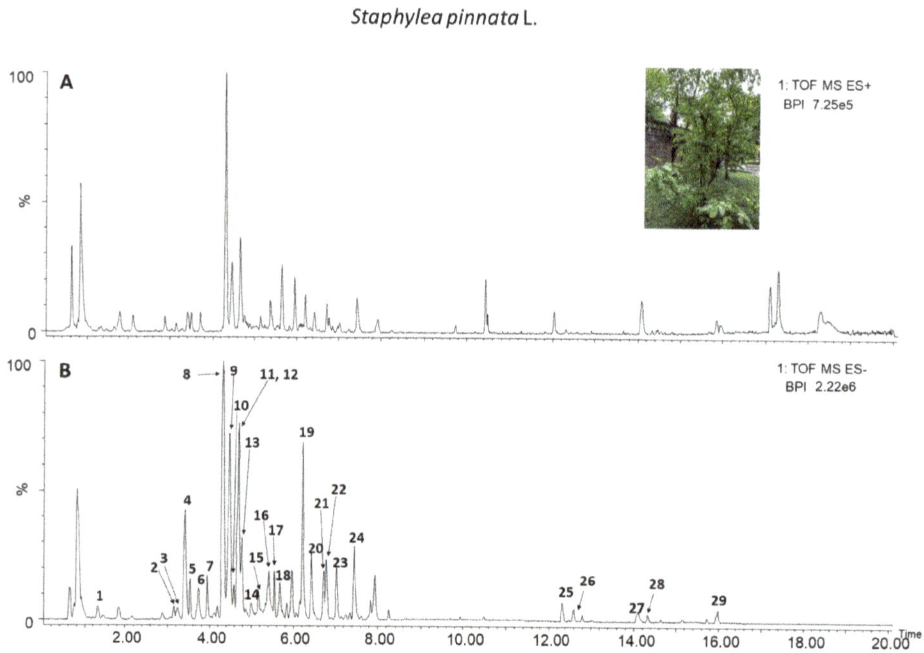

Figure 1. *S. pinnata* L. methanolic extract chromatograms. (**A**) Positive ionization mode; (**B**) negative ionization mode. The chromatograms show the main metabolites in *S. pinnata* methanol extracts.

As shown in Table 1, the highest value peak areas of the presented metabolites were for the procyanidin dimer 1, procyanidin trimer 3, procyanidin dimer 2, catechin, quercetin-o-rutinoside, and procyanidin tetramer 1, with peak areas of 84,368.54 ± 22,545.16, 76,013.76 ± 19,848.71, 40,872.21 ± 11,866.67, 32,270.00 ± 7888.78, and 28,883.55 ± 7503.36, respectively.

The phytochemical profile of *S. pinnata* leaf was strongly characterized by different B-type procyanidins and prodelphinidins. As shown in Figure 1, in the negative ionization mode, we identified three B-type procyanidins dimers (*m*/*z* 577.1346), four B-type procyanidins trimers (*m*/*z* 865.1980), one tetramer (*m*/*z* 1153.2614), and one B-type procyanidin pentamer (*m*/*z* 720.1582, double charges). The two prodelphinidins found in the extracts were a dimer (*m*/*z* 593.1295) and a trimer (*m*/*z* 881.1929). We also detected the presence of different O-glycosylated forms of two flavonols, quercetin and (iso)rhamnetin. The ending part of the chromatogram, where the chromatographic gradient reaches the highest acetonitrile levels, was characterized by the presence of different compounds belonging to the triterpenoid class. This group usually includes highly hydrophobic metabolites, mainly due to their chemical structure and elemental composition. In detail, we identified two isomers of maslinic acid (*m*/*z* 471.3474), asiatic acid and formic acid adducts (533.3478), and an oleanolic acid (*m*/*z* 455.3525). Other compounds putatively identified in *S. pinnata* extracts were citric acid, a dimeric lignin compound, piscidic acid ((p-Hydroxybenzyl) tartaric acid), a dihydroxybenzoic acid derivative, a benzyl alcohol-O-hexoside-pentoside, and esterification of caffeic acid with malic acid. Also, five unidentified compounds were observed. Information on the key characteristics of each numbered peak, including retention time, proposed annotation, experimental m/z, molecular formula, main adducts in both negative and positive acquisition modes, mass error in negative mode, and observed fragments in ms/ms are shown in Supplementary Table S1.

Table 1. Abundances (peak area—AU: arbitrary units).

Peak Number	Compound Name	Pool 1	Pool 2	Pool 3	Mean	STD
1	Citric acid	511.00	1116.00	574.00	733.67	297.49
2	Dimeric lignin (coniferyl alcohol derivative)	2428.54	1725.64	2077.01	2077.07	314.35
3	Prodelphinidin trimer 1	6983.25	7073.67	3991.68	6016.20	1568.71
4	Procyanidin trimer 1	32,232.76	28,374.76	16,314.96	25,640.83	7426.97
5	Piscidic acid	1922.00	3984.00	1991.00	2632.33	1047.45
6	Prodelphinidin dimer 1	9874.15	10,441.22	5545.04	8620.14	2395.42
7	Dihydroxybenzoic acid derivative	5822.43	4847.15	5390.92	5353.50	437.12
8	Procyanidin dimer 1	105,135.88	91,645.37	56,324.36	84,368.54	22,545.16
9	Procyanidin dimer 2	50,754.83	46,068.60	25,793.22	40,872.21	11,866.67
10	Procyanidin trimer 2	3683.00	3535.00	1618.00	2945.33	1030.28
11	Catechin	38,654.00	35,950.00	22,206.00	32,270.00	7888.78
12	Procyanidin trimer 3	91,430.89	86,030.83	50,579.56	76,013.76	19,848.71
13	Procyanidin tetramer 1	34,292.34	33,138.63	19,219.69	28,883.55	7503.36
14	Procyanidin pentamer 1	26,054.97	25,956.82	14,713.18	22,241.66	5831.70
15	Benzyl alcohol-O-hexoside-pentoside	6665.07	6029.68	6905.57	6533.44	404.76
16	Procyanidin trimer 4	22,770.24	21,524.11	12,379.30	18,891.22	5074.80
17	Caffeoyl malic acid	153.00	121.00	68.00	114.00	38.40
18	Procyanidin dimer 3	10,106.00	8839.00	4776.00	7907.00	2490.57
19	Quercetin-O-rutinoside	38,144.35	30,459.67	29,199.06	32,601.03	4330.69
20	Quercetin-O-hexoside	15,264.34	13,449.42	11,052.62	13,255.46	1889.52
21	Quercetin-O-malonylhexoside	11,796.04	9409.35	9075.77	10,093.72	1327.02
22	Isorhamnetin-O-rutinoside	18,728.16	16,116.04	15,764.19	16,869.47	1448.31
23	Isorhamnetin-O-hexoside	13,694.87	12,686.62	9526.32	11,969.27	1945.28
24	Isorhamnetin-O-malonylhexoside	22,373.07	18,442.05	16,459.41	19,091.51	2692.09
25	Asiatic acid	5178.89	4480.20	4386.19	4681.76	387.36
26	Asiatic acid isomer	4763.01	4240.51	4323.26	4442.26	251.19
27	Maslinic acid	21,692.20	20,709.16	20,638.08	21,013.14	526.95
28	Maslinic acid isomer	12,830.20	12,069.47	12,129.90	12,343.19	378.21
29	Oleanolic acid	2769.13	2637.53	2565.13	2657.26	92.50

3.2. Cytotoxicity Evaluation of S. pinnata Extract

Cytotoxicity is defined as the capacity of a substance to induce cell death [26]. To assess the cytotoxic impact of crude *S. pinnata* methanolic extract on CHO-K1 cells, the MTT cell viability assay was performed, with results illustrated in (Figure 2A,B). CHO-K1 cells incubated with *S. pinnata* extract for 24 h at concentrations below 125 µg/mL did not demonstrate significant cytotoxic effects when compared to both negative control (NC) and DMSO control cells. In contrast, at concentrations of 125, 250, 500, and 1000 µg/mL, a significant reduction in the viability of CHO-K1 cells was observed in comparison to the DMSO control cells. At the highest concentration tested, 1000 µg/mL, cell viability was reduced to 25% relative to both NCs and DMSO controls. The concentration of *S. pinnata* extract could not be increased beyond 1000 µg/mL due to its low solubility and its precipitation with the media. The extract exhibited an IC50 value of 325 µg/mL.

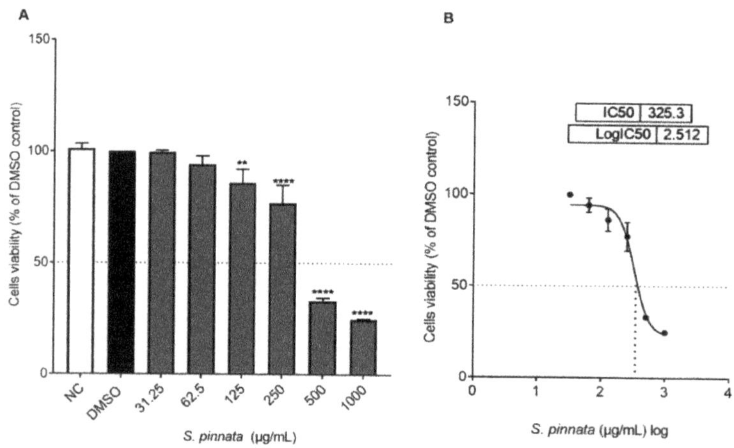

Figure 2. (**A**) The impact of different concentrations of *S. pinnata* extract (ranging from 0 to 1000 µg/mL) on the viability of CHO-K1 cells (% relative to DMSO control) after a 24 h treatment using the MTT assay. The figures illustrate the mean ± standard deviation derived from three independent experiments. Different symbols show statistically significant differences from the DMSO control ($p < 0.0001$). (**B**) A nonlinear regression analysis of log-transformed concentration values (curve fitting) was utilized to ascertain the IC_{50} value. Statistical evaluations were conducted using GraphPad Prism Version 7, employing ANOVA followed by the Tukey multiple comparison post-test (** $p = 0.0027$, **** $p < 0.0001$).

3.3. Genotoxicity Assessment Using CBMN Assay

3.3.1. Cytotoxicity and Binucleated Cells Evaluation

In our investigation, three different concentrations of *S. pinnata* extracts were utilized: 75, 150, and 300 µg/mL. This selection was based on the results of the MTT assay, which indicated an IC_{50} value of 325 µg/mL. Notably, the highest concentration of 300 µg/mL exhibited cytotoxicity levels below 55 ± 5% when compared to untreated control cells (DMSO control cells). To determine the genotoxic potential of *S. pinnata* extract, the initial focus was on assessing cytotoxicity. This evaluation was conducted as part of the MN experiment, utilizing the same cell lines for scoring the MN. Our results demonstrated that MMC, a well-established genotoxic clastogenic agent, led to an increase in cytotoxicity percentages, measured as % cytotoxicity CN, in a dose-dependent manner. Specifically, the percentages recorded were 15.67 ± 2.08% and 50.21 ± 4.65% at concentrations of 0.025 and 0.5 µg/mL, respectively, in comparison to NC cells (cells treated with DMSO at 0.4%) (Figure 3A).

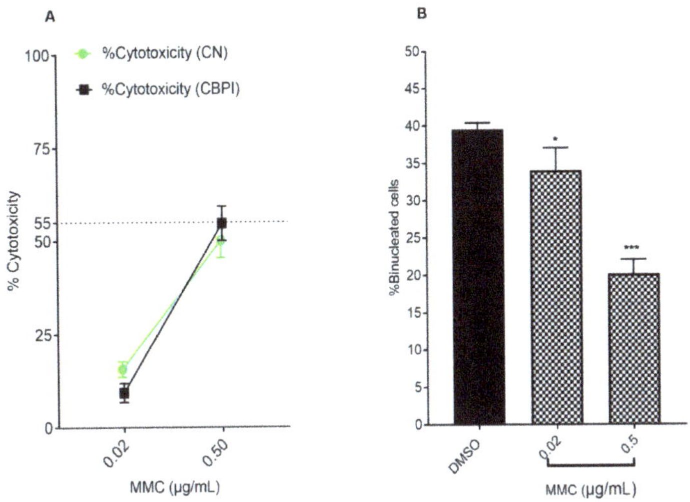

Figure 3. (**A**) The Cytotoxic effect of MMC on CHO-K1 cells [cytotoxicity (%) CN (green) and CBPI (black)] after treatment for 24 h, and additionally cells were incubated with Cyto B for another 24 h. (**B**) % of binucleated cells in CHO-K1 cells after 24 h incubation with DMSO control and different concentrations of MMC. The figures illustrate the mean ± standard deviation derived from three independent experiments. Different symbols show statistically significant differences from the NC using ANOVA followed by Tukey's multiple comparisons test in GraphPad Prism, with comparisons made to the NC group. (* $p = 0.0439$, *** $p < 0.001$).

With the CBPI method, MMC caused a significant increase in CBPI cytotoxicity percentages, ranging from 9.33 ± 2.51% to 54.69 ± 4.64% at concentrations of 0.025 and 0.5 μg/mL, respectively, compared to the NC cells (Figure 3A). To inhibit cytokinesis and promote the formation of binucleated post-mitotic cells, Cyto B was used to block actin assembly. In the NC group, CHO-K1 cells treated with 3 μg/mL of Cyto B for 24 h showed a mean binucleated cell percentage of 40.75 ± 2.29% (Figure 3B). In contrast, MMC-treated cells exhibited a significant reduction in binucleated cell formation, with percentages of 35.37 ± 1.53% and 20.16 ± 1.89% at MMC concentrations of 0.025 and 0.5 μg/mL, respectively, compared to the NC cells. Therefore, the MMC concentration of 0.025 μg/mL was selected for further experiments to assess the genotoxicity and antigenotoxicity of *S. pinnata* extracts, as it displayed lower cytotoxicity compared to the MMC at 0.5 μg/mL (Figure 3B).

The concentrations of *S. pinnata* applied, specifically 75, 150, and 300 μg/mL, resulted in cytotoxicity levels of 6.67 ± 1.53%, 12.67 ± 2.52%, and 32.33 ± 3.22%, respectively, as measured by the %CN method (Figure 4A). Conversely, the cytotoxicity assessed using the CBPI method indicated lower levels of cytotoxicity at 4.33 ± 1.53%, 11.33 ± 1.16%, and 24.15 ± 2.02% for the same concentrations of *S. pinnata* when compared to the NC cells (Figure 4A). When cells were treated with *S. pinnata* at a concentration of 75 μg/mL in the presence of MMC, a slight reduction in cytotoxicity was observed, yielding a value of 11.94 ± 1.0% as opposed to 15.65 ± 2.08% for MMC alone. Notably, an increase in cytotoxicity was recorded for cells treated with *S. pinnata* at concentrations of 150 and 300 μg/mL, showing values of 21.97 ± 4.55% and 39.54 ± 1.29%, respectively, in comparison to MMC-treated cells, which exhibited 15.67 ± 2.08% (Figure 4B). This indicates a significant cytotoxic effect of *S. pinnata* at higher concentrations, particularly at 300 μg/mL. In the presence of MMC, the cytotoxicity measured by CBPI revealed values of 10.34 ± 1.39%, 14.17 ± 2.05%, and 33.28 ± 3.23% for cells treated with MMC and *S. pinnata* at 75, 150, and 300 µg/mL, respectively, when compared to MMC-treated cells that showed 9.33 ± 2.52% (Figure 4B).

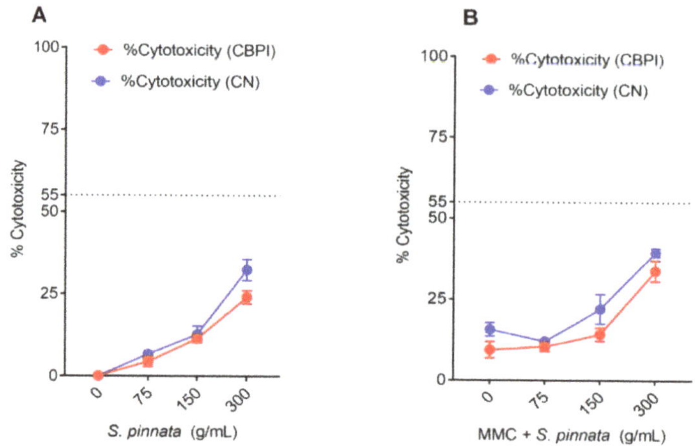

Figure 4. (**A**,**B**) The cytotoxic effect of *S. pinnata* extracts on CHO-K1 cells [cytotoxicity (%) CN (blue) and CBPI (red)] with or without 0.025 µg/mL of MMC for 24 h, with cells additionally incubated with Cyto B for another 24 h. Data were collected from three independent experiments.

No notable differences in the formation of binucleated cells were observed when treated with *S. pinnata* extracts at concentrations of 75 and 150 µg/mL compared to NC cells. However, at a concentration of 300 µg/mL, the formation of binucleated cells was significantly reduced compared to the NC cells. Binucleated cells formed were 41 ± 2.92%, 38 ± 1.44%, and 35 ± 2.10% for *S. pinnata* treated cells at 75, 150, and 300 µg/mL, respectively, while the NC exhibited a percentage of 40 ± 1.00% (Figure 5A).

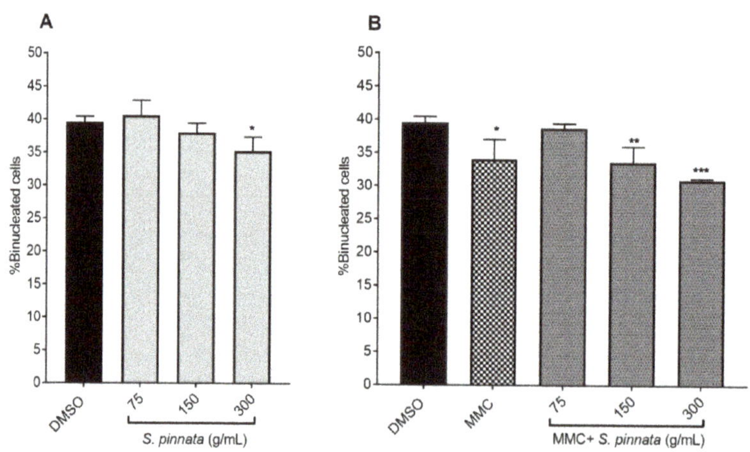

Figure 5. Formation of binucleated cells (%) in CHO-K1 cells treated with *S. pinnata* extract at three different concentrations without (**A**) or with (**B**) MMC at 0.025 µg/mL for 24 h and the cells subsequently incubated with 3 µg/mL Cyto B for 24 h. Data were collected from three independent experiments. Different symbols show statistically significant differences from the NC using ANOVA followed by Tukey's multiple comparisons test in GraphPad Prism, with comparisons made to the NC group. (* $p = 0.0118$, ** $p = 0.0078$, *** $p = 0.0006$).

Conversely, when cells were treated with *S. pinnata* extract alongside MMC, the percentage of binucleated cells was significantly lower in those treated with *S. pinnata* at concentrations of 150 and 300 µg/mL compared to the negative control group. The percentages of binucleated cells formed were 39 ± 1.01%, 34 ± 2.36%, and 30 ± 0.2% for *S. pinnata* treated cells at 75, 150, and 300 µg/mL, respectively, compared to the NC cells, which had a percentage of 40 ± 1.00% (Figure 5B).

3.3.2. Micronuclei Analysis

In our findings, the application of the MMC at a concentration of 0.025 µg/mL for 24 h on CHO-K1 cells led to a notable rise in the frequency of binucleated cells with micronuclei (MN frequency), measuring 4.15 ± 0.68%, compared to 0.72 ± 0.09% observed in NC cells (Figure 6). This concentration served as a positive control in this study, with the percentage of binucleated cells being close to that of NC cells.

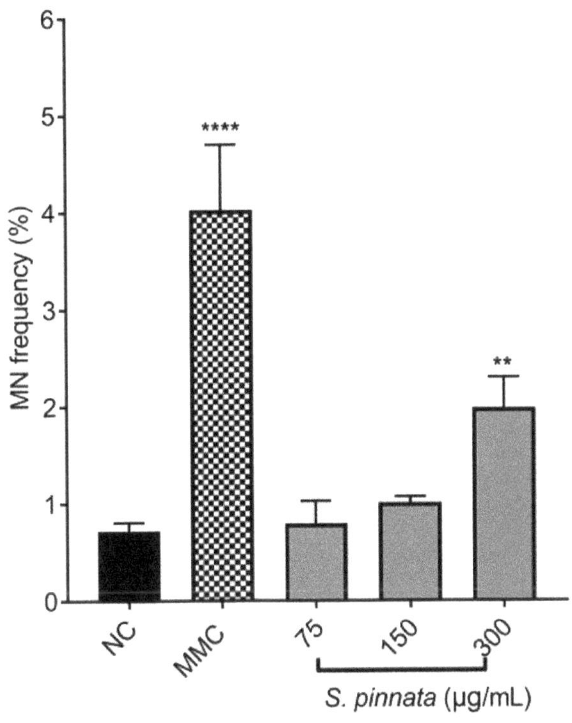

Figure 6. The effect of different *S. pinnata* extract treatments on the MN frequency (%) in CHO-K1 cells. Cells were treated for 24 h with 3 different concentrations of *S. pinnata* extract, then followed by 24 h incubation with 3 µg/mL of Cyto B. Graphs represent data collected from three independent experiments. One-way ANOVA and Tukey's multiple comparisons test using GraphPad Prism 7 software, were applied to calculate statistical significance in comparison with NC. (** $p = 0.0046$, **** $p < 0.0001$). MN frequency (%) was calculated from the following equation: micronuclei frequency (%) = binucleated cells with MN /binucleated cells *100.

Upon treatment with *S. pinnata* at concentrations of 75 and 150 µg/mL, no significant difference in the MN frequency was observed compared to NC cells (Figure 6). The MN frequency remained at 0.79 ± 0.2% and 1.01 ± 0.06% for 75 and 150 µg/mL *S. pinnata* treatments, respectively, in comparison to the NC value of 0.72 ± 0.09%. However, cells treated with *S. pinnata* at a concentration of 300 µg/mL showed a significant increase in MN frequency (1.99 ± 0.32%) compared to NC (0.72 ± 0.09%), indicating the genotoxic potential of *S. pinnata* extract treatment at the highest concentration and under the experimental conditions used. However, the MN frequency of *S. pinnata* extract (1.99 ± 0.32%) at 300 µg/mL was significantly lower than the MN frequency of MMC at 0.025 µg/mL (4.03 ± 0.68%), suggesting a weak genotoxic effect. Therefore, our study demonstrates a relationship between the induction of cytotoxicity and genotoxicity caused by *S. pinnata* extract at a concentration of 300 µg/mL. Images depicting the formation of micronuclei in binucleated CHO-K1 cells following a 24 h incubation with MMC at a concentration of 0.025 µg/mL and *S. pinnata* at the tested concentrations are presented in Figure 7.

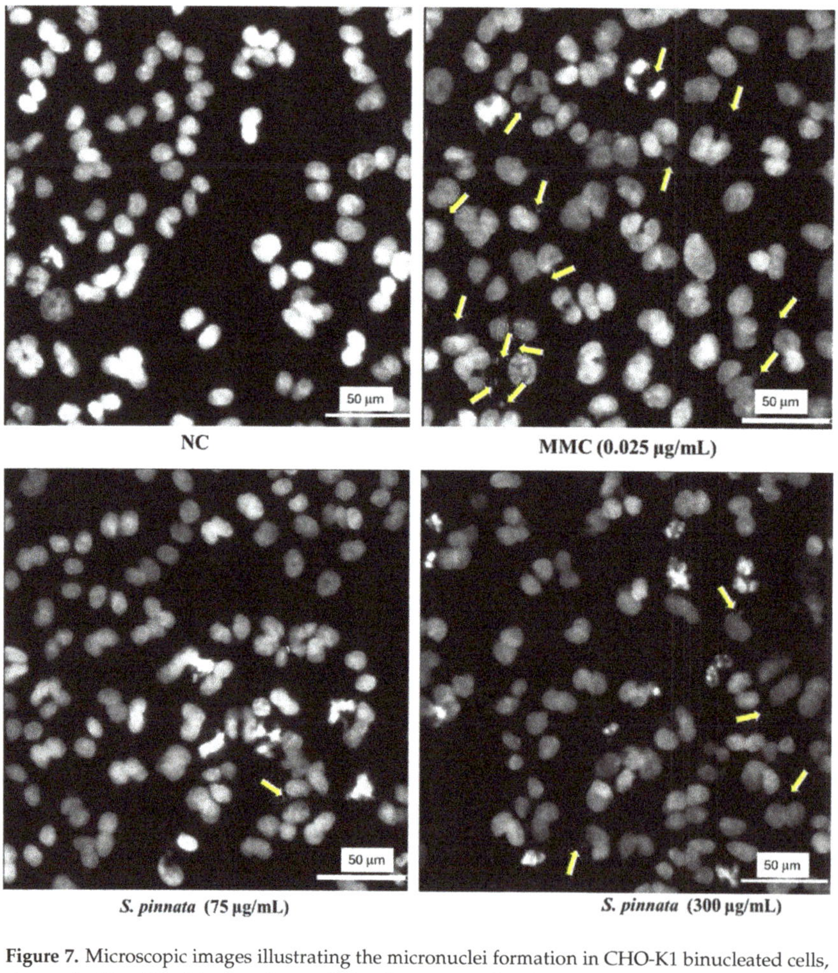

Figure 7. Microscopic images illustrating the micronuclei formation in CHO-K1 binucleated cells, observed with a 40× objective after a 24 h incubation period without treatment (NC) or treated with 0.025 µg/mL MMC or *S. pinnata* extract at 75 and 300 µg/mL. The DNA of CHO-K1 cells was stained using Hoechst dye. Yellow arrows indicate the micronuclei and the white line with a value of 50 µm in the image represents the scale bar.

3.3.3. Antigenotoxicity Determination

In a separate phase of the experiment, cells were treated with *S. pinnata* extract in combination with MMC. The MMC concentration of 0.025 μg/mL resulted in a significant increase in the MN frequency compared to the NC cells under the experimental conditions. The results showed a substantial reduction in MN frequency in cells treated with *S. pinnata* extract at a lower concentration of 75 μg/mL alongside MMC (2.58 ± 0.43%) when compared to cells treated with MMC alone (4.03 ± 0.68%) (Figure 8). In contrast, no significant differences in MN frequency were observed in cells treated with *S. pinnata* extract at 150 μg/mL in the presence of MMC (4.36 ± 0.32%) compared to those treated with MMC alone (4.03 ± 0.68%) (Figure 8).

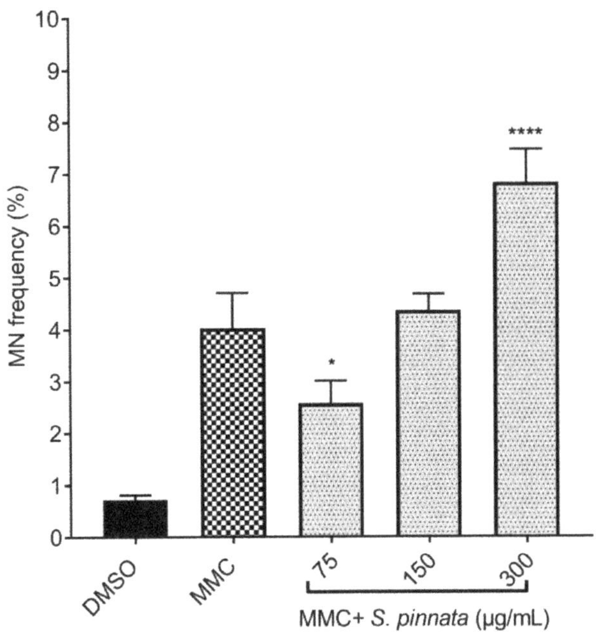

Figure 8. MN frequency (%) in CHO-K1 cells treated with MMC at 0.02 μg/mL alone or MMC + *S. pinnata* extract at three different concentrations of 57, 150, and 300 μg/mL for 24 h, followed by incubation with Cyto B at 3 μg/mL for another 24 h. The graphs display data from three independent experiments. Statistical significance was determined using one-way ANOVA followed by Tukey's multiple comparisons test, conducted with GraphPad Prism 7 software, comparing the results to the MMC control (* $p = 0.0135$, **** $p < 0.0001$).

Notably, cells treated with *S. pinnata* at 300 μg/mL in combination with MMC displayed a significant increase in MN frequency (6.83 ± 0.64%) compared to those treated with MMC alone (4.03 ± 0.68%) (Figure 8), suggesting that the *S. pinnata* extract enhances MMC-induced genotoxicity at this highest concentration. As a result, a correlation between the induction of both cytotoxicity and genotoxicity was observed in cells treated with *S. pinnata* extract at 300 μg/mL. Representative images of MN formation in binucleated CHO-K1 cells treated with *S. pinnata* extract and MMC at 0.025 μg/mL are presented in Figure 9.

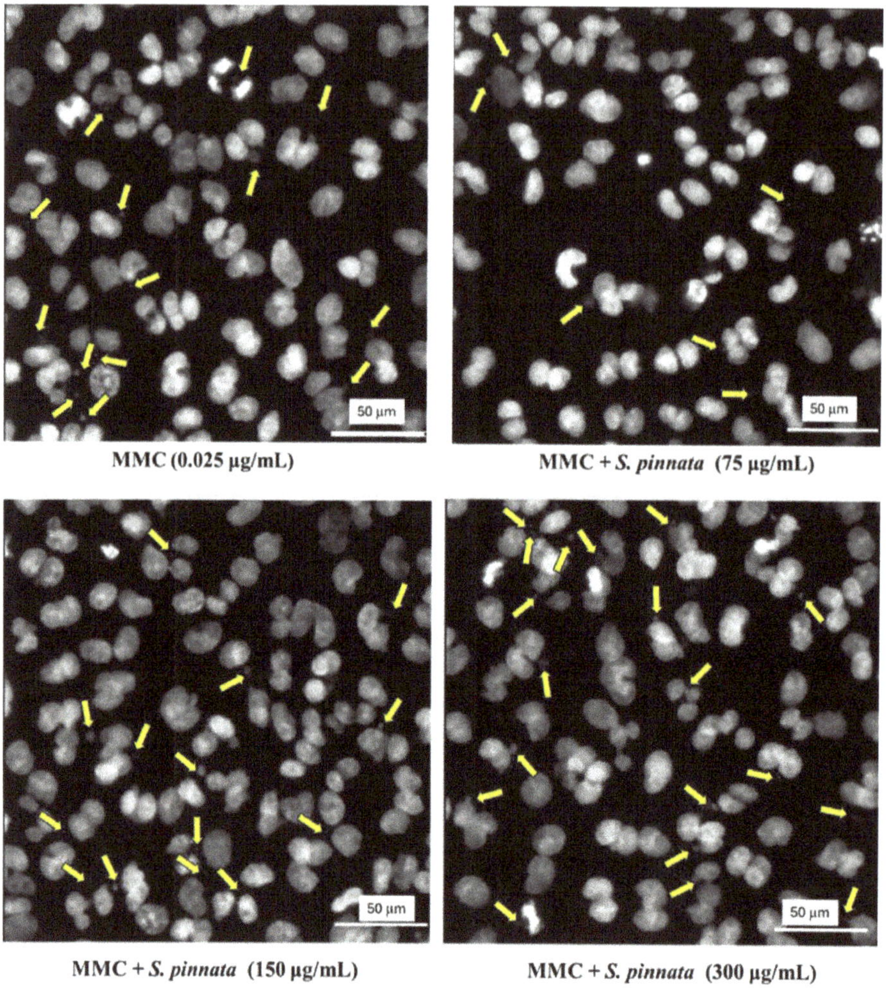

Figure 9. Representative microscopic images of MN formation in binucleated CHO-K1 cells with 40× objectives after 24 h incubation with 0.025 μg/mL MMC alone or MMC + *S. pinnata* extract at 75, 150, and 300 μg/mL. CHO-K1 cell DNA was stained with Hoechst dye. Yellow arrows indicate micronuclei and the white line with a value of 50 μm in the image represents the scale bar.

4. Discussion

While the safety and effectiveness of many plants utilized as traditional medicines for human consumption have not been comprehensively studied, individuals continue to rely on them due to their accessibility. To ensure the safety of these traditional medicinal plants, it is crucial to assess their potential cytotoxic and genotoxic effects, as this evaluation is a key factor in determining the efficacy of these remedies. Genotoxicity testing in vitro has gained increasing importance as a method for evaluating the safety of different drugs and chemicals. The formation of micronuclei acts as a biomarker for assessing the DNA-damaging potential of the substances under investigation. In this regard, the CBMN assay has become a widely used tool for detecting genotoxicity and identifying compounds that may reduce genotoxic and cytotoxic effects [27].

The biological activities exhibited by plant extracts, which consist of a variety of natural compounds, result from the unique roles of individual components and their interactions. In toxicological research, the chemical characterization of these extracts is essential, as it establishes a foundation for understanding the interactions among various compounds within biological systems. By determining the chemical properties and structures of a compound, researchers can evaluate its potential toxicity, which may include cytotoxic, genotoxic, and antigenotoxic effects. Therefore, this study aimed to characterize the metabolomic profile of the methanolic extract obtained from *S. pinnata* leaves using UPLC-HRMS and to examine its cytotoxic, genotoxic, and antigenotoxic effects on CHO-K1 cells through the CBMN assay.

The metabolomic investigation of *S. pinnata* indicated that the most abundant compounds are proanthocyanidins or condensed tannins, including various B-type procyanidins and prodelphinidins. Additionally, O-glycosylated variants of two flavonols, namely quercetin and (iso)rhamnetin, were also detected. Numerous tannins are acknowledged for their protective characteristics, which include anti-inflammatory, anti-fibrotic, antimicrobial, and anti-diabetic properties [28]. Procyanidins are generally accumulated in flowers, fruits, bark, and the seeds of various plants [29,30] and may exert significant health-promoting effects, such as antioxidant [31], antibacterial [32], anti-inflammatory [33], antineoplastic anti-allergic [34,35], lipid-lowering, and anti-obesity effects [36]. Because of these properties, they are widely recognized and employed in the healthcare industry [37].

In our initial experiments, we evaluated the cytotoxic effects of *S. pinnata* leaf extract at concentrations varying from 0 to 1000 μg/mL on CHO-K1 cells utilizing the MTT assay. The existing literature on the in vitro cytotoxicity of *S. pinnata* is limited, with a notable study by [8] demonstrating that water infusions of *S. pinnata* leaves administered to A431 cells yielded ED_{50} values of 35.3 ± 3.8 and 12.6 ± 1.9 μg/mL after 24 and 72 h of incubation, respectively, as assessed by the MTT assay. The cytotoxic effects observed were linked to the presence of various phytochemicals, including flavonoids, polyphenol triterpenes, and steroids [8]. Additionally, the crude hydroalcoholic extract of *S. pinnata* aerial parts, along with its fractions obtained through liquid–liquid extraction (LLE) and solid-phase extraction (SPE), were evaluated for their cytotoxicity against HeLa cells using the MTT assay. The findings indicated that the crude hydroalcoholic extract of *S. pinnata* at a concentration of 500 μg/mL for 24 h did not exhibit cytotoxicity towards HeLa and HaCaT cells. Conversely, both fractions of *S. pinnata* demonstrated significant toxicity against both cell lines at the same concentration [1]. As a result, the IC_{50} value obtained from this extract aligns with the range reported for other *S. pinnata* extracts in the literature.

The genotoxic effect of *S. pinnata* extract in the presence of MMC was evaluated. To our knowledge, this study is the first to explore the genotoxic and antigenotoxic characteristics of the methanolic extract derived from *S. pinnata* leaves. There is limited research on the genotoxic and antigenotoxic effects of *Staphylea* species, making it challenging to compare our findings with those of other *Staphylea* species. MMC has been recognized as genotoxic in both in vitro and in vivo testing involving mammalian cells and animal models, and it is widely acknowledged as a carcinogenic substance [38]. In our study, MMC was utilized as a positive control, validating the sensitivity of the CBMN assay and demonstrating a clear positive response at the concentration used. According to OECD guidelines, a definitive positive result for the substance tested is indicated when at least one concentration shows a statistically significant increase compared to the concurrent negative control under any experimental conditions. Consequently, our findings indicate that the methanolic leaf extract of *S. pinnata* exhibits genotoxic effects in CHO-K1 cells at the highest concentration tested, as confirmed by the in vitro CBMN assay.

It was noted that the extract from *S. pinnata* leaves exhibited contrasting effects that were dependent on the concentration applied, demonstrating a hormetic concentration-dependent response, characterized by the most pronounced effect observed at 300 μg/mL. Hormesis, as previously defined, refers to a phenomenon where a stressor agent yields beneficial effects at lower doses while producing adverse effects at higher doses, thus exhibiting varying effects based on the concentration tested [39]. Similar concentration-dependent antigenotoxic hormetic effects have been observed in various plant extracts. For instance, the stem extract of *Dendrobium speciosum* demonstrated protective properties against DNA damage induced by 4NQO in HePG2 cells at lower concentrations, as evaluated through the comet assay [40]. Furthermore, the lowest concentration of wastewater extracts from olive mill demonstrated an antigenotoxic hormetic effect when tested on HepaRG™ cells [41]. In addition, extract from *Gentiana lutea* at a lower tested dose reduced the genotoxicity induced by 2-amino-3-methyl-3H-imidazo[4,5-f]quinolone and 2-amino-1-methyl-6-phenylimidazo[4,5-b]pyridine (PhIP) in genetically modified Salmonella typhimurium TA1535, [42].

Genotoxicity investigations involving various plant extracts indicate that the observed genotoxic effects are contingent upon employed concentrations. For instance, *Hemidesmus indicus* (L.) R.Br. root ethanolic extract exhibited genotoxic effects at higher concentrations while demonstrating antigenotoxic properties at lower concentrations [43]. Conversely, both oil and water extracts from *Eugenia uniflora* L. showed mutagenic effects at lower concentrations but were classified as antimutagenic at elevated concentrations [44]. Moreover, manool extracted from the dried leaves of *Salvia officinal* L. was genotoxic at high employed doses, yet it conferred a protective effect against chromosome damage induced by methyl methanesulphonate at lower doses used [45]. Additionally, certain plant extracts, including *Artemisia vulgaris* L. and *Artemisia alba* Turra., displayed both genotoxic and antigenotoxic phytochemical effects at the same concentrations tested [46]. This phenomenon can be explained by the presence of complex mixtures of phytochemicals within the plant extracts, which may interact synergistically, while some phytochemicals may exhibit genotoxicity and others may mitigate these effects.

Plant extracts consist of a variety of metabolomic compounds, and their diverse biological activities arise from the distinct functions of individual components or their interactions [47]. Many studies have shown that various phenolic compounds display antioxidant or pro-oxidant properties that depend on their concentration [48]. Consequently, compounds may exhibit genotoxic or antigenotoxic protective effects when presented alone or in combination [46]. In our investigation, the genotoxic effect observed in treated cells with the highest concentration of *S. pinnata* may be attributed to the presence of certain specialized metabolites in the phytocomplex. Our results revealed the presence of O-glycosylated forms of two flavonols, namely quercetin and (iso)rhamnetin. Previous studies have shown that quercetin and its natural glycosides may exert various mutagenic and genotoxic effects in vitro [17,49] and in vivo [50].

The antigenotoxic effect observed for the extract of *S. pinnata* at its lowest concentration might be attributed to proanthocyanidins. In particular, the B-type procyanidins and prodelphinidines, as shown in Table 1, are metabolites reported to exert antioxidant activity, activate antioxidant cell machinery, and form complexes with other macromolecules. Previous studies, indeed, reported that proanthocyanidin-rich extracts derived from grape seeds did not exhibit any genotoxic effects in vivo [51,52] and significantly reduced DNA damages induced by H_2O_2 in Fao cells, as evidenced by the Comet assay [53]. Furthermore, proanthocyanidin-rich extracts from grape seeds resulted in a notable reduction in the frequency of micronuclei in vivo [19]. A rich fraction of proanthocyanidin obtained from the stem bark of *Stryphnodendron adstringens* was found to be non-genotoxic and showed an antigenotoxic effect against damages caused by cyclophosphamide [54]. Procyanidins from grape seeds activate antioxidative enzymes and the glutathione cycle in both in vivo and in vitro studies [55,56], thus contributing to the enhancement of cellular antioxidant capacity. Finally, the capacity of tannins to form complexes with various macromolecules might also contribute to protective activity through its antioxidants effect, as suggested in previous studies [57].

5. Conclusions

This study provides important insights into the metabolomic profile and potential cytotoxic, genotoxic, and antigenotoxic effects of the *S. pinnata* leaf extract belonging to Italian flora. Our findings show that the methanolic extract of *S. pinnata* leaf contains a wide variety of metabolites from several chemical classes, with the most abundant metabolites being different B-type procyanidins and prodelphinidins in addition to quercetin and (iso)rhamnetin (Table 1). To the best of our knowledge, this study is the first to investigate the genotoxic and antigenotoxic properties of the methanolic extract from *S. pinnata* leaves in vitro. Under the experimental conditions used in this study, our results indicate that the genotoxic effect of *S. pinnata* leaf extract is influenced by the concentration used. *S. pinnata* leaf methanolic extract demonstrates cytotoxic properties and shows both genotoxic and antigenotoxic effects. The levels of genotoxicity and cytotoxicity of *S. pinnata* extract were found to increase with higher concentrations used. The genotoxic and antigenotoxic effects of *S. pinnata* leaf extract are suggested to be linked to the phytochemical presence in the extracts, such as procyanidins and prodelphinidins, in addition to quercetin and (iso)rhamnetin derivatives as shown in Table 1. This study utilized crude extracts of *S. pinnata* leaves, with complex mixtures of biologically active compounds. Some of these compounds may exhibit cytotoxic and/or genotoxic effects, while others may possess cytoprotective and/or antigenotoxic properties. Consequently, this study indicates that despite the natural origin of the *S. pinnata* leaf extract, its indiscriminate use could lead to genotoxic and toxic effects, particularly at elevated concentrations. Therefore, careful consideration of dosage is essential for its safe consumption. Further investigations are necessary to assess the safety of *S. pinnata* leaf extracts using animal models and to elucidate the mechanisms underlying these effects.

Supplementary Materials: The following supporting information can be downloaded at: https://www.mdpi.com/article/10.3390/biom15091212/s1, Table S1: Putative identification of the metabolites presented in *S. pinnata* leaf methanol extracts.

Author Contributions: Conceptualization, G.A.-N., R.D.G. and H.C.; Methodology, G.A.-N., F.P. and M.C.; Software, G.A.-N., M.C. and A.O.; Validation, G.A.-N., M.C., R.D.G. and H.C; Formal Analysis, G.A.-N., M.C. and A.O.; Investigation, G.A.-N., F.P. and M.C.; Resources, H.C., M.C. and R.D.G.; Data Curation, G.A.-N., F.P. and M.C.; Writing—Original Draft Preparation, G.A.-N. and M.C. Writing—Review and Editing, G.A.-N., F.P., M.C., A.K., A.O., S.B., B.M., R.D.G. and H.C.; Visualization, G.A.-N., F.P., M.C., R.D.G. and H.C.; Project Administration, R.D.G. and H.C.; Funding Acquisition, H.C. All authors have read and agreed to the published version of the manuscript.

Funding: Project funded under the National Recovery and Resilience Plan (NRRP), Mission 4 Component 2 Investment 1.4—Call for tender No. 3138 of 16 December 2021, rectified by Decree n.3175 of 18 December 2021 of Italian Ministry of University and Research funded by the European Union—NextGenerationEU; Project code CN_00000033, Concession Decree No. 1034 of 17 June 2022 adopted by the Italian Ministry of University and Research, CUP F13C22000720007, Project title "National Biodiversity Future Center (NBFC)".

Institutional Review Board Statement: Not applicable.

Informed Consent Statement: Not applicable.

Data Availability Statement: Data are contained within the article or Supplementary Materials. The original contributions presented in this study are included in the article/Supplementary Materials. Further inquiries can be directed at the corresponding author.

Acknowledgments: The authors acknowledge Flavia Guzzo, Department of Biotechnology, University of Verona for providing the frozen plant materials, and Alessia Pascale Department of Drug Sciences, Section of Pharmacology, University of Pavia for providing the facilities to perform the in vitro cell work. The authors gratefully acknowledge Centro Grandi Strumenti and the microscope Facility for their assistance in this work.

Conflicts of Interest: The authors declare no conflicts of interest.

References

1. Paolillo, I.; Roscigno, G.; Innangi, M.; Zorrilla, J.G.; Petraglia, G.; Russo, M.T.; Carraturo, F.; Guida, M.; Pollice, A.; Cimmino, A.; et al. Health-Promoting Properties of Natural Flavonol Glycosides Isolated from *Staphylea pinnata* L. *Int. J. Mol. Sci.* **2024**, *25*, 5582. [CrossRef]
2. Madeja, J.; Harmata, K.; Kołaczek, P.; Karpińska-Kołaczek, M.; Piątek, K.; Naks, P. Bracken (*Pteridium aquilinum* (L.) Kuhn), mistletoe (*Viscum album* (L.)) and bladder-nut (*Staphylea pinnata* (L.))-mysterious plants with unusual applications. Culturaland ethnobotanical studies. In *Plants and Culture: Seeds of the Cultural Heritage of Europe*; Morel, J.-P., Mercuri, A.M., Eds.; Edipuglia: Santo Spirito, Italy, 2009; pp. 207–215.
3. Heiss, A.G.; FilipoviĆ, D.; Nedelcheva, A.; Ruß-Popa, G.; Wanninger, K.; Schramayr, G.; Perego, R.; Jacomet, S. A Fistful of Bladdernuts: The Shifting Uses of *Staphylea pinnata* L. as Documented by Archaeology, History, and Ethnology. *Folk Life* **2014**, *52*, 95–136. [CrossRef]
4. Redžić, S.; Ferrier, J. The Use of Wild Plants for Human Nutrition During a War: Eastern Bosnia (Western Balkans). In *Ethnobotany and Biocultural Diversities in the Balkans*; Pieroni, A., Quave, C., Eds.; Springer: New York, NY, USA, 2014; pp. 149–182.
5. Sircelj, H.; Vidrih, R.; Veberič, R.; Mikulič-Petkovšek, M. Evaluation of bioactive constituents in European bladdernut (*Staphylea pinnata* L.) seed kernels. *J. Food Compos. Anal.* **2019**, *78*, 33–41. [CrossRef]
6. Maia, A. Medicinal Remedies is Plant are Species of *Staphylea pinnata* L. and Family *Staphyleaceae* Lindl. in the Georgia (South Caucasus). *Agric. Res. Technol.* **2020**, *24*, 556255.
7. Lacikova, L.; Muselik, J.; Masterova, I.; Grancai, D. Antioxidant activity and total phenols in different extracts of four *Staphylea* L. Species. *Molecules* **2007**, *12*, 98–102. [CrossRef] [PubMed]
8. Lacikova, L.; Jancova, M.; Muselik, J.; Masterova, I.; Grancai, D.; Fickova, M. Antiproliferative, cytotoxic, antioxidant activity and polyphenols contents in leaves of four *Staphylea* L. species. *Molecules* **2009**, *14*, 3259–3267. [CrossRef]
9. Rutkowska, J.; Pasqualone, A. Plant Extracts as Functional Food Ingredients. *Foods* **2025**, *14*, 374. [CrossRef]
10. Al-Naqeb, G.; Kalmpourtzidou, A.; Giampieri, F.; De Giuseppe, R.; Cena, H. Genotoxic and antigenotoxic medicinal plant extracts and their main phytochemicals: "A review". *Front. Pharmacol.* **2024**, *15*, 1448731. [CrossRef]
11. De Quadros, A.P.O.; Oshiiwa, B.; Petreanu, M.; Niero, R.; Rosa, P.C.P.; Sawaya, A.C.H.F.; Mantovani, M.S.; O'Neill De Mascarenhas Gaivão, I.; Maistro, E.L. *Rubus rosifolius* (Rosaceae) stem extract induces cell injury and apoptosis in human hepatoma cell line. *Toxicol In Vitro* **2023**, *86*, e105485. [CrossRef]
12. Quadros Gomes, A.R.; da Rocha Galucio, N.C.; de Albuquerque, K.C.O.; Brígido, H.P.C.; Varela, E.L.P.; Castro, A.L.G.; Vale, V.V.; Bahia, M.O.; Rodriguez Burbano, R.M.; de Molfeta, F.A.; et al. Toxicity evaluation of *Eleutherine plicata* Herb. extracts and possible cell death mechanism. *Toxicol. Rep.* **2021**, *8*, 1480–1487. [CrossRef]
13. Jantová, S.; Nagy, M.; Ruzeková, L.; Grancai, D. Antibacterial activity of plant extracts from the families Fabaceae, Oleaceae, Philadelphaceae, Rosaceae and Staphyleaceae. *Phytother. Res.* **2000**, *14*, 601–603. [CrossRef] [PubMed]
14. Kim, M.; Jee, S.C.; Shin, M.K.; Han, D.H.; Bu, K.B.; Lee, S.C.; Jang, B.Y.; Sung, J.S. Quercetin and Isorhamnetin Reduce Benzo[a]pyrene-Induced Genotoxicity by Inducing RAD51 Expression through Downregulation of miR-34a. *Int. J. Mol. Sci.* **2022**, *23*, 13125. [CrossRef]
15. Musonda, C.A.; Chipman, J.K. Quercetin inhibits hydrogen peroxide (H2O2)-induced NF-kappaB DNA binding activity and DNA damage in HepG2 cells. *Carcinogenesis* **1998**, *19*, 1583–1589. [CrossRef] [PubMed]
16. Undeğer, U.; Aydin, S.; Başaran, A.A.; Başaran, N. The modulating effects of quercetin and rutin on the mitomycin C induced DNA damage. *Toxicol. Lett.* **2004**, *151*, 143–149. [CrossRef]

17. Engen, A.; Maeda, J.; Wozniak, D.E.; Brents, C.A.; Bell, J.J.; Uesaka, M.; Aizawa, Y.; Kato, T.A. Induction of cytotoxic and genotoxic responses by natural and novel quercetin glycosides. *Mutat. Res. Genet. Toxicol. Environ. Mutagen.* **2015**, *784–785*, 15–22. [CrossRef]
18. Zorzi, G.; Gambini, S.; Negri, S.; Guzzo, F.; Commisso, M. Untargeted Metabolomics Analysis of the Orchid Species *Oncidium sotoanum* Reveals the Presence of Rare Bioactive C-Diglycosylated Chrysin Derivatives. *Plants* **2023**, *12*, 655. [CrossRef]
19. Santos, G.S.; Tsutsumi, S.; Vieira, D.P.; Bartolini, P.; Okazaki, K. Effect of Brazilian propolis (AF-08) on genotoxicity, cytotoxicity and clonogenic death of Chinese hamster ovary (CHO-K1) cells irradiated with ^{60}Co gamma-radiation. *Mutat. Res. Genet. Toxicol. Environ. Mutagen.* **2014**, *762*, 17–23. [CrossRef]
20. El Hosry, L.; Di Giorgio, C.; Birer, C.; Habib, J.; Tueni, M.; Bun, S.S.; Herbette, G.; De Meo, M.; Ollivier, E.; Elias, R. In vitro cytotoxic and anticlastogenic activities of saxifragifolin B and cyclamin isolated from *Cyclamen persicum* and *Cyclamen libanoticum*. *Pharm. Biol.* **2014**, *52*, 1134–1140. [CrossRef] [PubMed]
21. Kirkland, D.; Aardema, M.; Henderson, L.; Müller, L. Evaluation of the ability of a battery of three in vitro genotoxicity tests to discriminate rodent carcinogens and non-carcinogens I. Sensitivity, specificity and relative predictivity. *Mutat. Res.* **2005**, *584*, 1–256. [CrossRef]
22. Al-Naqeb, G.; Zorzi, G.; Oldani, A.; Azzalin, A.; Avesani, L.; Guzzo, F.; Pascale, A.; De Giuseppe, R.; Cena, H. Phytochemical Profile and In Vitro Cytotoxic, Genotoxic, and Antigenotoxic Evaluation of *Cistus monspeliensis* L. Leaf Extract. *Int. J. Mol. Sci.* **2024**, *25*, 13707. [CrossRef]
23. Organization for Economic Co-operation and Development (OECD). Test No. 487: In Vitro Mammalian Cell Micronucleus Test. In *OECD Guidelines for the Testing of Chemicals;* Section 4; OECD Publishing: Paris, France, 2023.
24. Phelps, J.B.; Garriott, M.L.; Hoffman, W.P. A protocol for the in vitro micronucleus test. II. Contributions to the validation of a protocol suitable for regulatory submissions from an examination of 10 chemicals with different mechanisms of action and different levels of activity. *Mutat. Res.* **2002**, *521*, 103–112. [CrossRef]
25. Ramadhani, D.; Purnami, S. Automated Detection of Binucleated Cell and Micronuclei using CellProfiler 2.0 Software. *Hayati J. Biosci.* **2013**, *20*, 151–156. [CrossRef]
26. Salunke, M.; Wakure, B.; Wakte, P. HR-LCMS assisted phytochemical screening and an assessment of anticancer activity of Sargassum Squarrossum and Dictyota Dichotoma using in vitro and molecular docking approaches. *J. Mol. Struct.* **2022**, *1270*, 133833. [CrossRef]
27. Hashimoto, K.; Nakajima, Y.; Matsumura, S.; Chatani, F. Comparison of four different treatment conditions of extended ex-posure in the in vitro micronucleus assay using TK6 lymphoblastoid cells. *Regul. Toxicol. Pharmacol.* **2011**, *59*, 28–36. [CrossRef]
28. Rajasekar, N.; Sivanantham, A.; Ravikumar, V.; Rajasekaran, S. An overview on the role of plant-derived tannins for the treatment of lung cancer. *Phytochemistry* **2021**, *188*, 112799. [CrossRef] [PubMed]
29. Li, W.; He, Y.; Zhao, H.; Peng, L.; Li, J.; Rui, R.; Ju, S. Grape Seed Proanthocyanidin Ameliorates FB_1-Induced Meiotic Defects in Porcine Oocytes. *Toxins* **2021**, *13*, 841. [CrossRef] [PubMed]
30. Enomoto, T.; Nagasako-Akazome, Y.; Kanda, T.; Ikeda, M.; Dake, Y. Clinical effects of apple polyphenols on persistent allergic rhinitis: A randomized double-blind placebo-controlled parallel arm study. *Investig. Allergol. Clin. Immunol.* **2006**, *16*, 283–289.
31. Grace, M.H.; Xiong, J.; Esposito, D.; Ehlenfeldt, M.; Lila, M.A. Simultaneous LC-MS quantification of anthocyanins and non-anthocyanin phenolics from blueberries with widely divergent profiles and biological activities. *Food. Chem.* **2019**, *277*, 336–346. [CrossRef]
32. Nawrot-Hadzik, I.; Matkowski, A.; Hadzik, J.; Dobrowolska-Czopor, B.; Olchowy, C.; Dominiak, M.; Kubasiewicz-Ross, P. Proanthocyanidins and Flavan-3-Ols in the Prevention and Treatment of Periodontitis-Antibacterial Effects. *Nutrients* **2021**, *13*, 165. [CrossRef]
33. Tian, Y.; Yang, C.; Yao, Q.; Qian, L.; Liu, J.; Xie, X.; Ma, W.; Nie, X.; Lai, B.; Xiao, L.; et al. Procyanidin B2 Activates PPARγ to Induce M2 Polarization in Mouse Macrophages. *Front. Immunol.* **2019**, *10*, 1895. [CrossRef]
34. Choy, Y.Y.; Fraga, M.; Mackenzie, G.G.; Waterhouse, A.L.; Cremonini, E.; Oteiza, P.I. The PI3K/Akt pathway is involved in procyanidin-mediated suppression of human colorectal cancer cell growth. *Mol. Carcinog.* **2016**, *55*, 2196–2209. [CrossRef] [PubMed]
35. Sazwi, N.N.; Nalina, T.; Abdul Rahim, Z.H. Antioxidant and cytoprotective activities of *Piper betle, Areca catechu, Uncaria gambir* and betel quid with and without calcium hydroxide. *BMC Complement. Altern. Med.* **2013**, *13*, 351.
36. Wang, Q.Q.; Gao, H.; Yuan, R.; Han, S.; Li, X.X.; Tang, M.; Dong, B.; Li, J.X.; Zhao, L.C.; Feng, J.; et al. Procyanidin A2, a polyphenolic compound, exerts anti-inflammatory and anti-oxidative activity in lipopolysaccharide-stimulated RAW264.7 cells. *PLoS ONE* **2020**, *15*, e0237017. [CrossRef]
37. Lee, Y.A.; Cho, E.J.; Tanaka, T.; Yokozawa, T. Inhibitory activities of proanthocyanidins from persimmon against oxidative stress and digestive enzymes related to diabetes. *J. Nutr. Sci. Vitaminol.* **2007**, *53*, 287–292. [CrossRef]

38. Tang, J.; Li, B.; Hong, S.; Liu, C.; Min, J.; Hu, M.; Li, Y.; Liu, Y.; Hong, L. Punicalagin suppresses the proliferation and invasion of cervical cancer cells through inhibition of the β-catenin pathway. *Mol. Med. Rep.* **2017**, *16*, 1439–1444. [CrossRef]
39. Carneiro, C.C.; da Costa Santos, S.; de Souza Lino, R.; Bara, M.T., Jr.; Chaibub, B.A.; de Melo Reis, P.R.; Chaves, D.A.; da Silva, A.J.; Silva, L.S.; de Melo e Silva, D.; et al. Chemopreventive effect and angiogenic activity of punicalagin isolated from leaves of *Lafoensia pacari* A. St.-Hil. *Toxicol. Appl. Pharmacol.* **2016**, *310*, 1–8. [CrossRef] [PubMed]
40. Moretti, M.; Cossignani, L.; Messina, F.; Dominici, L.; Villarini, M.; Curini, M.; Marcotullio, M.C. Antigenotoxic effect, composition and antioxidant activity of *Dendrobium speciosum*. *Food Chem.* **2013**, *140*, 660–665. [CrossRef]
41. Rondini, T.; Branciari, R.; Franceschini, E.; Acito, M.; Fatigoni, C.; Roila, R.; Ranucci, D.; Villarini, M.; Galarini, R.; Moretti, M. Olive Mill Wastewater Extract: In Vitro Genotoxicity/Antigenotoxicity Assessment on HepaRG Cells. *Int. J. Environ. Res. Public Health* **2024**, *21*, 1050. [CrossRef]
42. Cvetković, S.; Todorović, S.; Nastasijević, B.; Mitić-Ćulafić, D.; Đukanović, S.; Knežević-Vukčević, J.; Nikolić, B. Assessment of genoprotective effects of *Gentiana lutea* extracts prepared from plants grown in field and in vitro. *Ind. Crops Prod.* **2020**, *154*, 112690. [CrossRef]
43. Ananthi, R.; Chandra, N.; Santhiya, S.T.; Ramesh, A. Genotoxic and antigenotoxic effects of *Hemidesmus indicus* R. Br. root extract in cultured lymphocytes. *J. Ethnopharmacol.* **2010**, *127*, 558–560. [CrossRef] [PubMed]
44. Kuhn, A.W.; Tedesco, M.; Laughinghouse, H.D.; Flores, F.C.; Silva, C.D.; Canto-Dorow, T.S.; Tedesco, S.B. Mutagenic and antimutagenic effects of *Eugenia uniflora* L. by the *Allium cepa* L. test. *Caryologia* **2015**, *68*, 25–30. [CrossRef]
45. Nicolella, H.D.; de Oliveira, P.F.; Munari, C.C.; Costa, G.F.; Moreira, M.R.; Veneziani, R.C.; Tavares, D.C. Differential effect of manool a diterpene from *Salvia officinalis*, on genotoxicity induced by methyl methanesulfonate in V79 and HepG2 cells. *Food Chem. Toxicol.* **2014**, *72*, 8–12. [CrossRef] [PubMed]
46. Jakovljević, M.R.; Grujičić, D.; Vukajlović, J.T.; Markovic, A.; Milutinović, M.G.; Stanković, M.S.; Vukovic, N.L.; Vukić, M.D.; Milošević-Djordjević, O. In vitro study of genotoxic and cytotoxic activities of methanol extracts of *Artemisia vulgaris* L. and *Artemisia alba* Turra. *S. Afr. J. Bot.* **2020**, *132*, 117–126. [CrossRef]
47. Vaou, N.; Stavropoulou, E.; Voidarou, C.C.; Tsakris, Z.; Rozos, G.; Tsigalou, C.; Bezirtzoglou, E. Interactions between Medical Plant-Derived Bioactive Compounds: Focus on Antimicrobial Combination Effects. *Antibiotics* **2022**, *11*, 1014. [CrossRef]
48. Azqueta, A.; Collins, A. Polyphenols and DNA Damage: A Mixed Blessing. *Nutrients* **2016**, *8*, 785. [CrossRef]
49. Utesch, D.; Feige, K.; Dasenbrock, J.; Broschard, T.H.; Harwood, M.; Danielewska-Nikiel, B.; Lines, T.C. Evaluation of the potential in vivo genotoxicity of quercetin. *Mutat. Res.* **2008**, *654*, 38–44. [CrossRef]
50. Hobbs, C.A.; Koyanagi, M.; Swartz, C.; Davis, J.; Kasamoto, S.; Maronpot, R.; Recio, L.; Hayashi, S.M. Comprehensive evaluation of the flavonol anti-oxidants, alpha-glycosyl isoquercitrin and isoquercitrin, for genotoxic potential. *Food Chem. Toxicol.* **2018**, *113*, 218–227. [CrossRef]
51. Yamakoshi, J.; Saito, M.; Kataoka, S.; Kikuchi, M. Safety evaluation of proanthocyanidin-rich extract from grape seeds. *Food Chem. Toxicol.* **2002**, *40*, 599–607. [CrossRef]
52. Erexson, G.L. Lack of in vivo clastogenic activity of grape seed and grape skin extracts in a mouse micronucleus assay. *Food Chem. Toxicol.* **2003**, *41*, 347–350. [CrossRef] [PubMed]
53. Llópiz, N.; Puiggròs, F.; Céspedes, E.; Arola, L.; Ardévol, A.; Bladé, C.; Salvadó, M.J. Antigenotoxic effect of grape seed procyanidin extract in Fao cells submitted to oxidative stress. *J. Agric. Food Chem.* **2004**, *52*, 1083–1087. [CrossRef] [PubMed]
54. Costa, M.A.; Ishida, K.; Kaplum, V.; Koslyk, E.D.; de Mello, J.C.; Ueda-Nakamura, T.; Dias Filho, B.P.; Nakamura, C.V. Safety evaluation of proanthocyanidin polymer-rich fraction obtained from stem bark of *Stryphnodendron adstringens* (BARBATIMAO) for use as a pharmacological agent. *Regul. Toxicol. Pharmacol.* **2010**, *58*, 330–335. [CrossRef]
55. Roig, R.; Cascón, E.; Arola, L.; Bladé, C.; Salvadó, M.J. Moderate wine consumption protects the rat against oxidation in vivo. *Life Sci.* **1999**, *64*, 1517–1524. [CrossRef] [PubMed]
56. Roig, R.; Cascón, E.; Arola, L.; Bladé, C.; Salvadó, M.J. Procyanidins protect Fao cells against hydrogen peroxide-induced oxidative stress. *Biochim. Biophys. Acta.* **2002**, *1572*, 25–30. [CrossRef]
57. Haslam, E. Natural polyphenols (vegetable tannins) as drugs: Possible modes of action. *J. Nat. Prod.* **1996**, *59*, 205–215. [CrossRef]

biomolecules

MDPI

Article

Biological Activities of Essential Oils and Hydrolates from Different Parts of Croatian Sea Fennel (*Crithmum maritimum* L.)

Livia Slišković [1], Nikolina Režić Mužinić [2], Olivera Politeo [3], Petra Brzović [4], Josip Tomaš [4], Ivana Generalić Mekinić [4,*] and Marijana Popović [5,*]

1 University Department for Forensic Sciences, University of Split, R. Boškovića 33, 21000 Split, Croatia; lsliskovic@forenzika.unist.hr

2 Department of Medical Chemistry and Biochemistry, School of Medicine, University of Split, Šoltanska 2a, 21000 Split, Croatia; nikolina.rezic.muzinic@mefst.hr

3 Department of Biochemistry, Faculty of Chemistry and Technology, University of Split, R. Boškovića 35, 21000 Split, Croatia; olivera@ktf-split.hr

4 Department of Food Technology and Biotechnology, Faculty of Chemistry and Technology, University of Split, R. Boškovića 35, 21000 Split, Croatia; petra.brzovic@ktf-split.hr (P.B.); jtomas@unist.hr (J.T.)

5 Department of Applied Science, Institute for Adriatic Crops and Karst Reclamation, Put Duilova 11, 21000 Split, Croatia

* Correspondence: gene@ktf-split.hr (I.G.M.); mpopovic@krs.hr (M.P.); Tel.: +385-21-329458 (I.G.M.); +385-21-434450 (M.P.)

Abstract: The traditional nutritional use of sea fennel (*Crithmum maritimum* L.) has been rediscovered and reestablished, making this halophyte plant a prominent ingredient in coastal cuisine and a subject of interest in various scientific disciplines, including pharmacy and medicine. The first objective of this study was to identify the volatile profiles of essential oils (EOs) and hydrolates derived from the leaves, flowers, and fruits of sea fennel using gas chromatography–mass spectrometry. A total of 25 different volatiles were identified in the EOs and 63 were identified in the hydrolates. Limonene was the most abundant component in the EOs (74.85%, 74.30%, and 67.41%, respectively), while in the hydrolates, it was terpinen-4-ol in the leaves (27.8%) and the flowers (36.7%) and (Z)-carveol in the fruits (11.4%). The second objective was to investigate the biological activities of the samples. The antioxidant and choline inhibitory activities of hydrolates were generally low, with the flower hydrolate providing the inhibition of both enzymes and the leaf hydrolate with the highest antiradical activity. The cytotoxic activities of the EOs and hydrolates were also investigated. The human breast adenocarcinoma cell line MDA-MB-23 was the most sensitive against EOs from the flowers and fruits, reaching the IC_{50} after 48 and 72 h, respectively. The leaf hydrolate exhibited cytotoxic activity after 72 h, while the flower hydrolate was effective after 48 h. The MCF-7 cell line was sensitive to the flower and fruit EOs, and the IC_{50} was reached at all the tested periods. Overall, the results highlight sea fennel as a rich source of bioactive compounds that have significant potential for greater utilization in the nutraceutical and pharmaceutical industries.

Keywords: *Crithmum maritimum* L.; essential oils; hydrolates; GC-MS; HS-SPME; antioxidant activities; acetylcholine inhibitory activity; butyrylcholine inhibitory activity; cytotoxic activities; multivariate analysis

1. Introduction

The increasing salinization of soils and waters due to climate change is driving interest in salt-tolerant crops like sea fennel. This emerging halophyte, with an exceptional resilience to harsh environmental conditions, has developed unique adaptations to saline environments, including the production of biologically active compounds that boost both its survival and nutritional profile [1,2]. Due to its potential as a "cash crop", sea fennel is gaining increasing attention as a sustainable agricultural option in coastal regions affected by soil deterioration and water pollution [3–5].

The use of sea fennel (*Crithmum maritimum* L.) dates back to ancient times, when coastal communities consumed it mainly out of necessity. Sea fennel leaves were particularly valued for their role in human nutrition, as they are rich in vitamin C and were used to prevent scurvy, which was a significant concern among sailors and maritime populations [6,7]. Today, however, it has been rediscovered and has rightfully earned its status as a star of coastal cuisine, but its application is also expanding into other sectors, such as cosmetics, food supplements, and the pharmaceutical industry [8]. The aerial parts of sea fennel contain a wealth of bioactive compounds, including essential oils (EOs), vitamins C and E, phenolics, carotenoids and essential fatty acids, essential amino acids, dietary fibers, and minerals, especially iodine. The distribution of metabolites varies among the different plant parts (leaves, flowers, stems, and fruits) [7,9–12].

The most significant difference among the sea fennel populations in the Mediterranean basin lies in the composition of their essential oils. The key distinguishing factor is the presence of the volatile compound dillapiole [13]. The Croatian sea fennel chemotype is completely free of dillapiole, which has been confirmed by various studies that have analyzed different parts of the plant, collected from various locations along the Adriatic coast and over different time periods [10,14,15].

Nowadays, we are witnessing a growing consumer demand for natural products and their derivatives. Due to their richness in bioactive compounds, essential oils (EOs) are increasingly being investigated for their potential as preservatives, flavorings, and therapeutic agents [16,17]. Hydrolates (aromatic waters, hydrosols) are another highly valued natural product obtained as a by-product of the EO distillation process, and they are increasingly attracting interest in fields such as aromatherapy, perfumery, and food flavoring [18,19]. Hydrolates are highly diluted mixtures of volatile compounds generated during the distillation and condensation of the oil and water vapors. Due to the immiscibility and differences in the densities of these two components, essential oils separate into the upper layer, while hydrolates, being denser, form the lower layer [20]. However, polar volatiles from EOs, mainly oxygenated compounds that are capable of forming hydrogen bonds, often migrate from the EOs into hydrolates, contributing to their aroma and properties. This is often the reason why the chemistry of EOs differs from that of their respective hydrolates [21,22]. In addition to the chemistry of EOs and hydrolates, the bioactivity assessments of these fractions, including their antioxidant, antimicrobial, and anti-inflammatory properties, have attracted the interest of scientists over the last few decades, especially due to their more frequent application in food/pharmaceutical/cosmetic products [23–26].

Although the EOs from sea fennel have been well characterized, there is little or no information on the corresponding hydrolates [21,22].

Due to the increasing demand for natural products and their wide range of applications, this study conducted a comprehensive chemical and biological analysis of sea fennel essential oils and hydrolates. To the best of our knowledge, the chemical composition of hydrolates derived from sea fennel flowers and fruits has not been previously reported. Likewise, no existing literature was found on the cytotoxic effects of sea fennel EOs or hydrolates against the MDA-MB-231, MCF-7, or OVCAR-3 cell lines, or on the acetylcholinesterase (AChE) and butyrylcholinesterase (BuChE) inhibitory activities of sea fennel hydrolates. To expand the knowledge on the potential health benefits of sea fennel, the authors investigated the volatile profiles of the leaf, flower, and fruit EOs and hydrolates, along with their antioxidant, cholinesterase-inhibitory, and cytotoxic effects.

2. Materials and Methods

2.1. Plant Material

Sea fennel (*Crithmum maritimum* L.) aerial parts were collected from the same location and increments in June (leaves), August (flowers), and November (fruits) 2023 in Split, Croatia (43.5147° N, 16.4435° E). The plant material was air-dried for two weeks at room temperature in a shaded and aerated place, and later used for the isolation of EOs. A voucher specimen of plant material was deposed in an herbarium at the Faculty of Chemistry and Technology (University of Split, Split, Croatia).

2.2. Extraction

2.2.1. Hydrodistillation of EOs

The EOs and hydrolates of sea fennel were extracted from 100 g of plant material by hydrodistillation (HD) in a modified Clevenger-type apparatus, as detailed by Popović et al. [27]. The distillation lasted 3 h and a mixture of pentane/diethyl ether (2:1, v/v) was used as a solvent trap for the EO's lipophilic components. The extracted EOs were dried over anhydrous sodium sulphate (Kemika, Zagreb, Croatia) and stored at −20 °C until further analysis. Likewise, the hydrolates, the water layer under the solvent trap, were separated and stored at −20 °C for subsequent analysis, as described by Politeo et al. [22].

2.2.2. Isolation of VOCs from Hydrolates

The isolation of VOCs from hydrolates was performed according to Politeo et al. [22] by solid-phase microextraction (SPME) fibers with a 50/30 μm DVB/CAR/PDMS coating purchased from Supelco (Sigma Aldrich, Bellefonte, PA, USA). Before use, the fibers were conditioned at 250 °C for 1 h. For sample preparation, the hydrolates and NaCl were placed in 20 mL glass vials, which were then sealed using magnetic caps with silicone septa. The SPME fibers were inserted into the vial headspace, and the samples were equilibrated at 40 °C for 30 min. Subsequently, the fibers were transferred to the GC injection port for thermal desorption, which was conducted at 250 °C for 3 min. Each sample analysis was performed in duplicate.

2.3. Identification of the Volatiles from EOs and Hydrolates by GC-MS

The identification and quantification of volatiles in the EOs and hydrolates of sea fennel were conducted using GC-MS. A gas chromatograph (model 8890 GC) equipped with an automatic liquid injector (model 7693A) and a tandem mass spectrometer (MS, model 7000D GC/TQ, Agilent Inc., Santa Clara, CA, USA) were used for the analysis. A non-polar HP-5MS UI column (30 m × 0.25 mm × 0.25 μm, Agilent Inc., Santa Clara, CA, USA) was used for chromatographic separation. Helium served as the carrier gas with a flow rate of 1.0 mL/min, with a sample injection volume of 1 μL. The analyses were performed in MS full-scan mode (33–350 m/z), the ion source temperature was set to 200 °C, the interface temperature to 250 °C, and the ionization voltage to 70 eV.

For the EO analysis, the column temperature was held at 60 °C for 3 min and then increased at a rate of 3 °C/min to 246 °C, where it was maintained isothermally for 25 min. For the hydrolate analysis, the oven temperature was initially held at 40 °C for 3 min and then increased to 80 °C at 3 °C/min, followed by a further increase to 220 °C at a rate of 10 °C/min, where it was held for 5 min.

Compound identification was based on comparing the retention indices with a series of n-hydrocarbons and matching the mass spectra with commercial databases (Wiley 7 MS library, Wiley, NY, USA, and NIST02, Gaithersburg, MD, USA) and the literature [28]. Quantification was performed by calculating the relative peak areas (%) using the normalization method (without correction factors). All the analyses were conducted in duplicate, and the percentages shown in Tables 1 and 2 represent the mean values.

Table 1. GC-MS analysis of sea fennel leaf, flower, and fruit essential oils.

No.	Compound	LRI	Leaves	Flowers	Fruits	Mode of Identification
1.	Hexanal	837	0.70 ± 0.03	nd	nd	GC, MS
2.	Acetic acid ethyl ester	842	1.41 ± 0.08	nd	nd	GC, MS
3.	Heptanal	898	0.41 ± 0.01	nd	nd	GC, MS
4.	α-Pinene	927	3.20 ± 0.02 a	1.01 ± 0.02 b	9.07 ± 0.04 c	GC, MS
5.	Camphene	941	nd	nd	0.18 ± 0.01	GC, MS
6.	Sabinene	966	nd	15.72 ± 0.22 a	12.37 ± 0.06 a	GC, MS
7.	β-Pinene	969	nd	nd	0.74 ± 0.00	GC, MS
8.	β-Myrcene	985	0.42 ± 0.01 a	1.04 ± 0.00 b	1.18 ± 0.02 b	GC, MS
9.	Octanal	999	0.49 ± 0.032 a	nd	0.15 ± 0.01 a	GC, MS
10.	α-Terpinene	1011	nd	0.52 ± 0.04 a	0.20 ± 0.00 a	GC, MS
11.	*o*-Cymene	1018	nd	0.35 ± 0.01 a	0.14 ± 0.01 b	GC, MS
12.	Limonene	1022	74.85 ± 0.03 a	74.30 ± 0.39 ab	68.41 ± 0.06 b	GC, MS
13.	β-Ocimene	1032	6.54 ± 0.04 a	3.72 ± 0.04 b	3.96 ± 0.02 b	GC, MS
14.	γ-Terpinene	1052	nd	1.23 ± 0.05 a	0.51 ± 0.03 b	GC, MS
15.	(*Z*)-Sabinene hydrate	1060	nd	nd	0.30 ± 0.01	GC, MS
16.	(*E*)-Sabinene hydrate	1092	nd	nd	0.20 ± 0.00	GC, MS
17.	(*Z*)-*p*-Mentha-2,8-dien-1-ol	1114	1.32 ± 0.01	nd	nd	GC, MS
18.	(*Z*)-Limonene oxide	1127	nd	nd	0.23 ± 0.00	GC, MS
19.	(*E*)-Limonene oxide	1131	nd	nd	0.26 ± 0.00	GC, MS
20.	Terpinen-4-ol	1171	nd	1.56 ± 0.01 a	0.66 ± 0.00 b	GC, MS
21.	Dodecane	1181	nd	nd	nd	GC, MS
22.	α-Terpineol	1184	nd	nd	0.37 ± 0.00	GC, MS
23.	Dodecane	1200	1.07 ± 0.03 a	0.30 ± 0.02 a	0.31 ± 0.01 a	GC, MS
24.	Tetradecane	1400	4.28 ± 0.25 a	0.25 ± 0.00 a	nd	GC, MS
25.	Hexadecane	1600	4.11 ± 0.08	nd	nd	GC, MS
		Aldehydes	1.60	0.00	0.15	
		Esters	1.41	0.00	0.00	
		Monoterpene hydrocarbons	85.01	97.89	96.76	
		Monoterpenoids	1.32	1.56	20.02	
		Alkanes	9.46	0.55	0.31	
	Total chromatogram identified (%)		98.80	100	99.24	

LRI—linear retention indices, nd—not detected. Identified compounds are expressed as mean ± SD. Different letters (a–c) denote statistically significant difference ($p < 0.05$).

Table 2. HS-SPME-GC-MS analysis of sea fennel leaf, flower, and fruit hydrolates.

No.	Compound	LRI	Leaves	Flowers	Fruits	Mode of Identification
1	α-Thujene	926	0.2 ± 0.00 a	nd	0.2 ± 0.05 a	GC, MS
2	(*E*)-2-Heptenal	949	nd	nd	0.1 ± 0.00	GC, MS
3	Benzaldehyde	952	1.00 ± 0.00 a	0.7 ± 0.00 b	0.1 ± 0.05 c	GC, MS
4	Sabinene	971	0.2 ± 0.00 a	0.3 ± 0.00 a	0.2 ± 0.00 a	GC, MS
5	α-Phellandrene	1001	1.00 ± 0.00 a	0.8 ± 0.00 b	1.1 ± 0.05 a	GC, MS
6	α-Terpinene	1012	0.45 ± 0.05 a	0.7 ± 0.00 b	0.4 ± 0.00 a	GC, MS
7	*p*-Cymene	1021	0.55 ± 0.05 a	nd	1.3 ± 0.05 b	GC, MS
8	β-Phellandrene	1025	0.85 ± 0.05 a	1 ± 0.05 a	0.4 ± 0.00 b	GC, MS
9	1,8-Cineole	1027	0.5 ± 0.4 a	nd	0.3 ± 0.1 a	GC, MS

Table 2. *Cont.*

No.	Compound	LRI	Leaves	Flowers	Fruits	Mode of Identification
10	Benzeneacetaldehyde	1040	0.2 ± 0.00 a	0.9 ± 0.05 b	0.1 ± 0.15 a	GC, MS
11	γ-Terpinene	1057	0.35 ± 0.05 a	0.6 ± 0.00 b	0.4 ± 0.00 a	GC, MS
12	(*Z*)-Sabinene hydrate	1067	2.45 ± 0.05 a	5.9 ± 0.15 b	2.5 ± 0.3 a	GC, MS
13	*p*-Cymenene	1090	0.35 ± 0.05 a	nd	0.2 ± 0.05 b	GC, MS
14	Methyl benzoate	1092	0.25 ± 0.05	nd	nd	GC, MS
15	(*E*)-Sabinene hydrate	1096	3.15 ± 0.05 a	7.4 ± 0.15 b	3 ± 0.15 a	GC, MS
16	Linalool	1098	0.6 ± 0.00 a	1.2 ± 0.00 a	4.6 ± 2.15 a	GC, MS
17	(*E*)-*p*-Mentha-2,8-dien-1-ol	1120	4.8 ± 0.00 a	5.1 ± 0.05 a	9 ± 0.05 b	GC, MS
18	α-Campholenal	1124	nd	nd	0.1 ± 0.05	GC, MS
19	Cosmene	1130	nd	nd	0.6 ± 0.00	GC, MS
20	(*Z*)-*p*-Mentha-2,8-dien-1-ol	1135	1.7 ± 0.00 a	1 ± 0.00 b	5.1 ± 0.25 c	GC, MS
21	(*E*)-Limonene oxide	1136	0.7 ± 0.00	nd	nd	GC, MS
22	(*Z*)-*p*-Menth-2-en-1-ol	1140	2.3 ± 0.00 a	2.7 ± 0.05 a	0.6 ± 0.05 b	GC, MS
23	Camphor	1141	nd	nd	0.1 ± 0.00	GC, MS
24	(*E*)-Verbenol	1145	0.25 ± 0.05 a	nd	0.3 ± 0.05 a	GC, MS
25	Sabina ketone	1157	0.5 ± 0.00 a	0.5 ± 0.00 a	0.4 ± 0.1 a	GC, MS
26	Borneol	1165	nd	0.4 ± 0.1 a	0.3 ± 0.00 a	GC, MS
27	(*Z*)-Chrysanthenol	1168	nd	nd	0.8 ± 0.1	GC, MS
28	Terpinen-4-ol	1178	27.8 ± 0.1 a	36.7 ± 0.1 b	10.3 ± 1.5 c	GC, MS
29	*p*-Methyl acetophenone	1183	0.4 ± 0.00	nd	nd	GC, MS
30	*p*-Cymen-8-ol	1184	0.4 ± 0.1	nd	nd	GC, MS
31	Cryptone	1185	nd	nd	0.3 ± 0.05	GC, MS
32	(*E*)-Isocarveol	1188	0.65 ± 0.05 a	nd	10.6 ± 0.35 b	GC, MS
33	α-Terpineol	1190	4.2 ± 0.2 a	5.8 ± 0.85 a	6.9 ± 0.35 a	GC, MS
34	Myrtenol	1194	1.3 ± 0.2 a	nd	2.4 ± 0.00 b	GC, MS
35	(*Z*)-Piperitol	1195	nd	1 ± 0.1 a	0.2 ± 0.00 b	GC, MS
36	Dihydrocarvone	1197	1.4 ± 0.2	nd	nd	GC, MS
37	(*E*)-Isopiperitenol	1199	nd	0.7 ± 0.00 a	1.9 ± 0.25 b	GC, MS
38	(*E*)-Piperitol	1207	1.6 ± 0.00 a	1.4 ± 0.05 a	0.4 ± 0.05 b	GC, MS
39	(*Z*)-Isopiperitenol	1218	nd	nd	1.6 ± 0.65	GC, MS
40	(*E*)-Carveol	1220	7.3 ± 0.2 a	3.6 ± 0.00 b	4.2 ± 0.3 b	GC, MS
41	(*Z*)-Carveol	1232	3.7 ± 0.1 a	1.00 ± 0.00 b	11.4 ± 0.4 c	GC, MS
42	Carvone	1244	4.95 ± 0.05 a	0.5 ± 0.00 b	3.7 ± 0.15 c	GC, MS
43	Carvotanacetone	1251	0.4 ± 0.1	nd	nd	GC, MS
44	(*Z*)-Chrysanthenyl acetate	1256	nd	0.2 ± 0.05 a	0.1 ± 0.05 a	GC, MS
45	Ionene	1257	0.25 ± 0.15	nd	nd	GC, MS
46	Geraniol	1258	nd	nd	1.2 ± 0.45	GC, MS
47	Perilla aldehyde	1273	nd	nd	0.2 ± 0.00	GC, MS
48	Isobornyl acetate	1285	nd	nd	0.1 ± 0.00	GC, MS
49	*p*-Cymen-7-ol	1290	1.25 ± 0.05 a	0.8 ± 0.05 b	nd	GC, MS
50	Perilla alcohol	1290	nd	0.2 ± 0.00 a	1.3 ± 0.15 b	GC, MS
51	*p*-Menth-1-en-9-ol	1296	nd	nd	0.5 ± 0.00	GC, MS
52	Carvacrol	1299	0.4 ± 0.00	nd	nd	GC, MS
53	(*E*)-Cinnamyl alcohol	1303	0.25 ± 0.05	nd	nd	GC, MS
54	*p*-Mentha-1,4-dien-7-ol	1330	nd	0.9 ± 0.00 a	0.3 ± 0.15 a	GC, MS
55	Myrtenyl acetate	1334	0.95 ± 0.15	nd	nd	GC, MS
56	*β*-Damascenone	1384	1.7 ± 0.1 a	0.3 ± 0.05 b	nd	GC, MS
57	Geranyl acetate	1385	nd	nd	0.2 ± 0.05	GC, MS
58	10-(Acetylmethyl)-3-carene	1387	8.6 ± 0.5	nd	nd	GC, MS
59	(*E*)-*β*-ionone	1389	0.2 ± 0.00	nd	nd	GC, MS
60	Spathulenol	1578	0.95 ± 0.75 a	8.3 ± 0.9 b	1.8 ± 0.1 a	GC, MS
61	Dillapiol	1626	0.2 ± 0.00 a	0.5 ± 0.1 a	0.3 ± 0.00 a	GC, MS
62	Isospathulenol	1640	nd	2 ± 0.25	nd	GC, MS

Table 2. *Cont.*

No.	Compound	LRI	Leaves	Flowers	Fruits	Mode of Identification
63	α-Bisabolol	1684	nd	0.4 ± 0.05 a	1.1 ± 0.45 a	GC, MS
	Aldehydes		1.2	1.6	0.3	
	Ketones		0.4	0	0	
	Esters		0.25	0	0	
	Phenols		0.4	0	0	
	Phenylpropanoids		0.2	0.5	0.3	
	Monoterpene hydrocarbons		12.55	3.4	4.2	
	Monoterpenoids		75.25	77.3	87.4	
	Sesquiterpenes		0	0	0.6	
	Sesquiterpenoid hydrocarbons		0.95	10.7	2.9	
	Total chromatogram identified (%)		91.2	93.5	95.7	

LRI—linear retention indices, nd—not detected. Identified compounds are expressed as mean ± SD. Different letters (a–c) denote statistically significant difference ($p < 0.05$).

2.4. Antioxidant Activities

The antioxidant assays were performed in 96-well plates using a Tecan MicroPlate Reader, model Sunrise (Tecan Group Ltd., Männedorf, Switzerland), and a Synergy HTX Multi-Mode Reader (BioTek Instruments, Inc., Winooski, VT, USA). All the measurements were carried out in triplicate.

The reducing activity of the samples was evaluated using the Ferric reducing antioxidant power (FRAP) assay, following the protocol by Čagalj et al. [29]. The procedure involves adding samples (10 µL) to a FRAP reagent mixture (300 µL, acetate buffer (0.3 M)/TPTZ (10 mM) in HCl (40 mM)/$FeCl_3$ (20 mM) = 10:1:1) and measuring the absorbance at 593 nm after 4 min. The results were compared with a calibration curve prepared using different concentrations of Fe^{2+} and expressed as µM of Fe^{2+}.

An oxygen radical absorbance capacity (ORAC) assay was conducted as described by Čagalj et al. [29]. The reagent mixture of fluorescein (150 µL, 0.08 mM in 0.075 M phosphate buffer) and the sample (25 µL) (or Trolox for the calibration curve/buffer for blank measurements) was added into microplate wells and thermostated (30 min at 37 °C). The reaction, initiated by adding 2,2′-azobis(2-methylpropionamidine) dihydrochloride (25 µL), was monitored for 80 min at the excitation and emission wavelengths of 485 and 520 nm, with measurements taken every minute. The results are expressed in micromolar Trolox equivalents (µM TE), calculated from the calibration curve obtained by testing different concentrations of Trolox.

The 2,2-diphenyl-1-picrylhydrazyl (DPPH) radical scavenging activity of the samples was assessed by following the method of Čagalj et al. [29]. In this assay, the decrease in absorbance of a DPPH ethanolic solution (290 µL, initial absorbance of 1.2 ± 0.02) after the addition of the sample (10 µL) was measured at 517 nm after 1 h. The results are expressed as the percentage inhibition of the DPPH radical (inhibition %), calculated according to the following equation:

$$\text{Inhibition } \% = [(Ac - As)/Ac] \times 100,$$

where Ac is the initial absorbance of the DPPH solution and As is the absorbance of the DPPH solution containing the sample 1 h after its addition.

2.5. Cholinesterase Inhibitory Activity

The samples' inhibition capacity of AChE and BuChE was examined by modified Ellman's assays, as described by Bektašević et al. [30]. The reaction mixture consisted of phosphate buffer (0.1 M, pH = 8), 10 μL of DTNB (2-nitro-5-thiobenzoate, 6.6 mM, dissolved in phosphate buffer, pH = 7), 10 μL of the sample solution or distilled water as a negative control, and 10 μL of enzyme solutions (AChE, BuChE, 0.66 U/mL). Acetylthiocholine (ATChI) or butyrylthiocholine iodide (BTChI) (10 μL, 11 mM, prepared in phosphate buffer, pH = 8), was used to initialize the reaction. Non-enzymatic hydrolysis was also monitored by measuring blank runs. All the spectrophotometric measurements were carried out at 405 nm during 6 min and performed in quadruplicate. Finally, the inhibition, expressed as a percentage, was calculated by the following formula:

$$\text{Inhibition (\%)} = ((\text{Ae} - \text{Abe}) - (\text{Au} - \text{Abu}))/((\text{Ae} - \text{Abe})) \times 100,$$

where Ae represents the absorbance of the enzyme solution without the sample (inhibitor), Abe is the absorbance of the blank sample of the enzyme solution without the enzyme, Au is the absorbance of the enzyme solution with the sample (inhibitor), and Abu is the absorbance of the blank sample of the enzyme solution without the sample (inhibitor).

2.6. Cytotoxic Activities

The cytotoxic activity of sea fennel EOs and hydrolates was determined by a cell viability assay (gold standard) using 3-(4,5-dimethylthiazol-2-yl)-2,5-diphenyltetrazolium bromide (MTT). The MTT assay was performed on four cell lines: the human breast adenocarcinoma MDA-MB-231 cell line, the human breast metastatic adenocarcinoma MCF-7 cell line, the human ovarian carcinoma OVCAR-3 cell line (LGC standards), and the human embryonic kidney HEK-293 cell line (ATCC Cell Lines), according to Režić Mužinić et al. [31]. The MDA-MB-231, MCF-7, OVCAR-3, and HEK-293 cell lines were seeded in 96-well plates at a density of 9500, 7000, 8000, and 9000 cells/well, respectively, and incubated overnight for adherence. The cells were then treated with the medium or serial dilutions of extracts (1–250 μg/mL) for 4, 24, 48, and 72 h (in triplicate). After treatment, 100 μL of a 0.5 g/L MTT solution was added, and the cells were incubated at 37 °C for 2 h. MTT was replaced with 10% DMSO to dissolve the formazan crystals for ten minutes. Purpuric-colored products were measured at 570 nm using a microplate reader (BioSan, Riga, Latvia). The IC_{50} values were calculated in the IC_{50} calculator (AAT Bioquest, Sunnyvale, CA, USA).

2.7. Statistical Analysis

2.7.1. ANOVA

To determine the percentage differences in the volatile compounds of sea fennel EOs and hydrolates from the leaves, flowers, and fruits, a one-way ANOVA with a Dunnett T3 post hoc significance test for unequal variances was performed at the significance level of $p \leq 0.05$ by the SPSS software, version 25.0, IBM Corporation, New York, NY, USA.

Additionally, to evaluate the differences in the concentrations of sea fennel EOs from the leaves, flowers, and fruits on different cancer cell lines, a one-way ANOVA was performed by the Past 3.X software (version 3.14, University of Oslo, Oslo, Norway), with the significance level at $p < 0.05$.

2.7.2. Linear Regression

To determine the impact of various factors on the cytotoxic activity of the essential oils and hydrolates from sea fennel leaves, flowers, and fruits, a linear regression analysis was conducted using the SPSS software, version 25.0 (IBM Corporation, New York, NY, USA). The dependent variable in the analysis was the IC_{50} value, while the independent variables included the volatile compounds, essential oil concentration, plant part, and exposure time (4, 24, 48, and 72 h). The aim of the analysis was to determine the most significant predictors of cytotoxicity. Prior to the regression analysis, the dataset was checked for linearity, normality, and multicollinearity to ensure the validity of the model.

3. Results and Discussion

3.1. Volatile Composition of Sea Fennel Leaf, Flower, and Fruit EOs and Hydrolates

EOs from sea fennel leaves, flowers, and fruits (Figure 1) were obtained by hydrodistillation and analyzed by GC-MS (Table 1), along with their corresponding hydrolates (Table 2). While the EOs from sea fennel leaves and flowers of the Croatian chemotype have been extensively studied and well documented [10,14,15], there have been no reports focusing on the chemistry of the fruit EO. Likewise, while some studies have investigated hydrolates from sea fennel leaves [18,19], this study is, to the best of the authors' knowledge, the first to report the volatile composition of hydrolates obtained from other plant parts.

(a) (b) (c)

Figure 1. Sea fennel (**a**) leaves, (**b**) flowers, and (**c**) fruits.

The classification of sea fennel chemotypes remains a topic of debate. Pateira et al. [32] identified two chemotypes distinguished by their dillapiole content: chemotype I contains 15–17%, while chemotype II has 0–6%. In contrast, due to significant variations in essential oil composition depending on the geographical origin, Renna [33] proposed a broader classification consisting of four chemotypes: the aromatic monoterpene type, the monoterpene hydrocarbon type, the phenylpropanoid type, and an intermediate type.

Previous studies on Croatian sea fennel [10,14,15] have classified its EOs as chemotype II and of the monoterpene hydrocarbon type. The most abundant compounds in all the plant parts were monoterpene hydrocarbons, reaching 97.87% in the flowers, 96.76% in the fruits, and 85.01% in the leaves (Table 1). The GC-MS analysis of sea fennel leaf, flower, and fruit EOs conducted in this study supports this classification, as dillapiole was completely absent, while limonene (a monoterpene hydrocarbon) was the predominant compound, ranging from 68.41% in the fruits to 74.85% in the leaves. Sabinene, another abundant monoterpene hydrocarbon, was absent in the leaf essential oil, but reached 15.72% in the flowers and 12.37% in the fruits. The limonene content in the leaf (74.85%) and flower (74.30%) EOs was consistent with the literature data, according to which the EOs from summer leaves from the Split region contained limonene levels ranging from 57.5 [15] to 79.13% [10] and up to 84.53% [27]. In flower EOs, limonene ranged from 62.2 [15] to as high as 93.39% (based on our unpublished data). Previous studies have reported sabinene levels in leaves ranging from 0.15 [27] to 51.47% [21], and in flowers from 1.65 (unpublished data) to 12% [15]. Additionally, two other monoterpene hydrocarbons were found in notable amounts: α-pinene was the most abundant in the fruit EO (9.07%), while β-ocimene was the highest in the leaves (6.54%), slightly exceeding previously published values (0.42–4.1%) [10,21]. To evaluate the differences among EOs from different plant parts, a one-way ANOVA was performed. The level of α-pinene showed significant differences across all the samples. Although there was no significant variation in the limonene content between the leaf and flower essential oils, a notable difference was observed between the fruit and leaf essential oils. In contrast, the β-ocimene levels differed significantly between the leaf and flower essential oils, as well as between the leaf and fruit essential oils.

The most abundant compounds in sea fennel EOs are of significant value to human health. Limonene, an aromatic and volatile monoterpene, exhibits a wide range of biological activities, including anticancer, antidiabetic, gastroprotective, anti-inflammatory, antioxidant, antinociceptive, antihyperalgesic, and antiviral properties [34]. Another notable compound, α-pinene, has been shown to possess cytoprotective, gastroprotective, neuroprotective, anxiolytic, and anticonvulsant effects [35]. Although sabinene has been less extensively researched, an available study suggest that it exhibits angiostatic and antiangiogenic effects, as well as antimicrobial activity [36].

To examine the differences in sea fennel leaf, flower, and fruit hydrolates, an HS-SPME-GC-MS analysis was performed (Table 2). As expected, the hydrolates contained more polar compounds than those found in the EOs, since their constituents are generally water-soluble. Monoterpenes, the predominant constituents of the EOs, were typically absent in the hydrolates due to their higher affinity for lipophilic environments [37]. In total, 63 volatile compounds were identified in the sea fennel hydrolates. The greatest diversity was observed in the fruit hydrolate, with 48 compounds identified, followed by the leaf hydrolate, with 43, and the flower hydrolate, with 32 compounds. Monoterpenoids represented the most abundant class of compounds across all three plant parts, with the highest proportion observed in the fruits (87.4%) and the lowest in the leaves (75.25%). The second most prevalent group was monoterpene hydrocarbons, accounting for 12.55% in the leaves, while sesquiterpenoids were notably present in the flowers (10.7%). Ketones, esters, and phenol were detected only in the leaves (Table 2). Considering that the volatiles in the hydrosols predominantly originated from oxygenated constituents of the essential oils, the high content of monoterpenoids in the hydrosols was an expected outcome.

Terpinen-4-ol was identified as the most abundant compound in the leaf and flower hydrolates (27.8% and 36.7%, respectively) and was also present in significant amounts in the fruit hydrolate (10.6%). Sabinene, one of the dominant compounds in the flower and fruit EOs, appeared in the hydrolates in its oxygenated forms, (*Z*)- and (*E*)-sabinene hydrate. (*Z*)-Sabinene hydrate was detected at 5.9% in the flowers and 2.5% in the fruits, while (*E*)-sabinene hydrate was found at 7.4% in the flowers and 3.0% in the fruits. Similarly, (*Z*)- and (*E*)-carveol, the hydroxylated derivatives of limonene, were detected in high concentrations across all samples. (*Z*)-Carveol was most abundant in the fruit hydrolate (11.4%), while (*E*)-carveol was the highest in the leaf hydrolate (7.3%). The fruit hydrolate was also rich in (*E*)-isocarveol (10.6%) and (*E*)-*p*-mentha-2,8-dien-1-ol (9.0%). The flower hydrolate contained a considerable amount of spathulenol (8.3%), while the leaf hydrolate was characterized by 10-(acetylmethyl)-3-carene (8.6%), which was absent in the flower and fruit hydrolates. Previous studies on sea fennel leaf hydrolates collected in Central Dalmatia in August and October similarly identified terpinen-4-ol as the most abundant volatile compound (13.86% and 41.93%, respectively), followed by 10-(acetylmethyl)-3-carene (13.45% and 13.80%, respectively). Comparable levels of sabinene hydrates (1.54–4.73%) and (*E*)-carveol (3.36%–4.92%) were also reported [21,22]. α-Terpineol was present in similar concentrations in all the samples, and increased slightly from the leaves to the flowers to the fruits (4.2% < 5.8% < 6.9%). A one-way ANOVA revealed no significant differences in the α-terpineol levels between the samples. However, significant differences were found for terpinen-4-ol, (*Z*)-carveol, and carvone in all the hydrolates. The flower hydrolate exhibited significantly different levels of (*Z*)- and (*E*)-sabinene hydrate and spathulenol compared to the other hydrolates, while the fruit hydrolate differed significantly in its content of (*E*)-*p*-mentha-2,8-dien-1-ol.

Terpinen-4-ol, the most abundant volatile compound in all the sea fennel hydrolates, exhibits a wide range of pharmacological properties, including strong antimicrobial and antivirulent activity as well as antioxidant, anti-inflammatory, antihypertensive, and anticancer effects. Isolated mainly from tea tree oil, terpinen-4-ol can be synthesized by the hydration of sabinene or obtained via the oxidation, hydrogenation, and selective hydrogenation of terpinolene [38]. Terpineol occurs in four isomeric monoterpenoid forms, including terpine-4-ol and α-terpineol. Due to its pleasant fragrance, it is widely used in the perfume and cosmetics industry. Additionally, it has shown a range of biological activities, such as antioxidant, anticancer, antihypertensive, anticonvulsant, antiulcer, and antinociceptive properties [39]. Carvone and carveol, known for their aromatic properties, are extensively used in the flavor and fragrance industry. However, their potential extends beyond these applications, as they also exhibit antioxidant, antimicrobial, anti-inflammatory, antidiabetic, antitumor, and insecticidal activities [40]. Spathulenol, a tricyclic sesquiterpene alcohol, possesses antioxidant, antiseptic, immunomodulatory, and antinociceptive properties. Its potential use for the treatment of cardiac fibrosis, as an adjunct in chemotherapy, and in the therapy of neurodegenerative diseases have also been explored [41].

3.2. Biological Activities of Sea Fennel Leaf, Flower, and Fruit EOs and Hydrolates

Sea fennel extracts have been extensively studied for their bioactive constituents and properties, among which the antioxidant activity has attracted the most interest. Previous studies on Croatian sea fennel populations have confirmed the good antioxidant activity of the non-volatile fractions (polar/phenolic extracts) [14,15], while the activity of the tested EOs was extremely low [15,42] due to a high content of monoterpenic hydrocarbons that are known as weak radical scavengers and reducing agents. Therefore, in order to provide novelty in the research, this study focused solely on the antioxidant properties of the corresponding hydrolates (from different plant parts, Table 3), which were characterized by the presence of polar, water-soluble, and, in most cases, oxygenated compounds. To the best of the authors' knowledge, this is first report to compare the antioxidant potential of hydrolates from different parts of the sea fennel plant, namely flowers, leaves, and fruits.

As can be seen from the results shown in Table 3, the reducing activity of the samples, assessed using the FRAP method (which measures the ability of antioxidants to reduce Fe^{3+} to Fe^{2+} under acidic conditions), was not observed. This finding aligns with our previous study [21], where the reducing activity was negligible compared to more potent antioxidants. The DPPH and ORAC method, on the other hand, provided information about the free radical scavenging activity of the samples. The difference was in the fact that the DPPH uses stabile, synthetic radicals, whereas the ORAC method monitors the inhibition of peroxyl radicals, which are common in biological systems. Based on the obtained results, it can been seen that the DPPH method also did not show significant activity, with inhibition percentages ranging from 5.67 for the fruit hydrolate to 6.30 for the leaf hydrolate, respectively. However, for the ORAC method, slight differences were seen between the tested results. The highest activity was detected for the leaf hydrolate (123 μmTE/L), which was close to previously reported values [21]. The relative weak antioxidant activity of the fruit hydrolate can be potentially attributed to this hydrolate having the lowest amount of terpinen-4-ol compared to the other samples, while the leaf hydrolate had higher levels of (*E*)-carveol and carvone, as well as 10-(acetylmethyl)-3-carene (8.6%), which was not detected in the other two samples.

Table 3. Antioxidant activity and inhibition of acetyl- (AChE) and butyryl- (BuChE) cholinesterase by hydrolates from different sea fennel plant parts.

Plant Part	Antioxidant Activity			Enzyme Inhibition	
	FRAP (mM Fe^{2+}/L)	DPPH (Inhibition %)	ORAC (μmTE/L)	AChE (Inhibition %)	BuChE (Inhibition %)
Leaves	n.a.	6.30 ± 1.48	122.98 ± 0.50	n.a.	4.59 ± 2.56
Flowers	n.a.	6.11 ± 1.68	103.22 ± 0.44	5.05 ± 0.75	3.04 ± 1.73
Fruits	n.a.	5.67 ± 0.38	98.88 ± 1.00	n.a.	n.a.

n.a.—not active, detected. All results are expressed as mean ± SD.

Similar to the antioxidant activity, the enzyme inhibitory activity of sea fennel EOs has been previously studied by various authors [15,43–47], but there is limited information on the activity of sea fennel hydrolates. In our previous study, we tested the antityrosinase and anticollegenase inhibitory activity of sea fennel hydrolates [21], whereas this research focused on their activity on AChE and BuChE. The obtained results are presented in Table 3.

Although several studies have investigated the AChE and/or BuChE inhibitory activity of sea fennel's polar, non-volatile fractions/extracts [44,48], only two studies, including one from our research group, have investigated the inhibitory activity of EOs [15,47]. To the authors' knowledge, no study has examined the AChE and/or BuChE activity of hydrolates.

Ismail et al. [47] investigated the enzyme inhibition activities of a sea fennel EO (from the coastal region of Libya, North Africa) on AChE and detected an IC_{50} value of 34.43 mg/mL. However, their EO was characterized by a different volatile profile, with the major constituents being thymyl methyl ether (56.86%), γ-terpinene (16.17%), and ledene oxide (4.32%). In our previous study, EOs from sea fennel flowers and leaves were tested at a concentration of 1 mg/mL, and the IC_{50} values were not detected. However, according to the obtained results, the flower EO was more effective at inhibiting AChE, while the leaf EO provided greater inhibition against BuChE. The EOs from flowers and leaves exerted better inhibition against AChE rather than BuChE [15]. The results for the AChE and BuChE inhibitory activity of hydrolates are presented in Table 3, and they also indicate better activity of the flower hydrolate against AChE than BuChE (two-fold lower inhibition). This sample was recognized for the highest content of terpinen-4-ol, which could be the reason for its inhibition activity, as previously reported. The activity of the hydrolates on the inhibition of AChE and BuChe could be explained by terpinene-4-ol, 1,8-cineole, and linalool. In the case of the leaf hydrolate, activity only against BuChE was recorded (3.31% inhibition).

The cytotoxic effects of sea fennel EOs from the leaves, flowers, and fruits at different concentrations, along with the hydrolates, were tested against three human cancer cell lines (MDA-MB-231, MCF-7, and OVCAR-3) after 4, 24, 48, and 72 h of incubation. The IC_{50} values are presented in Figure 2. Additionally, healthy human embryonic kidney HEK-293 cells were treated with fruit and flower EOs, and the results are shown in Figure S3.

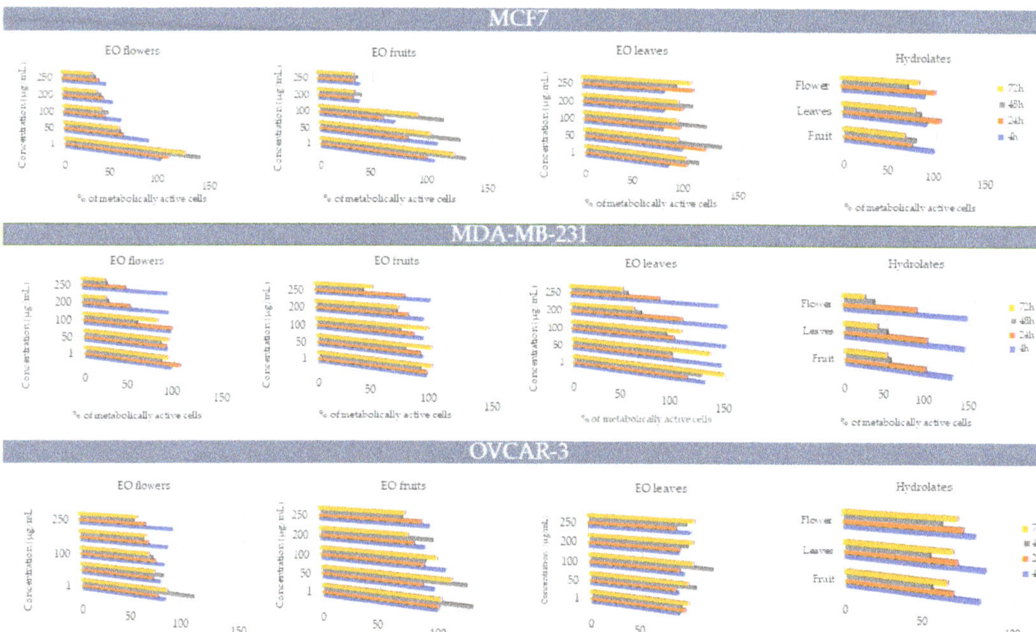

Figure 2. Percentage of metabolically active human breast adenocarcinoma MDA-MB-231, human breast metastatic adenocarcinoma MCF-7, and human ovarian carcinoma OVCAR-3 cell lines after 4, 24, 48, and 72 h of incubation with different concentrations of sea fennel essential oils and hydrolates from leaves, flowers, and fruits.

Of the three tested cell lines, cytotoxic activity was observed against two, namely MDA-MB-231 and MCF-7, while none of the EOs or hydrolates reached the IC_{50} value for any tested concentration on the OVCAR-3 cell line (Figure 2). The human breast adenocarcinoma MDA-MB-231 cell line was the most sensitive to the sea fennel flower EO at concentrations of 200 and 250 μg/mL, with 28.25 and 27.36% cell survival after 48 h and 26.46% and 25.71% cell survival after 72 h, respectively. The fruit EO reached the IC_{50} (250 μg/mL) after 48 h with 43.39% cell survival. The flower hydrolate reached the IC_{50}

after 48 h (with 37.42% cell survival) and after 72 h (with 26.75% cell survival), while the leaf hydrolate reached the IC_{50} (with 41.92% cell survival) only after 72 h. The human breast metastatic adenocarcinoma MCF-7 cell line was sensitive to flower and fruit EOs for all the tested times, but the best results were observed after 72 h of incubation (IC_{50} = 250 μg/mL), with 31.97% and 34.22% cell survival.

No reduction in the amount of healthy human embryonic kidney HEK-293 cells (Figure S3) was observed when they were treated with different concentrations of fruit and flower essential oils (the EOs that had the highest cytotoxic activity on cancer cell lines). Interestingly, at concentrations <200 μg/mL, the cell count doubled after 48 and 72 h, while at a concentration of 250 μg/mL, some cytotoxic effects were observed, but the IC_{50} was not reached. These results suggest the need for further research on sea fennel EOs regarding the potential mechanism of their healing properties [49].

Since essential oils and hydrolates contain a wide range of diverse components, and their cytotoxic activity was tested at different concentrations and exposure times, a linear regression analysis was performed to evaluate the impact of various factors on the cytotoxic activity of EOs and hydrolates from sea fennel leaves, flowers, and fruits (Tables 4 and 5). The concentration of the EOs had the greatest effect on the MCF-7 cell line (B = −0.226, $p < 0.0001$), with a lower IC_{50} value observed at higher concentrations. The concentration was a significant factor for all the cell lines tested. For the MCF-7 and MDA-MB-232 cell lines, octanal showed a significant effect (B = 75.86, $p < 0.0001$ and B = 74.10, $p < 0.0001$, respectively).

Table 4. Results of linear regression for the cytotoxicity of sea fennel leaf, flower, and fruit essential oils on cancer (MDA-MB-231, MCF-7, and OVCAR-3) cell lines.

Cell Line	Predictor	Coefficient	Standard Error	F-Value	R^2	p-Value
MCF-7	Concentration (μg/mL)	−0.226	0.03			<0.0001
	Octanal	75.86	13.01			<0.0001
	Intercept	91.10	5.153	48.00	0.627	<0.0001
MDA-MB-232	Concentration (μg/mL)	−0.17	0.03			<0.0001
	Octanal	74.10	11.81			<0.0001
	Time (h)	−0.43	0.09			<0.0001
	Intercept	110.66	5.83	33.00	0.645	<0.0001
OVCAR-3	Concentration (μg/mL)	−0.064	0.016			<0.0001
	Dodecane	36.695	7.174			<0.0001
	Intercept	85.833	3.018	21.56	0.481	<0.0001

Table 5. Results of linear regression for the cytotoxicity of sea fennel leaf, flower, and fruit hydrolates on cancer (MDA-MB-231, MCF-7, and OVCAR-3) cell lines.

Cell Line	Predictor	Coefficient	Standard Error	F-Value	R^2	p-Value
MCF-7	Time (h)	−0.28	0.11			0.025
	Intercept	96.48	4.853	6.92	0.409	<0.0001
MDA-MB-232	Time (h)	−1.51	0.17			<0.0001
	Intercept	137.88	7.54	80.93	0.89	<0.0001
OVCAR-3	Time (h)	−0.254	0.085			0.014
	Intercept	77.477	3.835	8.83	0.47	<0.0001

The only factor that had a significant effect on all the hydrolate-treated cell lines was time (Table 5). For all three cell lines (MCF-7, MDA-MB-232, and OVCAR-3), the coefficients were negative (B = −0.28, B = −1.51, and B = −0.254, respectively), indicating that the IC_{50} values decreased with increasing exposure time.

Regarding the chemistry of essential oils, it can be suggested that the high concentration of limonene plays a key role in the strong cytotoxic activity of sea fennel EOs. Limonene is well known for its chemopreventive properties against various types of cancer and possesses no mutagenic, carcinogenic, or nephrotoxic risks to humans [50]. However, the presence of α-pinene [51] and several other biologically active compounds suggests that the cytotoxic effects observed for sea fennel EOs may be due to their synergistic effect. Since hydrolates are highly diluted, they generally exhibit less biological activity than their corresponding essential oils. However, due to the presence of a wide variety of compounds with anticancer properties, such as terpinen-4-ol, α-terpineol, carvone, and carveol [38–40], as well as their probable synergistic effects, the sea fennel hydrolates also demonstrated cytotoxic activity.

Previous studies have investigated the cytotoxicity of sea fennel aqueous and chloroform extracts against tumor cell lines [52,53], and one study examined the cytotoxic effects of a sea fennel EO. Beeby et al. also performed studies on the MCF-7 and HER-293 cell lines; however, due to the dominance of γ-terpinene, thymol methyl ether, *o*-cymene, and β-phellandrene in their EO sample, compounds that were minor or absent in our EOs, a direct comparison of the results is not possible [54].

To the best of the authors' knowledge, this is the first report on the cytotoxic activities of sea fennel essential oils (chemotype II, monoterpene hydrocarbon type) and hydrolates against the MDA-MB-231, MCF-7, and OVCAR-3 cell lines.

4. Conclusions

This study provides new insights into the chemical composition and biological activities of essential oils and hydrolates from different parts of sea fennel (*Crithmum maritimum* L.). The volatile profiles of EOs from the leaves, flowers, and fruits were dominated by monoterpene hydrocarbons, with limonene as the most abundant constituent (up to 74.85%). The hydrolates were more chemically diverse than the EOs and were dominated by monoterpenoids, with terpinen-4-ol found in the highest amount in the leaves (27.8%). The results demonstrate weak antioxidant and enzyme-inhibitory and significant cytotoxic activities, with variations depending on the plant part and extract type. The essential oils showed significant effects against certain cancer cell lines, while the hydrolates exhibited moderate activity, without any toxicity to healthy cells for all the tested extracts.

These findings highlight the potential of sea fennel extracts for applications in the nutraceutical and pharmaceutical industries, particularly due to their bioactive properties. Further investigations are needed to explore their mechanisms of action, safety profiles, and possible therapeutic applications.

Supplementary Materials: The following supporting information can be downloaded at: https://www.mdpi.com/article/10.3390/biom15050666/s1, Figure S1: Total ion chromatogram of sea fennel essential oils from (a) leaves, (b) flowers, and (c) fruits; Figure S2: Total ion chromatogram of sea fennel hydrolates from (a) leaves, (b) flowers, and (c) fruits; Figure S3: Percentage of metabolically active human embryonic kidney HEK-293 cell line after 24, 48 and 72 h of incubation with different concentrations of sea fennel flower and fruit essential oils; Table S1: Results of linear regression for the cytotoxicity of sea fennel flower and fruit essential oils on healthy HEK-293 cell line.

Author Contributions: Conceptualization, M.P.; methodology, M.P., N.R.M., O.P. and I.G.M.; validation, M.P., N.R.M., O.P. and I.G.M.; formal analysis, L.S., O.P., N.R.M., P.B., J.T., I.G.M. and M.P.; investigation, M.P. and I.G.M.; data curation, L.S., O.P., N.R.M., P.B., J.T., I.G.M. and M.P.; writing—

original draft preparation, M.P. and I.G.M.; writing—review and editing, L.S., N.R.M., O.P., P.B., J.T., I.G.M. and M.P.; visualization, N.R.M. and M.P.; funding acquisition, I.G.M. and M.P. All authors have read and agreed to the published version of the manuscript.

Funding: This work is part of the PRIMA programme supported by the European Union. Project title: "Innovative sustainable organic sea fennel (*Crithmum maritimum* L.)-based cropping systems to boost agrobiodiversity, profitability, circularity, and resilience to climate changes in Mediterranean small farms" (acronym: SEAFENNEL4MED) (URL: https://seafennel4med.com/, accessed on 19 March 2025).

Institutional Review Board Statement: Not applicable.

Informed Consent Statement: Not applicable.

Data Availability Statement: The original contributions presented in this study are included in the article/Supplementary Material; further inquiries can be directed to the corresponding author/s.

Acknowledgments: The authors would like to extend thanks for the scientific research equipment financed by the EU grant "Functional integration of the University of Split, PMFST, PFST and KTFST through the development of the scientific and research infrastructure" (KK.01.1.1.02.0018).

Conflicts of Interest: The authors declare no conflicts of interest. The funders had no role in the design of the study; in the collection, analyses, or interpretation of data; in the writing of the manuscript; or in the decision to publish the results.

Abbreviations

The following abbreviations are used in this manuscript:

EO	essential oil
HD	hydrodistillation
VOC	volatile organic compound
GC-MS	gas chromatography–mass spectrometry
HS-SPME	headspace solid-phase microextraction
FRAP	ferric-reducing antioxidant power
DPPH	2,2-diphenyl-1-picrylhydrazyl
ORAC	oxygen radical absorbance capacity assay
AChE	acetylcholine
BuChE	butyrylcholine
MDA-MB-231	human breast adenocarcinoma
MCF-7	human breast metastatic adenocarcinoma
OVCAR-3	human ovarian carcinoma
HER-293	healthy human embryonic kidney
ANOVA	analysis of variance

References

1. Renna, M.; Gonnella, M.; Caretto, S.; Mita, G.; Serio, F. Sea Fennel (*Crithmum maritimum* L.): From Underutilized Crop to New Dried Product for Food Use. *Genet Resour. Crop. Evol* **2017**, *64*, 205–216. [CrossRef]
2. Ventura, Y.; Myrzabayeva, M.; Alikulov, Z.; Omarov, R.; Khozin-Goldberg, I.; Sagi, M. Effects of Salinity on Flowering, Morphology, Biomass Accumulation and Leaf Metabolites in an Edible Halophyte. *AoB Plants* **2014**, *6*, plu053. [CrossRef] [PubMed]
3. Jiménez-Becker, S.; Ramírez, M.; Plaza, B.M. The Influence of Salinity on the Vegetative Growth, Osmolytes and Chloride Concentration of Four Halophytic Species. *J. Plant Nutr.* **2019**, *42*, 1838–1849. [CrossRef]
4. Martins-Noguerol, R.; Matías, L.; Pérez-Ramos, I.M.; Moreira, X.; Francisco, M.; Pedroche, J.; DeAndrés-Gil, C.; Gutiérrez, E.; Salas, J.J.; Moreno-Pérez, A.J.; et al. Soil Physicochemical Properties Associated with the Yield and Phytochemical Composition of the Edible Halophyte *Crithmum maritimum*. *Sci. Total Environ.* **2023**, *869*, 161806. [CrossRef] [PubMed]
5. Abdallah, A.; Zouhaier, B.; Rabhi, M.; Abdelly, C.; Smaoui, A. Environmental Eco-Physiology and Economical Potential of the Halophyte *Crithmum maritimum* L. (Apiaceae). *J. Med. Plants Res.* **2011**, *5*, 3564–3571.

6. Tardío, J.; Sánchez-Mata, M.; Morales, R.; Molina, M.; García-Herrera, P.; Morales, P.; Díez-Marqués, C.; Fernández-Ruiz, V.; Cámara, M.; Pardo-de-Santayana, M.; et al. *Mediterranean Wild Edible Plants*; Sánchez-Mata, M.d.C., Tardío, J., Eds.; Springer: New York, NY, USA, 2016; pp. 273–470. [CrossRef]
7. Franke, W. Vitamin C in Sea Fennel (*Crithmum maritimum*), an Edible Wild Plant. *Econ. Bot.* **1982**, *36*, 163–165. [CrossRef]
8. Kraouia, M.; Nartea, A.; Maoloni, A.; Osimani, A.; Garofalo, C.; Fanesi, B.; Ismaiel, L.; Aquilanti, L.; Pacetti, D. Sea Fennel (*Crithmum maritimum* L.) as an Emerging Crop for the Manufacturing of Innovative Foods and Nutraceuticals. *Molecules* **2023**, *28*, 4741. [CrossRef]
9. Sarrou, E.; Siomos, A.S.; Riccadona, S.; Aktsoglou, D.-C.; Tsouvaltzis, P.; Angeli, A.; Franceschi, P.; Chatzopoulou, P.; Vrhovsek, U.; Martens, S. Improvement of Sea Fennel (*Crithmum maritimum* L.) Nutritional Value through Iodine Biofortification in a Hydroponic Floating System. *Food Chem.* **2019**, *296*, 150–159. [CrossRef] [PubMed]
10. Generalić Mekinić, I.; Politeo, O.; Ljubenkov, I.; Mastelić, L.; Popović, M.; Veršić Bratinčević, M.; Šimat, V.; Radman, S.; Skroza, D.; Ninčević Runjić, T.; et al. The Alphabet of Sea Fennel: Comprehensive Phytochemical Characterisation of Croatian Populations of *Crithmum maritimum* L. *Food Chem. X* **2024**, *22*, 101386. [CrossRef]
11. Maoloni, A.; Milanović, V.; Osimani, A.; Cardinali, F.; Garofalo, C.; Belleggia, L.; Foligni, R.; Mannozzi, C.; Mozzon, M.; Cirlini, M.; et al. Exploitation of Sea Fennel (*Crithmum maritimum* L.) for Manufacturing of Novel High-Value Fermented Preserves. *Food Bioprod. Process.* **2021**, *127*, 174–197. [CrossRef]
12. Martins-Noguerol, R.; Matías, L.; Pérez-Ramos, I.M.; Moreira, X.; Muñoz-Vallés, S.; Mancilla-Leytón, J.M.; Francisco, M.; García-González, A.; DeAndrés-Gil, C.; Martínez-Force, E.; et al. Differences in Nutrient Composition of Sea Fennel (*Crithmum maritimum*) Grown in Different Habitats and Optimally Controlled Growing Conditions. *J. Food Compos. Anal.* **2022**, *106*, 104266. [CrossRef]
13. Porrello, A.; Vaglica, A.; Savoca, D.; Bruno, M.; Sottile, F. Variability in *Crithmum maritimum* L. Essential Oils' Chemical Composition: PCA Analysis, Food Safety, and Sustainability. *Sustainability* **2024**, *16*, 2541. [CrossRef]
14. Radman, S.; Mastelić, L.; Ljubenkov, I.; Lazarevski, S.; Politeo, O.; Podrug, R.; Prga, I.; Čorić, I.; Popović, M.; Veršić Bratinčević, M.; et al. Sea Fennel (*Crithmum maritimum* L.) Flowers as an Emerging Source of Bioactive Compounds. *Pol. J. Food Nutr. Sci.* **2024**, *74*, 221–231. [CrossRef]
15. Generalić Mekinić, I.; Blažević, I.; Mudnić, I.; Burčul, F.; Grga, M.; Skroza, D.; Jerčić, I.; Ljubenkov, I.; Boban, M.; Miloš, M.; et al. Sea Fennel (*Crithmum maritimum* L.): Phytochemical Profile, Antioxidative, Cholinesterase Inhibitory and Vasodilatory Activity. *J. Food Sci. Technol.* **2016**, *53*, 3104–3112. [CrossRef] [PubMed]
16. Ni, Z.-J.; Wang, X.; Shen, Y.; Thakur, K.; Han, J.; Zhang, J.-G.; Hu, F.; Wei, Z.-J. Recent Updates on the Chemistry, Bioactivities, Mode of Action, and Industrial Applications of Plant Essential Oils. *Trends Food Sci. Technol.* **2021**, *110*, 78–89. [CrossRef]
17. Jugreet, B.S.; Suroowan, S.; Rengasamy, R.R.K.; Mahomoodally, M.F. Chemistry, Bioactivities, Mode of Action and Industrial Applications of Essential Oils. *Trends Food Sci. Technol.* **2020**, *101*, 89–105. [CrossRef]
18. Almeida, H.H.S.; Fernandes, I.P.; Amaral, J.S.; Rodrigues, A.E.; Barreiro, M.-F. Unlocking the Potential of Hydrosols: Transforming Essential Oil Byproducts into Valuable Resources. *Molecules* **2024**, *29*, 4660. [CrossRef]
19. D'Amato, S.; Serio, A.; López, C.C.; Paparella, A. Hydrosols: Biological Activity and Potential as Antimicrobials for Food Applications. *Food Control* **2018**, *86*, 126–137. [CrossRef]
20. De Elguea-Culebras, G.O.; Bravo, E.M.; Sánchez-Vioque, R. Potential Sources and Methodologies for the Recovery of Phenolic Compounds from Distillation Residues of Mediterranean Aromatic Plants. An Approach to the Valuation of by-Products of the Essential Oil Market—A Review. *Ind. Crops Prod.* **2022**, *175*, 114261. [CrossRef]
21. Politeo, O.; Ćurlin, P.; Brzović, P.; Auzende, K.; Magné, C.; Generalić Mekinić, I. Volatiles from French and Croatian Sea Fennel Ecotypes: Chemical Profiles and the Antioxidant, Antimicrobial and Antiageing Activity of Essential Oils and Hydrolates. *Foods* **2024**, *13*, 695. [CrossRef]
22. Politeo, O.; Popović, M.; Veršić Bratinčević, M.; Koceić, P.; Ninčević Runjić, T.; Mekinić, I.G. Conventional vs. Microwave-Assisted Hydrodistillation: Influence on the Chemistry of Sea Fennel Essential Oil and Its By-Products. *Plants* **2023**, *12*, 1466. [CrossRef]
23. Kaloustian, J.; Mikail, C.; Abou, L.; Vergnes, M.F.; Nicolay, A.; Portugal, H. Nouvelles perspectives industrielles pour les hydrolats. *Acta Bot. Gall.* **2008**, *155*, 367–373. [CrossRef]
24. Ainseba, N.; Soulimane, A.; El Amine Dib, M.; Djabou, N.; Muselli, A. Comparative Study of the Antioxidant, Antimicrobial and Anti-Inflammatory Activity between Essential Oil and Hydrosol Extract of the Aerial Parts of *Inula viscosa* L. *J. Appl. Biotechnol. Rep.* **2023**, *10*, 1169–1175.
25. Miguel, M.G. Antioxidant and Anti-Inflammatory Activities of Essential Oils: A Short Review. *Molecules* **2010**, *15*, 9252–9287. [CrossRef] [PubMed]
26. Hay, Y.-O.; Abril Sierra, M.A.; Sequeda-Castañeda, L.G.; Bonnafous, C.; Delgado Raynaud, C. Evaluation of Combinations of Essential Oils and Essential Oils Withhydrosols on Antimicrobial and Antioxidant Activities. *J. Pharm. Pharmacogn. Res.* **2018**, *6*, 216–230. [CrossRef]

27. Popović, M.; Radman, S.; Generalić Mekinić, I.; Ninčević Runjić, T.; Urlić, B.; Veršić Bratinčević, M. A Year in the Life of Sea Fennel: Annual Phytochemical Variations of Major Bioactive Secondary Metabolites. *Appl. Sci.* **2024**, *14*, 3440. [CrossRef]
28. Adams, R.P. *Identification of Essential Oil Components by Gas Chromatography/Quadrupole Mass Spectrometry*, 4th ed.; Allured Publishing Corporation: Carol Stream, IL, USA, 2007.
29. Čagalj, M.; Skroza, D.; Razola-Díaz, M.d.C.; Verardo, V.; Bassi, D.; Frleta, R.; Generalić Mekinić, I.; Tabanelli, G.; Šimat, V. Variations in the Composition, Antioxidant and Antimicrobial Activities of Cystoseira Compressa during Seasonal Growth. *Mar. Drugs* **2022**, *20*, 64. [CrossRef]
30. Bektašević, M.; Politeo, O.; Carev, I. Comparative Study of Chemical Composition, Cholinesterase Inhibition and Antioxidant Potential of *Mentha pulegium* L. Essential Oil. *Chem. Biodivers.* **2021**, *18*, e2000935. [CrossRef]
31. Režić Mužinić, N.; Veršić Bratinčević, M.; Grubić, M.; Frleta Matas, R.; Čagalj, M.; Visković, T.; Popović, M. Golden Chanterelle or a Gold Mine? Metabolites from Aqueous Extracts of Golden Chanterelle (*Cantharellus cibarius*) and Their Antioxidant and Cytotoxic Activities. *Molecules* **2023**, *28*, 2110. [CrossRef]
32. Pateira, L.; Nogueira, T.; Antunes, A.; Venâncio, F.; Tavares, R.; Capelo, J. Two Chemotypes of *Crithmum maritimum* L. from Portugal. *Flavour. Fragr. J.* **1999**, *14*, 333–343. [CrossRef]
33. Renna, M. Reviewing the Prospects of Sea Fennel (*Crithmum maritimum* L.) as Emerging Vegetable Crop. *Plants* **2018**, *7*, 92. [CrossRef]
34. Vieira, A.J.; Beserra, F.P.; Souza, M.C.; Totti, B.M.; Rozza, A.L. Limonene: Aroma of Innovation in Health and Disease. *Chem. Biol. Interact.* **2018**, *283*, 97–106. [CrossRef] [PubMed]
35. Salehi, B.; Upadhyay, S.; Erdogan Orhan, I.; Kumar Jugran, A.L.D.; Jayaweera, S.; Dias, D.A.; Sharopov, F.; Taheri, Y.; Martins, N.; Baghalpour, N.; et al. Therapeutic Potential of α- and β-Pinene: A Miracle Gift of Nature. *Biomolecules* **2019**, *9*, 738. [CrossRef]
36. Sharma, S.; Gupta, J.; Prabhakar, P.K.; Gupta, P.; Solanki, P.; Rajput, A. Phytochemical Repurposing of Natural Molecule: Sabinene for Identification of Novel Therapeutic Benefits Using In Silico and In Vitro Approaches. *Assay. Drug Dev. Technol.* **2019**, *17*, 339–351. [CrossRef] [PubMed]
37. Aćimović, M.; Tešević, V.; Smiljanić, K.; Cvetković, M.; Stanković, J.; Kiprovski, B.; Sikora, V. Hydrolates: By-Products of Essential Oil Distillation: Chemical Composition, Biological Activity and Potential Uses. *Adv. Technol.* **2020**, *9*, 54–70. [CrossRef]
38. Prerna; Chadha, J.; Khullar, L.; Mudgil, U.; Harjai, K. A Comprehensive Review on the Pharmacological Prospects of Terpinen-4-Ol: From Nature to Medicine and Beyond. *Fitoterapia* **2024**, *176*, 106051. [CrossRef]
39. Khaleel, C.; Tabanca, N.; Buchbauer, G. α-Terpineol, a Natural Monoterpene: A Review of Its Biological Properties. *Open Chem.* **2018**, *16*, 349–361. [CrossRef]
40. Zhang, L.-L.; Chen, Y.; Li, Z.-J.; Li, X.; Fan, G. Bioactive Properties of the Aromatic Molecules of Spearmint (*Mentha Spicata* L.) Essent. Oil: A Review. *Food Funct.* **2022**, *13*, 3110–3132. [CrossRef]
41. Manjima, R.B.; Ramya, S.; Kavithaa, K.; Paulpandi, M.; Saranya, T.; Harysh Winster, S.B.; Balachandar, V.; Arul, N. Spathulenol Attenuates 6-Hydroxydopamine Induced Neurotoxicity in SH-SY5Y Neuroblastoma Cells. *Gene Rep.* **2021**, *25*, 101396. [CrossRef]
42. Kulisic-Bilusic, T.; Blažević, I.; Dejanović, B.; Miloš, M.; Pifat, G. Evaluation of the antioxidant activity of essential oils from caper (*Capparis spinosa*) and sea fennel (*Crithmum maritimum*) by different methods. *J. Food Biochem.* **2010**, *34*, 286–302. [CrossRef]
43. Nguir, A.; Besbes, M.; Ben Jannet, H.; Flamini, G.; Harzallah-Skhiri, F.; Hamza, M.A. Chemical Composition, Antioxidant and Anti-Acetylcholinesterase Activities of Tunisian *Crithmum maritimum* L. Essential Oils. *Mediterr. J. Chem.* **2011**, *1*, 173–179. [CrossRef]
44. Hulkko, L.S.S.; Chaturvedi, T.; Custódio, L.; Thomsen, M.H. Harnessing the Value of Tripolium Pannonicum and Crithmum Maritimum Halophyte Biomass through Integrated Green Biorefinery. *Mar. Drugs* **2023**, *21*, 380. [CrossRef]
45. Pedreiro, S.; Figueirinha, A.; Cavaleiro, C.; Cardoso, O.; Donato, M.M.; Salgueiro, L.; Ramos, F. Exploiting the *Crithmum maritimum* L. Aqueous Extracts and Essential Oil as Potential Preservatives in Food, Feed, Pharmaceutical and Cosmetic Industries. *Antioxidants* **2023**, *12*, 252. [CrossRef] [PubMed]
46. Lemoine, C.; Rodrigues, M.J.; Dauvergne, X.; Cérantola, S.; Custódio, L.; Magné, C. A Characterization of Biological Activities and Bioactive Phenolics from the Non-Volatile Fraction of the Edible and Medicinal Halophyte Sea Fennel (*Crithmum maritimum* L.). *Foods* **2024**, *13*, 1294. [CrossRef] [PubMed]
47. Ismail, A.; El-Shibani, F.A.; Mohammed, H.A.; Al-Najjar, B.O.; Korkor, A.M.; Abdulkarim, A.K.; Said, R.; Almahmoud, S.A.; Sulaiman, G.M. Chemical Composition, Antioxidant, and Enzyme Inhibition Activities of Crithmum Maritimum Essential Oils: The First Chemo-Biological Study for Species Grown in North Africa. *Sci. Rep.* **2024**, *14*, 25318. [CrossRef]
48. Generalić Mekinić, I.; Šimat, V.; Ljubenkov, I.; Burčul, F.; Grga, M.; Mihajlovski, M.; Lončar, R.; Katalinić, V.; Skroza, D. Influence of the Vegetation Period on Sea Fennel, *Crithmum maritimum* L. (Apiaceae), Phenolic Composition, Antioxidant and Anticholinesterase Activities. *Ind. Crops Prod.* **2018**, *124*, 947–953. [CrossRef]

49. Salas-Oropeza, J.; Rodriguez-Monroy, M.A.; Jimenez-Estrada, M.; Perez-Torres, A.; Castell-Rodriguez, A.E.; Becerril-Millan, R.; Jarquin-Yanez, K.; Canales-Martinez, M.M. Essential Oil of Bursera Morelensis Promotes Cell Migration on Fibroblasts: In Vitro Assays. *Molecules* **2023**, *28*, 6258. [CrossRef]
50. Sun, J. D-Limonene: Safety and Clinical Applications. *Altern. Med. Rev.* **2007**, *12*, 259–264.
51. Hou, J.; Zhang, Y.; Zhu, Y.; Zhou, B.; Ren, C.; Liang, S.; Guo, Y. α-Pinene Induces Apoptotic Cell Death via Caspase Activation in Human Ovarian Cancer Cells. *Med. Sci. Monit.* **2019**, *25*, 6631–6638. [CrossRef]
52. Alemán, A.; Marín-Peñalver, D.; de Palencia, P.F.; Gómez-Guillén, M.d.C.; Montero, P. Anti-Inflammatory Properties, Bioaccessibility and Intestinal Absorption of Sea Fennel (*Crithmum maritimum*) Extract Encapsulated in Soy Phosphatidylcholine Liposomes. *Nutrients* **2022**, *14*, 210. [CrossRef]
53. Meot-Duros, L.; Cérantola, S.; Talarmin, H.; Le Meur, C.; Le Floch, G.; Magné, C. New Antibacterial and Cytotoxic Activities of Falcarindiol Isolated in *Crithmum maritimum* L. Leaf Extract. *Food Chem. Toxicol.* **2010**, *48*, 553–557. [CrossRef] [PubMed]
54. Beeby, E.; Magalhães, M.; Poças, J.; Collins, T.; Lemos, M.F.L.; Barros, L.; Ferreira, I.C.F.R.; Cabral, C.; Pires, I.M. Secondary Metabolites (Essential Oils) from Sand-Dune Plants Induce Cytotoxic Effects in Cancer Cells. *J. Ethnopharmacol.* **2020**, *258*, 112803. [CrossRef] [PubMed]

biomolecules

MDPI

Article

Effect of *Callistemon citrinus* Phytosomes on Oxidative Stress in the Brains of Rats Fed a High-Fat–Fructose Diet

Oliver Rafid Magaña-Rodríguez [1], Luis Gerardo Ortega-Pérez [1], Aram Josué García-Calderón [1], Luis Alberto Ayala-Ruiz [1], Jonathan Saúl Piñón-Simental [1], Asdrubal Aguilera-Méndez [2], Daniel Godínez-Hernández [2] and Patricia Rios-Chavez [1,*]

[1] Facultad de Biología, Universidad Michoacana de San Nicolás de Hidalgo, Morelia 58030, Mexico; oliver.rodriguez@umich.mx (O.R.M.-R.); gerardo.ortega@umich.mx (L.G.O.-P.); 1208935x@umich.mx (A.J.G.-C.); 1232816g@umich.mx (L.A.A.-R.); jonathan.saul.simental@umich.mx (J.S.P.-S.)

[2] Instituto de Investigaciones Químico Biológicas, Universidad Michoacana de San Nicolás de Hidalgo, Morelia 58030, Mexico; amendez@umich.mx (A.A.-M.); daniel.godinez@umich.mx (D.G.-H.)

* Correspondence: prchavez17@gmail.com or patricia.rios@umich.mx

Abstract: *Callistemon citrinus* has shown antioxidant and anti-inflammatory properties in certain tissues. However, its impact on the brain remains unproven. This study investigates the effect of *C. citrinus* extract and phytosomes on the oxidative status of the brains of rats fed a high-fat–fructose diet (HFD). Fifty-four male Wistar rats were randomly divided into nine groups ($n = 6$). Groups 1, 2, and 3 received a standard chow diet; Group 2 also received the vehicle, and Group 3 was supplemented with *C. citrinus* extract (200 mg/kg). Groups 4, 5, 6, 7, 8, and 9 received a high-fat diet (HFD). Additionally, groups 5, 6, 7, 8, and 9 were supplemented with orlistat at 5 mg/kg, *C. citrinus* extract at 200 mg/kg, and phytosomes loaded with *C. citrinus* at doses of 50, 100, and 200 mg/kg, respectively. Administration was oral for 16 weeks. Antioxidant enzymes, biomarkers of oxidative stress, and fatty acid content in the brain were determined. A parallel artificial membrane permeability assay (PAMPA) was employed to identify compounds that can cross the intestinal and blood–brain barriers. The HFD group (group 4) increased body weight and adipose tissue, unlike the other groups. The brain fatty acid profile showed slight variations in all of the groups. On the other hand, group 4 showed a decrease in the activities of antioxidant enzymes SOD, CAT, and PON. It reduced GSH level, while increasing GPx activity as well as MDA, 4-HNE, and AOPP levels. *C. citrinus* extract and phytosomes restore the antioxidant enzyme activities and mitigate oxidative stress in the brain. *C. citrinus* modulates oxidative stress in brain tissue through 1.8-cineole and α-terpineol, which possess antioxidant and anti-inflammatory properties.

Keywords: antioxidant; brain; *Callistemon citrinus*; oxidative stress; phytosomes

1. Introduction

From 1990 to 2021, the number of men and women with overweight and obesity increased by 1.0 billion and 1.1 billion, respectively. It is projected that by 2050, 3.8 billion people will have this condition pathology [1].

Obesity and excessive fat accumulation in the body significantly generate reactive oxygen species (ROS). A decrease in antioxidant defenses and an increase in ROS indicate a state of oxidative stress that damages biomolecules such as carbohydrates, lipids, proteins, and nucleic acids [2].

High-fat, high-sugar diets are widely consumed worldwide, especially in the United States. This diet encourages an increase in reactive oxygen species and free radicals, ultimately leading to a state of oxidative stress [3]. The harmful effects of a high-fat diet (HFD) on cognitive function are mainly caused by increased adiposity and the subsequent pro-inflammatory and pro-oxidative impact on the brain [4]. These effects result from decreased activity of antioxidant enzymes such as superoxide dismutase (SOD), catalase (CAT), glutathione peroxidase (GPx), and paraoxonase-1 (PON-1) [5]. Additionally, HFD increases the levels of lipoperoxidation products in the brain (MDA and 4-HNE), which leads to protein damage [6,7].

Previous studies have demonstrated that the types of fatty acids in the diet play a key role in the pathophysiological changes observed in rodent models [8]. Pistell et al. [9] showed that a diet high in saturated fatty acids for rodents increases brain inflammatory markers. Similarly, Chen et al. [10] reported that mice fed a high-fat diet for 24 weeks showed increased expression of brain senescence markers. A high-fat diet also caused mitochondrial changes in microglial cells [11]. Conversely, consuming a high-sucrose or high-fructose diet triggers inflammation in the hippocampus, leading to memory impairments [12].

Orlistat (Xenical) is a lipase inhibitor approved by the FDA (Food and Drug Administration) for the treatment of obesity; however, it has also been reported to cause adverse effects, including difficulty controlling bowel movements, steatorrhea, flatulence, diarrhea, and stomach pain [13]. Natural products are widely used due to their diverse compounds, which can produce beneficial effects and often cause fewer side effects. Nonetheless, they must be present in high concentrations to achieve efficacy.

Phytosomes, also known as herbosomes, represent a novel technology utilizing vesicular systems for drug delivery. They consist of a phytoactive or natural product encapsulated within a phospholipid membrane, forming lipid-based nanoparticles that enhance the absorption and bioavailability of soluble compound drugs [14]. The lipid nature of the cell membrane improves its permeability and shields bioactive compounds from degradation in the gastrointestinal tract [15].

Callistemon citrinus belongs to the Myrtaceae family, is native to Australia, and was used by the Aboriginal people for the treatment of diarrhea, rheumatism, and cough. The first studies utilized leaves and flowers, reporting their antibacterial and antifungal properties [16]. *C. citrinus* has also been associated with anti-inflammatory, elastase inhibitory, molluscicidal, cardioprotective, antioxidant, and protective effects against type 2 diabetes [17]. Lopez-Mejia et al. reported the potential of *C. citrinus* leaf extract to inhibit colon cancer [18]. Piñón-Simental et al. found that the extract has a protective effect on the gastric mucosa against indomethacin [19]. Recently, we demonstrated that a high-fat, fructose-rich diet causes obesity and oxidative stress in the liver, heart, kidneys, and stomach of rats fed for 15 weeks [20]. Conversely, administering phytosomes loaded with *C. citrinus* leaf extract reduces oxidative stress and demonstrates anti-obesity, antioxidant, and anti-inflammatory effects [21].

We previously reported the phytochemical composition of the terpenes in *Callistemon citrinus* using gas chromatography–mass spectrometry (GC-MS) and two-dimensional gas chromatography (GC $\times$ GC) [22]. Additionally, a toxicology study conducted by Ortega-Pérez et al. [20] found that neither acute nor sub-chronic exposure exhibited toxicity. Finally, Ortega-Pérez et al. [23] identified p-coumaric acid, gallic acid, and ellagic acid in *C. citrinus* and characterized the phytosomes loaded with *C. citrinus*. The biological activity of *C. citrinus* is mainly attributed to the phenolic and terpenoid compounds previously reported, including several bioactive substances such as 1,8-cineole, limonene, α-terpineol, and terpinolene [24]. Numerous studies have investigated the role of these compounds as antioxidant and antiradical agents in mitigating oxidative stress caused by brain damage [25]. The *C. citrinus* phytosomes have been proven to be more stable and soluble than *C. citrinus* extract.

This formulation enhances the absorption of polar compounds and creates a hydrophobic bilayer for non-polar substances. This contrasts with the low bioavailability and absorption of terpenes, phenolic, and flavonoid compounds when used orally. The use of *C. citrinus* extract and phytosomes has the advantage of containing compounds that protect the brain from oxidative stress; however, this has not been studied. On the other hand, Rule et al. [26] reported the effect of diet on the fatty acid profile in the brain. Additionally, Ulmann et al. [27] showed that the concentration of arachidonic acid decreases in brain phosphatidylinositol and phosphatidylserine with age. This study investigates the impact of *C. citrinus* extract and phytosomes loaded with *C. citrinus* on oxidative stress in the brains of rats fed a high-fat–fructose diet and determines whether dietary fat–sugar induced differences in the brain fatty acids.

2. Materials and Methods

2.1. Chemical Materials

Boron trifluoride, Bradford reagent, dodecane, ferrous ammonium sulfate, o-dianisidine dihydrochloride, potassium iodide, 1-methyl-2-phenylindole, methanesulfonic acid, 5,5′-dithiobis 2-nitrobenzoic acid, reduced glutathione, cumene hydroperoxide, sodium carbonate, riboflavin, 4-nitrophenylacetate, and all other chemicals and solvents were purchased from Sigma-Aldrich (Carlsbad, CA, USA) (Merck/Mexico/Life Science Products & Service Solutions).

2.2. Biological Material and Preparation of Ethanolic Extract

Leaves from *Callistemon citrinus* plants were collected in Morelia, Michoacán, Mexico. Fresh leaves were crushed and macerated in ethanol 96% at a 1:10 ratio (*w*/*v*). The ethanolic extract was stored in the dark at 4 °C for 7 days. The leaf extract was dried using a rotary evaporator with a vacuum pump at 45 °C. The concentrated extract was stored in a refrigerator at 4 °C until further use. The Supplementary File S1 shows the chemical composition of *Callistemoon citrinus* and the chromatogram.

2.3. Preparation of Phytosomes

The preparation of phytosomes followed the protocol outlined by Ortega-Pérez et al. [23]. The mixture included 50 mL of hydration medium (0.01 M phosphate buffer solution, 150 mM NaCl, pH 7.4), 1:1 (*w*/*w*) of soybean phosphatidylcholine and *C. citrinus* pure extract, 1 g of *C. citrinus* extract, and 1 g of soybean phosphatidylcholine to achieve a final concentration of 200 mg/mL; 0.5 g of *C. citrinus* extract and 0.5 g of soybean phosphatidylcholine for a final concentration of 100 mg/mL; and 0.25 g of *C. citrinus* extract and 0.25 g of soybean phosphatidylcholine for a final concentration of 50 mg/mL—all in the same volume of hydration medium. Additionally, 0.72 g of Tween 80 and 1% ethyl acetate were added to improve solubility. The mixture was then emulsified using an ultrasonicator (VCX 500, SONICS, Oklahoma City, OK, USA) at 25% amplitude for 10 min at 10 °C. This phytosomal complex was stored in an amber-colored glass bottle at 20 °C.

2.3.1. Lyophilization and Scanning Electron Microscopy (SEM)

Phytosome samples were frozen at −80 °C for 8 h, then lyophilized in a high-vacuum environment of 34 Pa using a lyophilizer (Labconco Plus 12; Labconco, Kansas City, MO, USA) for 8 h with a condenser set at −43 °C. Lyophilized phytosomes were stored in a sealed glass ampoule at 4 °C. One drop of the lyophilized sample was placed on a brass electron microscope stub and coated with copper particles for sputtering. Representative images of the samples were taken, and particle diameters were calculated using scanning electron microscopy (JEOL JSM-7600F SEM, JEOL, Akishima, Japan) at an accelerating voltage of 20.0 kV and a working distance of 15.1 mm. Details of the morphological structure of the phytosomes were observed at an amplitude of up to 10,000× and a working distance that allowed for detailed observations at increasing depths of focus.

2.3.2. Particle Size and Polydispersity Index

The mean particle size and polydispersity index (PDI) (which refers to the width of a particle size distribution) were measured using a Nano Particle Analyzer SZ-100 (Horiba, Kyoto, Japan) based on the principle of dynamic light scattering; Ludox TM silica served as the reference material [23]. Ludox TM-50 was diluted to 10% with 0.01 M KCl. Approximately 10 milliliters of KCl/LUDOX solution were filtered through a 2.5 µm filter. The samples were equilibrated for 2 min at 25 °C and then placed in a plastic cuvette for analysis at 90° scattering angles. All batches were performed in triplicate, and the mean and SD were calculated.

2.3.3. Entrapment Efficiency

The entrapment efficiency of *C. citrinus* phytosomes was determined using a UV–visible spectrophotometer [23]. Approximately 1 mL of dialyzed vesicular suspension was taken and diluted with 0.1 mL of Triton X-100. The solution was centrifuged at 1350× g for 5 min, and the supernatant was diluted with ethanol. The amount of drug entrapped was analyzed spectrophotometrically at a maximum wavelength of 425 nm against ethanol containing Triton X-100 as a blank. Equation (1) computes the efficiency of entrapment (EE); Tdrug is the total amount of drug; Edrug is the extract entrapment in the formulation (phytosome); and Udrug is the extract not entrapped in phytosomal formulation.

$$\mathrm{EE} = \frac{\mathrm{Edrug}}{\mathrm{Tdrug}} \times 100\% = \frac{\mathrm{Edrug}}{\mathrm{Edrug} + \mathrm{Udrug}} \times 100\% = \left(1 - \frac{\mathrm{Udrug}}{\mathrm{Edrug} + \mathrm{Udrug}}\right) \times 100\% \quad (1)$$

2.3.4. Stability and Solubility

The stability analysis was conducted by storing the phytosomes at 20 ± 2 °C and 4 ± 1 °C, and the particle size was measured on days 1, 3, 5, and 10 after storage. It was then measured again after three and a half months.

Solubility analysis was performed by dissolving 2 mg of each formed complex (soybean phospholipid particles) and *C. citrinus* leaf extract in 5 mL of different solvents within small volumetric flasks. The solutions were stirred continuously for 1 h [23]. The experiments were conducted in triplicate.

2.4. Animals

White male Wistar rats were bred and maintained in a cage in the laboratory of the Institute of Chemical-Biological Research of UMSNH under bioterium conditions: a 12 h light/dark cycle, relative humidity of 60–70%, an average temperature of 20 °C, and food and water provided ad libitum. Young adults, aged 2 months and weighing between 200 and 250 g, were used in all treatments. The experimental process followed the guidelines for the care and use of laboratory animals as outlined in the Mexican Official Standard (NOM-062-ZOO-1999) [28] by the Mexican Secretary of Agriculture, Livestock, Rural Development, Fishing, and Food (published in the Official Gazette, Mexico City 2001). The Animal Bioethical Committee of UMSNH approved the study on 12 January 2024, with approval code ID IIQB-CIBE-06-2024. All efforts were made to minimize animal suffering and reduce the number of animals used in the study. The use of Wistar rats instead of other strains is based on their ease of handling, genetic uniformity, and ability to produce reproducible results.

2.5. Experimental Design, High-Fat–Fructose Diet

The model was conducted using a high-fat–fructose diet, following the protocol of Ortega-Pérez et al. [20]. Rats were given a diet consisting of 45% Rodent Diet® brand rat food, which provided 0.20 g of protein, 0.77 g of carbohydrates, 0.03 g of fat, 0.12 g of calcium, 0.012 g of phosphate, 0.021 g of sodium, and 0.007 g of potassium per gram, supplying 3.3 calories per gram. The HFD also included 14.8% pork lard and 14.8% vegetable shortening (INCA brand), and the food was supplemented with 25% fructose. The detailed fatty acid composition is presented in Table 1. All ingredients were mixed and formed into pellets for use; HFD was prepared daily to avoid oxidation. Table S2 shows the composition of the control diet (see Supplement Material). The animals were randomly divided into nine groups (n = 6 per group): Group 1 was given a chow diet and a saline solution (C); Group 2 followed a chow diet and the vehicle (C + V); Group 3 was provided a chow diet with *C. citrinus* leaf extract (200 mg/kg) (C + C.c); Group 4 received an HFD and a saline solution; Group 5 received an HFD with orlistat (5 mg/kg) (HFD + Orl); Group 6 received an HFD with *C. citrinus* leaf extract (200 mg/kg) (HFD + C.c); Groups 7, 8, and 9 received HFD with *C. citrinus* phytosomes at 50, 100, and 200 mg/kg, respectively (HFD + P50; P100 and P200). The dosage of phytosomes was chosen based on a previous study [23].

Table 1. Percentage of fatty acid composition in the fat diet.

Fatty Acid	Lard (%)	Vegetable Shortening (%)
C10:0, Capric	-	0.16
C12:0, Lauric	0.26	0.27
C13:0, Tridecanoic	-	0.067
C13:1, Tridecenoic	-	0.13
C14:0, Myristic	2.6	5.4
C14:1*n*-6	-	0.3
C14:2*n*-9	-	0.72
Tetradecanoic acid, 12-methyl	-	0.57
Pentadecanoic acid	0.14	1.01
C16:0, Palmitic	22.8	16.4
C16:1, Palmitoleic	5	2.27
Methyl 14-methylhexadecanoate	-	0.47
C17:0, Margaric	0.89	1.33
C17:1*n*-9	0.72	-
C18:0, Stearic	14.35	7.26
C18:1, Oleic	1.97	-
C18:1*n*-11t	23.29	14.07
Octadecanoic acid, 17-methyl	0.18	-
C18:2*n*-6, Linoleic	2.46	-
C20:0, arachidic	0.67	0.64
C20:1, Erucic	1.88	1.04
C20:2, Eicosadienoic	1.25	-
C20:4, Eicosatetranoic	0.55	-
C21:0, Heneicosanoic	-	0.045
C22:0, Behenic	0.037	0.093
C22:4*n*-6, Adrenic	0.255	0.03
C23:0, Tricosanoic	-	0.034
C24:0, Lignoceric	0.015	0.03
C26:0, Cerotic	-	0.01
Totaal	79.317	52.346

Treatments were given once daily by oral gavage at 7:00 a.m. throughout the 16 weeks. The weights of all groups were measured weekly to monitor weight gain throughout the study. The percentage of total weight gain (X) was calculated as X = [(Final BW − initial BW)/Initial BW × 100]. In addition to weight gain, the Lee index and adiposity index were also calculated to demonstrate the animals' obesity. The Lee index (g/cm), which is equivalent to BMI, was calculated as LI = $(3\sqrt{BW})/NAL \times 10$, and the equation to assess the percentage of body fat (%) is 0.73 (LI-280.8). In this way, the fat percentage depends only on the weight and the nasal–anus length of the rats. The adiposity index (AI) was calculated as AI = (total adipose tissue weight/final BW) × 100 [23] (Table 2). At the end of the experiment, the animals were anesthetized with an intraperitoneal injection of sodium pentobarbital (50 mg/kg). The entire brain was collected, washed with a 0.9% saline solution, weighed, and stored at −80 °C for further analysis.

Table 2. Characteristics of phytosomes containing *Callistemon citrinus* leaf extract.

Characteristic	Result (Mean ± SD)
Mean particle size (nm)	129.98 ± 18.30
Polydispersity index	1.13 ± 0.03
Entrapment efficiency (%)	80.49 ± 0.07
Temperature on the stability (106 days)	20 ± 2 °C
Total Solubility profile (%)	90.00

2.6. Preparation of Brain Tissue Homogenates

The whole brain was thoroughly homogenized on ice at a ratio of 1:4 (*w*/*v*) with 10 mM phosphate buffer (pH 7.4) in a mortar until the tissue was disintegrated. After that, it was centrifuged at 13,000 rpm for 20 min at 4 °C, and the supernatant was collected and stored at −20 °C. Total proteins were measured as follows: 1 µL of brain homogenate was mixed with 799 µL of distilled water, and then 1 mL of Bradford reagent was added. The mixture was incubated at room temperature for 5 min, and the absorbance was read at 595 nm. For the blank, only 800 µL of distilled water and Bradford reagent were used.

2.7. Total Lipid Extraction and Methyl Esterification

Whole brain tissue was homogenized in a cold mortar, with 0.1 g of this homogenate taken and mixed with 2 mL of a 2:1 (*v*/*v*) chloroform–methanol solution in a cold mortar. The homogenate was collected and centrifuged at 6000 rpm for 10 min at 4 °C. The supernatant was collected, and 100 µL of water was added. The mixture was centrifuged at 3000 rpm for 20 min at 4 °C, and the lower phase was recovered and stored at −20 °C. Methyl esterification was performed using 100 µL of the lower phase and 1 mL of 0.5 N NaOH in methanol. It was incubated at 100 °C for 60 min. Then, 1 mL of boron trifluoride (BF_3) dissolved in methanol was added and incubated for an additional 20 min at 100 °C. After cooling, the contents were transferred to a test tube, and 2 mL of water and 3 mL of hexane were added. The mixture was mixed for 20 s, and the upper (organic) phase was recovered and placed in a second test tube, where an additional 3 mL of hexane was added. The mixture continued for another 20 s, and the organic phase was collected once more. Finally, the organic phase was dried using a stream of nitrogen and resuspended in isooctane. The samples were analyzed using an Agilent 7890A gas chromatography system (Agilent Technologies, Santa Clara, CA, USA) equipped with an HP5MS30M column (5% phenyl methylsiloxane, 60 × 0.25 × 0.25 mm; Agilent Technologies, Santa Clara, CA, USA) and coupled to an electronic impact ionization quadrupole mass spectrometer.

Hewlett-Packard 5975C (Santa Clara, CA, USA). The oven temperature was initially set at 60 °C for 1 min, then increased to 280 °C at a rate of 8 °C/min. The injector, ionization source, and quadrupole temperature were maintained at 230 °C, 230 °C, and 150 °C, respectively. Helium served as the carrier gas at a constant flow of 1 mL/min. The mass spectrometer operated in EI mode at 70 eV, with a mass-to-charge (m/z) range of 50–500, and the voltage was set to −1737 V. Total ion chromatograms (TICs) were processed using the automated data processing software MSChem (Version E.02.01.1177, Agilent Technologies). To identify the different compounds, the mass spectrum of each detected compound was compared to those in mass spectral databases (Wiley 275 and US National Institute of Science and Technology (NIST) V. 2.0). The quantities of compounds were calculated from a standard calibration curve using 1,8-cineole in the range of 1–0.2 mg/mL.

2.8. Parallel Artificial Membrane Permeability Assay Blood–Brain Barrier (PAMPA BBB)

The PAMPA BBB assay kit was used as in vitro model to predict transcellular passive absorption. The assay is cost-effective, quick, and mimics the three barriers: skin, gastrointestinal, and blood–brain). The concentration and dose were selected based on the physicochemical properties of the active pharmaceutical ingredients, rather than on relevant in vivo dose data. The donor volume-to-membrane area ratio is also significantly higher compared to the in vivo environment. Permeability is determined and used for assay evaluation, which is normalized to the donor concentration [29].

Approximately 3 μL of PBL (porcine brain lipids) at a concentration of 5 mg/mL dissolved in dodecane was applied to the surface of the artificial membrane in the donor plate. Subsequently, 297 μL of 100 mM phosphate-buffered saline was added to the donor plate, along with 300 μL to the acceptor plate. Then, 3 μL of the compound, in this case, the *C. citrinus* extract at a concentration of 100 mg/mL dissolved in 5% DMSO, was added to the donor plate. The donor plate was carefully placed on the acceptor plate and incubated for 14 to 18 h at 37 °C. After the incubation period, the contents of the acceptor plate were collected in a microtube (Eppendorf, Hamburg, Germany). Then, 300 μL of chloroform was added to this content and vortexed for 5 min to extract the compounds in the buffer. The organic phase was recovered and placed in vials for analysis using gas chromatography coupled with mass spectrometry. The 1,8-cineole (Sigma-Aldrich) was used as an internal standard. The same protocol was followed for the intestinal barrier permeability test, with a slight modification in the membrane compounds, using soy lecithin (20 mg/mL) dissolved in chloroform.

2.9. Advanced Oxidation Protein Products (AOPP)

The reaction mixture consisted of 50 μL of tissue homogenate and 1 mL of 20 mM phosphate buffer at pH 7.4, then 50 μL of 1.16 M potassium iodide and 100 μL of acetic acid were added. After a 2 min incubation, the absorbance at 340 nm was measured [30].

2.10. Malondialdehyde (MDA) and 4-Hydroxynonenal (4-HNE)

The assay for malondialdehyde (MDA) and 4-hydroxynonenal (HNE) was based on using 1-methyl-2-phenylindole, as reported by Johnston et al. [31]. A mixture containing 0.5 g of tissue, 10 μL of 5 mM BHT, and 2 mL of 10 mM phosphate buffer at pH 7.4 was homogenized in a cold mortar with a pestle. The homogenate was centrifuged for 13 min at 13,000× *g*, and the supernatant was collected. For the assay, 200 μL of tissue homogenate was mixed with 650 μL of 10 mM 1-methyl-2-phenylindole dissolved in a 3:1 acetonitrile/methanol solution, and the sample was prepared in duplicate. One sample was treated with 150 μL of 37% HCl (*v*/*v*), while the other was treated with 150 μL of methanesulfonic acid. Both samples were incubated for 60 min at 45 °C. After incubation, the samples were placed on ice for 5 min, then centrifuged for 10 min at 13,000× *g*. The supernatant was read at 586 nm. For the blank, the 1-methyl-2-phenylindole solution was replaced with a 3:1 acetonitrile/methanol mixture (one blank for each acid).

2.11. Reduced Glutathione (GSH)

Reduced glutathione (GSH) levels in brain tissue were measured using the method described by Sedlak and Lindsay [32]. A mixture of 62.5 µL of tissue homogenate, 187.5 µL of 0.2 M Tris buffer at pH 8.2, and 12.5 µL of 0.01 M 5,5′-dithiobis(2-nitrobenzoic acid) (DNTB) was prepared, followed by the addition of 987.5 µL of absolute methanol. The mixture was then shaken at 240 rpm for 15 min. Afterwards, the samples were centrifuged at 3000 rpm at room temperature for 15 min. Finally, the yellow color of the reaction developed, and the absorbance was measured with a spectrophotometer at 412 nm.

2.12. Catalase (CAT) Activity

Catalase (CAT) activity was assessed by measuring the rate of hydrogen peroxide (H_2O_2) decomposition [33]. The reaction mixture contained 950 µL of 50 mM phosphate buffer at pH 7.0, 25 µL of brain homogenate, and 25 µL of 30 mM H_2O_2 substrate. Changes in absorbance at 240 nm were recorded every 30 s over 3 min. The blank consisted of 975 µL buffer and 25 µL H_2O_2. The molar extinction coefficient for peroxide was 43.6 $M^{-1}cm^{-1}$.

2.13. Glutathione Peroxidase (GPx) Activity

Glutathione peroxidase activity was measured following the protocol of Prabhu et al. [34]. A working solution included 10 µL of 0.5 mM NADPH, 10 µL of 100 mM reduced glutathione, 1 unit of glutathione reductase enzyme, 901 µL of 50 mM phosphate buffer at pH 7.0, 50 µL of 30 mM cumene hydroperoxide, and 25 µL of brain homogenate. The absorbance was measured at 340 nm every 30 s over 5 min against a blank containing all components except the sample and glutathione reductase.

2.14. Superoxide Dismutase (SOD) Activity

SOD activity was measured using the technique described by Bouhalit et al. [35], conducted in darkness or with minimal light exposure. For this method, the following were added to a microtube (Eppendorf): 542.5 µL of 67 mM phosphate buffer (pH 7.8), 25 µL of 100 mM EDTA, 50 µL of 50 mM sodium carbonate (Na_2CO_3) at pH 10, 7.5 µL of TEMED, 200 µL of 1.5 mM NTB, 125 µL of 20 mM methionine dissolved in 2% DMSO, 50 µL of brain homogenate, and finally 25 µL of 0.1 mM riboflavin. Two microtubes (Eppendorf) served as blanks; all the components were added except for the sample. All tubes were illuminated, except for one unilluminated blank tube, with a 100 W spotlight placed 15 cm away for 15 min. The measurement was then taken at 560 nm in a spectrophotometer.

2.15. Paraoxanase (PON) Activity

Assessing PON was performed following Dantoine's protocol [36]. A reaction mixture was prepared with 25 µL of methanol, 100 µL of 10 M $CaCl_2$, 562.5 µL of 25 mM Tris-HCl buffer at pH 8, and 312.5 µL of 1 M 4-nitrophenylacetate dissolved in methanol. A volume of 125 µL (1/20 dilution) of brain homogenate was added to a glass cell, followed by 1 mL of the prepared cocktail. The cell was shaken and read in a spectrophotometer at 402 nm every 30 s for 3 min against a blank of the cocktail.

2.16. Statistical Analysis

The entire sample ($n = 6$) was included in the treatment, and values were expressed as mean ± standard deviation (SD). Data were analyzed using JMP 8.0 and GraphPad Prism (version 7) with one-way analysis of variance (ANOVA) to identify statistical differences in parameters. Tukey's multiple comparison test (a, b, c) was performed. Tukey's Honestly Significant Difference (HSD) test is a post hoc test used in ANOVA to compare all possible pairs of means. When ANOVA reveals a significant difference among group means, a post hoc test, such as Tukey's, is necessary to determine which groups differ significantly. Values of * $p \leq 0.05$ were considered statistically significant.

3. Results

The use of scanning electron microscopy (SEM) has given significant insight into the solid-state properties and surface morphology of complex (Figure 1). Particles are small and adjoining; this size range can be used for oral absorption.

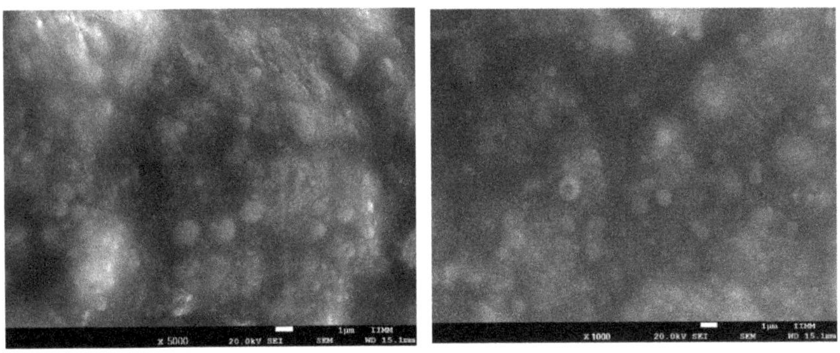

Figure 1. Scanning electron microscope images at 5000× (**left**) and 1000× (**right**) of phytosomes loaded with *Callistemon citrinus* leaf extract.

The size of the phytosomal formulation of *C. citrinus* prepared in this study was 129.98 nm ± 18.30 nm in the emulsion. The entrapment efficiency was 80.49 ± 0.07. The stability of the phytosomal formulation lasted 3.5 months at 20 °C and showed broad solubility (Table 2).

3.1. Effects of *C. citrinus* Phytosomes on Overall Weight Gain

Figure 2A,B shows that weight gain in the HFD group started to increase in the 4th week, and by the end of the 16th week, it was 267.33 ± 36.67 g, while the control group's gain was 171.0 ±36.4 g ($p < 0.0014$). This represents a 30% increase in body weight compared to the group that received the same high-fat–fructose diet supplemented with *C. citrinus* extract, phytosomes, and orlistat. The lower doses of phytosomes (50 mg/kg) displayed a pattern similar to that of the *C. citrinus* extract at 200 mg/kg. Table 3 presents some morphometric parameters that demonstrate the impact of a high-fat, high-fructose diet. The Adiposity index was higher in the HFD group, 9.43 ± 0.623, than the control group, 2.78 ± 0.55 ($p < 0.03$), while in the HFD supplement with *C. citrinus* at 200 mg/kg, phytosomes + *C. citrinus* at 200 mg/kg, and orlistat at 5 mg/kg, the AI was 6.19 ± 0.39, 4.02 ± 0.62, and 5.56 ± 0.7 (p rate 0.001 to 0.021), respectively. On the other hand, the only group with a Lee index of 0.33 ($p < 0.05$) was the HFD group; this value indicates obesity in the rodents, unlike the other groups. This finding suggests that a potential mechanism for managing weight gain in *C. citrinus* and the phytosomes is comparable to that of orlistat.

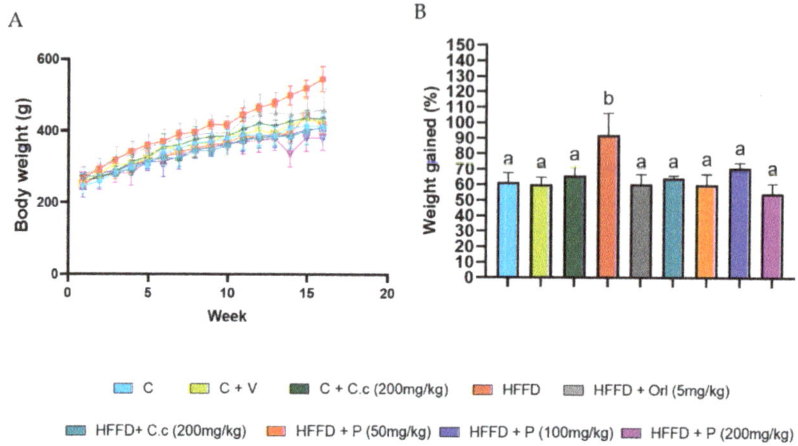

Figure 2. Weekly body weight in control rats and the different experimental treatments during the 16 weeks (**A**), and the percentage of total weight gained by control rats and various experimental treatments over 16 weeks is shown (**B**). Values are expressed as mean ± standard deviation (ANOVA followed by Tukey's test, $n = 6$). Different letters (a,b) indicate significant differences between groups.

Table 3. Impact of *C. citrinus* extract and phytosomes supplementation on brain and obesity indicators in rats consuming a high-fat–fructose diet.

Measurements	C	C + V	C + C.c (200 mg/kg)	HFD	HFD + Orl (5 mg/kg)	HFD + C.c (200 mg/kg)	HFD + P (50 mg/kg)	HFD + P (100 mg/kg)	HFD + P (200 mg/kg)
Brain (g)	1.93 ± 0.03 [a]	2.07 ± 0.02 [a]	2.08 ± 0.07 [a]	2.06 ± 0.04 [a]	1.95 ± 0.05 [a]	1.96 ± 0.04 [a]	2.04 ± 0.08 [a]	2.08 ± 0.03 [a]	2.17 ± 0.02 [a]
Body weight gain (g)	171.00 ± 36.42 [b]	168.66 ± 9.86 [b]	169.00 ± 35.36 [b]	267.33 ± 36.90 [a]	180.75 ± 40.42 [b]	146.25 ± 31.84 [b]	178.66 ± 27.22 [b]	167.33 ± 23.02 [b]	163.66 ± 3.21 [b]
Adiposity index	2.78 ± 0.55 [c]	2.77 ± 0.55 [c]	2.52 ± 0.62 [c]	9.43 ± 0.62 [a]	5.56 ± 0.71 [bc]	6.18 ± 0.39 [b]	5.98 ± 0.71 [b]	4.82 ± 0.71 [bc]	4.02 ± 0.62 [bc]
Lee index	0.30 ± 0.01 [b]	0.30 ± 0.02 [b]	0.30 ± 0.01 [b]	0.33 ± 0.01 [a]	0.30 ± 0.01 [b]	0.30 ± 0.01 [b]	0.29 ± 0.01 [b]	0.30 ± 0.01 [b]	0.31 ± 0.31 [b]

All values expressed as mean ± SEM ($n = 6$; ANOVA followed by Tukey's test, with statistically different values [a, b, c] between groups; $p < 0.05$).

3.2. Brain Lipid Profile

To evaluate the effect of *Callistemon citrinus* extract and the phytosome on the total brain lipid composition, we analyzed the whole brain. The lipid profile results show no significant differences in the percentages ($p < 0.05$) of saturated fatty acids (Figure 3A), monounsaturated fatty acids (Figure 3B), polyunsaturated fatty acids (Figure 3C), and dimethyl acetal fatty acids (Figure 3D) when comparing the control and HFD groups to other treatments after 16 weeks. Additionally, the main fatty acids found in brain tissue across various treatments include palmitic acid (16:0), stearic acid (18:0), arachidonic acid (ARA) (20:4), and 4,7,10,13,16,19-docosahexaenoic acid (DHA) (22:6) (Table 4). A high level of palmitic acid was observed in the HFD + Orl group, but it was not significantly different ($p < 0.05$) from the control group. However, the other groups did not differ substantially from one another. Stearic acid levels showed a significant ($p < 0.05$) increase in the HFD + Orl (5 mg/kg). Conversely, in the HFD + C.c (200 mg/kg), there was a significant ($p < 0.05$) reduction compared to the other groups. Palmitoleic acid levels in the HFD group increased significantly ($p < 0.05$) compared to the different groups. The high-fat–fructose diet did not produce significant changes in the percentages of ARA (20:4) and DHA (22:6) across all the treatments. On the other hand, no effects were observed in the profiles of DMA compounds.

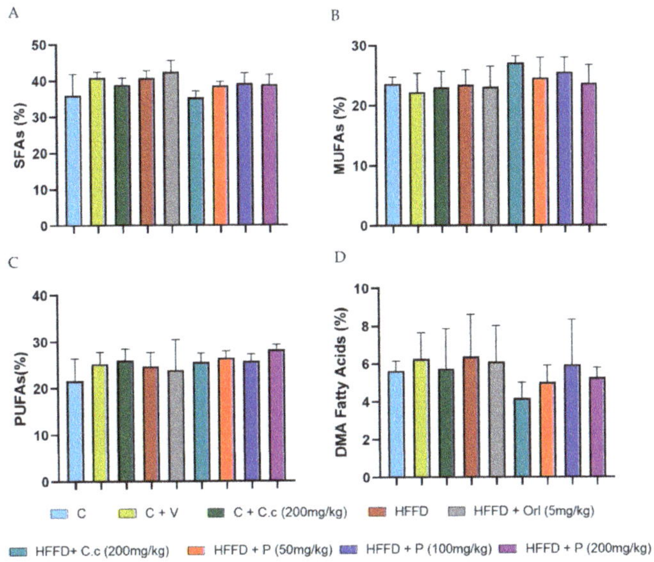

Figure 3. Effect of *Callistemon citrinus* extract and phytosomes on the lipid profile in brain tissue, (**A**) saturated fatty acids (SFAs), (**B**) monounsaturated fatty acids (MUFAs), (**C**) polyunsaturated fatty acids (PUFAs), and (**D**) dimethyl acetaldehyde (DMA) fatty acids. Control (C); control + vehicle (C + V); control + *C. citrinus* leaf extract (C + C.c, 200 mg/kg), high-fat diet (HFD), HFD + orlistat (HFD + Orl, 5 mg/kg), HFD + *C. citrinus* leaf extract (HFD + C.c, 250 mg/kg), HFD + phytosomes (HFD + P, 50 mg/kg, 100 mg/kg, and 200 mg/kg, respectively). Values are expressed as mean ± standard deviation (ANOVA followed by Tukey's test, $p < 0.05$, $n = 6$).

Table 4. Percentage of the major fatty acids in the brain of control rats and the different treatments.

Fatty Acid	C	C + V	C + C.c (200 mg/kg)	HFFD	HFFD + Orl (5 mg/kg)	HFFD+ C.c (200 mg/kg)	HFFD + P (50 mg/kg)	HFFD + P (100 mg/kg)	HFFD + P (200 mg/kg)
C14:0, Myristic	0.12 ± 0.04 c	0.24 ± 0.04 ab	0.18 ± 0.04 abc	0.24 ± 0.03 abc	0.18 ± 0.03 abc	0.29 ± 0.08 a	0.16 ± 0.01 bc	0.17 ± 0.02 abc	0.15 ± 0.05 bc
C15:0, Pentadecylic	0.07 ± 0.01	0.11 ± 0.01	0.10 ± 0.01	0.10 ± 0.01	0.09 ± 0.01	0.07 ± 0.02	0.09 ± 0.01	0.07 ± 0.03	0.10 ± 0.04
C16:0, Palmitic	17.40 ± 1.56	20.03 ± 1.60	18.95 ± 0.91	18.80 ± 0.30	20.42 ± 1.09	18.52 ± 2.39	19.29 ± 0.66	19.18 ± 1.85	19.27 ± 1.66
C17:0, Margaric	0.23 ± 0.01	0.30 ± 0.02	0.27 ± 0.04	0.33 ± 0.02	0.31 ± 0.09	0.23 ± 0.03	0.29 ± 0.03	0.29 ± 0.05	0.30 ± 0.02
C18:0, Stearic	15.08 ± 1.85 bc	18.74 ± 0.84 ab	18.15 ± 0.67 abc	18.24 ± 0.74 abc	19.91 ± 0.95 a	14.61 ± 3.18 c	17.66 ± 0.35 abc	17.88 ± 0.60 abc	18.20 ± 0.79 abc
C20:0, Arachidic	0.42 ± 0.21	0.22 ± 0.08	0.26 ± 0.06	0.42 ± 0.20	0.32 ± 0.09	0.38 ± 0.16	0.23 ± 0.04	0.31 ± 0.08	0.173 ± 0.040
C22:0, Behenic	0.37 ± 0.20	0.29 ± 0.04	0.31 ± 0.09	0.37 ± 0.20	0.34 ± 0.14	0.32 ± 0.19	0.16 ± 0.02	0.29 ± 0.09	0.16 ± 0.05
C16:1*n*-9, Palmitoleic	0.41 ± 0.06 b	0.45 ± 0.05 ab	0.54± 0.02 ab	0.56 ± 0.05 a	0.49 ± 0.05 ab	0.43 ± 0.01 ab	0.50 ± 0.04 ab	0.53 ± 0.05 ab	0.49 ± 0.01 ab
C18:1*n*-9, Oleic	12.34 ± 0.12	14.58 ± 1.35	14.868 ±1.00	14.340 ± 1.019	16.42 ± 2.23	14.33 ± 3.51	14.263 ± 0.348	15.987 ± 0.771	15.506 ± 1.872
C20:1*n*-9, Eicosenoic	2.48 ± 0.06 a	0.94 ± 0.56 b	1.12 ± 0.14 b	0.95 ± 0.21 b	1.23 ± 0.41 ab	2.03 ± 0.058 ab	1.26 ± 0.36 ab	1.62 ± 0.54 ab	0.89 ± 0.25 b
C22:1*n*-9, Erucic	0.21 ± 0.08	0.18 ± 0.10	0.13 ± 0.06	0.14 ± 0.04	0.20 ± 0.10	0.13 ± 0.06	0.13 ± 0.05	0.20 ± 0.21	0.17 ± 0.21
C24:1*n*-9, Nervonic	1.33 ± 0.25 a	0.25 ± 0.18 b	0.31 ± 0.03 b	0.24 ± 0.16 b	0.55 ± 0.35 b	0.45 ± 0.07 b	0.23 ± 0.07 b	0.39 ± 0.17 b	0.21 ± 0.18 b
C18:2*n*-6, LA	0.81 ± 0.10 ab	0.94 ± 0.19 a	0.88 ± 0.15 ab	0.73 ± 0.10 ab	0.64 ± 0.13 ab	0.53 ± 0.12 b	0.68 ± 0.08 ab	0.68 ± 0.10 ab	0.80 ± 0.16 ab
C20:4*n*-6. Arachidonic	7.92 ± 1.64	9.57 ± 0.67	9.77 ± 0.31	9.44 ± 0.75	9.60 ± 1.50	7.21 ± 1.35	9.96 ± 0.14	9.91 ± 0.38	10.48 ± 0.14
C22:4*n*-6, Adrenic	2.08 ± 0.37 bc	2.93 ± 0.15 abc	3.18 ± 0.45 a	2.92 ± 0.33 abc	3.01 ± 0.57 ab	1.98 ± 0.46 c	3.12 ± 0.15 a	3.34 ± 0.15 a	3.15 ± 0.25 a
C22:5*n*-3, DPAn-3	0.25 ± 0.15 c	0.63 ± 0.18 abc	1.05 ± 0.38 ab	1.05 ± 0.30 ab	0.79 ± 0.46 abc	0.36 ± 0.09 bc	0.93 ± 0.16 abc	1.20 ± 0.17 a	1.12 ± 0.07 a
C22:6*n*-3, DHA	10.09 ± 2.32	11.55 ± 1.39	10.54 ± 1.10	10.35 ± 1.51	9.52 ± 3.99	6.53 ± 1.55	10.74 ± 0.77	9.95 ± 0.65	12.09 ± 0.45
DMA C16:0	1.66 ± 0.06	1.59 ± 0.35	1.76 ± 0.18	1.67 ± 0.43	1.79 ± 0.56	0.95 ± 0.07	1.51 ± 0.14	1.49 ± 0.45	1.68 ± 0.22
DMA C18:0	3.36 ± 0.37	3.48 ± 0.62	2.81 ± 1.35	3.39 ± 0.83	3.73 ± 1.08	1.87 ± 0.41	2.56 ± 0.51	2.28 ± 0.88	3.29 ± 0.17
DMA C18:1 cis	1.430± 0.06	0.85 ± 0.51	0.82 ± 0.40	0.91 ± 0.37	0.96 ± 0.31	0.76 ± 0.84	0.65 ± 0.13	0.88 ± 0.15	0.84 ± 0.26
DMA C18:1 trans	1.68 ± 0.23 a	0.88 ± 0.51 ab	0.84 ± 0.40 ab	0.87 ± 0.34 ab	0.90 ± 0.30 ab	0.77 ± 0.01 ab	0.63 ± 0.11 b	0.86 ± 0.16 ab	0.75 ± 0.24 ab

All values expressed as mean ± SEM ($n = 6$); values statistically different (a, b, c) among groups ($p \leq 0.05$) according to Tukey's test.

3.3. Parallel Artificial Membrane Permeability Assay (PAMPA)

In the research to use natural products with neuroprotective effects, the first step is to demonstrate their permeability through the BBB using the PAMPA, and thus select the compounds that can access the brain. However, a pharmacokinetics study in vivo is required to demonstrate the compound's bioavailability. In this study, the PAMPA was used to identify components of *Callistemon citrinus* capable of crossing the intestinal and blood–brain barriers (BBBs). 1,8-Cineole exhibited partial permeability in the artificial intestinal barrier assay, with a concentration of 0.024 mg/mL. Meanwhile, α-terpineol and L-pinocarveol were detected at very low levels (Figure 4). Conversely, in the BBB permeability assay, 1,8-cineole and α-terpineol demonstrated higher permeability, with concentrations of 0.094 mg/mL and 0.016 mg/mL, respectively (Figure 5).

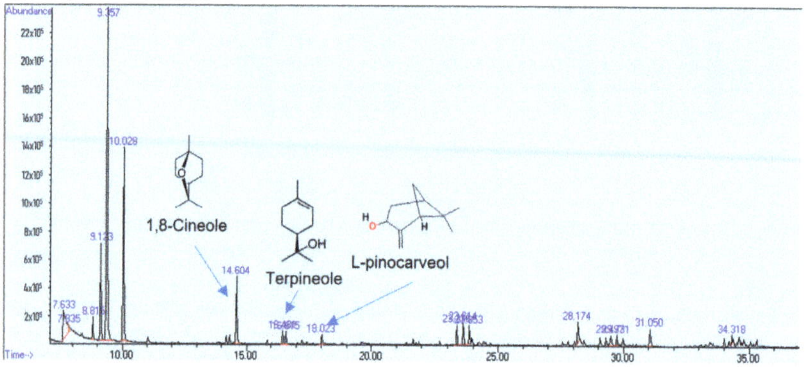

Figure 4. GC/MS chromatogram (arbitrary units) of *C. citrinus* leaf extract showing compounds that could cross the intestinal barrier in vitro.

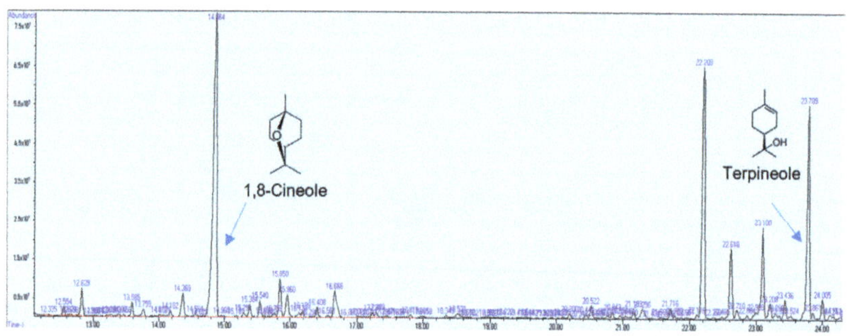

Figure 5. GC/MS chromatogram (arbitrary units) of *C. citrinus* leaf extract showing compounds that could cross the blood–brain barrier in vitro.

3.4. Biomarkers of Oxidative Stress

Figure 6A–C shows that the high-fat diet (HFD) group showed high levels of advanced oxidation protein products (AOPPs) and lipid peroxidation products, such as 4-hydroxynonenal (4-HNE) and malondialdehyde (MDA), as compared to the other groups ($p < 0.0001$). However, the level of 4-HNE in the control group was higher than in the treatment groups ($p < 0.0001$), which showed significantly lower levels of 4-HNE.

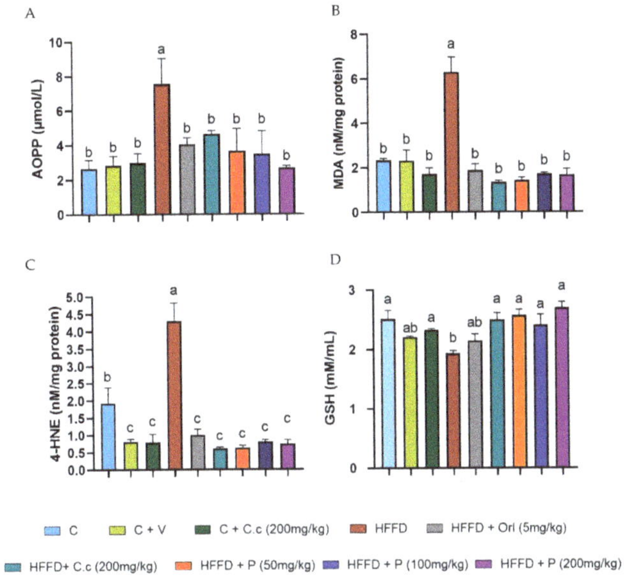

Figure 6. Effect of *Callistemon citrinus* extract and phytosomes on the generation of oxidative stress markers. Advanced oxidation protein products (AOPPs) (**A**), malondialdehyde (MDA) (**B**), 4-hydroxynonenal (4-HNE) (**C**), and reduced glutathione (GSH) (**D**). Control (C), control + vehicle (C + V), control + *C. citrinus* leaf extract (C + C.c, 200 mg/kg), high-fat diet (HFD), HFD + orlistat (HFD + Orl, 5 mg/kg), HFD + *C. citrinus* leaf extract (HFD + C.c, 250 mg/kg), HFD + phytosomes (HFD + P, 50, 100, and 200 mg/kg, respectively). Values are expressed as mean ± standard deviation (ANOVA followed by Tukey's test, $p < 0.001$, $n = 6$). Different letters (a, b, c) indicate statistically significant differences between groups.

Finally, the HFD group exhibited significantly lower levels of reduced glutathione (GSH) compared to the control, *C. citrinus*, HFD-treated *C. citrinus*, and phytosomes (p values ranging from 0.0001 to 0.016). Meanwhile, the groups supplemented with vehicle and HFD + Orlistat showed no statistically significant differences compared to the HFD-alone groups (Figure 6D). The groups that received a high-fat diet supplemented with the extract and phytosomes from *C. citrinus* exhibited a decrease in reactive oxygen species (ROS) production, thereby preventing oxidative stress in the brain.

3.5. Antioxidant Enzyme Activities

CAT activity in the HFD group showed a statistically significant 77% decrease ($p < 0.05$) compared to the control group ($p < 0.0001$). On the other hand, the HFD group supplemented with orlistat also showed a 55% reduction in CAT activity compared to the control group ($p < 0.05$). However, the groups supplemented with *Callistemon citrinus* extract and phytosones at the three different doses reduced the impact of the high-fat–fructose diet, maintaining CAT activity levels similar to the control group (Figure 7A). Additionally, as shown in Figure 7C, SOD activity in the brain of the HFD group was significantly reduced by 24% compared to both groups without HFD and those fed HFD but supplemented with *C. citrinus*, phytosome, and orlistat (p values ranging from 0.0003 to 0.007). Conversely, the PON activity of the HFD group revealed a statistically significant 20% decrease compared to the control and HFD groups supplemented with phytosomes at three doses (p values ranging from 0.0004 to 0.0017). In contrast, the groups with vehicle, *C. citrinus* extract without and with diet were slightly lower but not statistically significant compared to the control group. Analysis of the HFD group treated with orlistat again showed no significant difference between the HFD group and the other groups (Figure 7D).

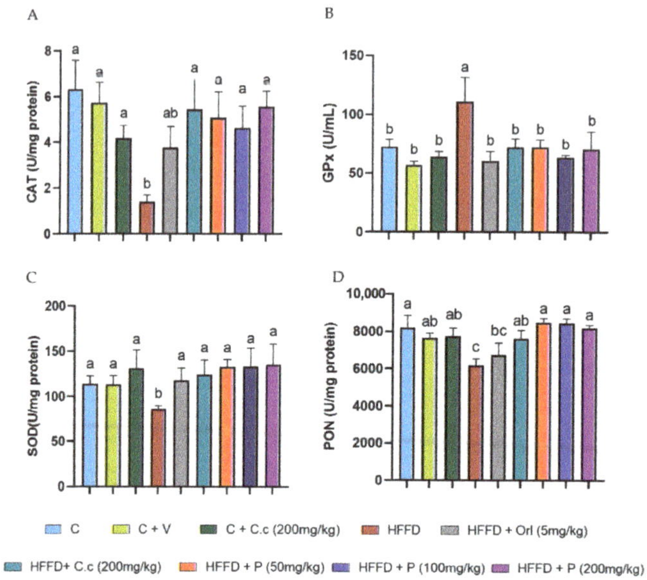

Figure 7. Effect of *C. citrinus* leaf extract and phytosomes on the activity of catalase (CAT) (**A**), glutathione peroxidase (GPx) (**B**), superoxide dismutase (SOD) (**C**), and paraoxonase (PON) (**D**) in brain tissue. Control (C), control + vehicle (C + V), control + *C. citrinus* leaf extract (C + C.c, 200 mg/kg), high-fat diet (HFD), HFD + orlistat (HFD + Orl, 5 mg/kg), HFD + *C. citrinus* leaf extract (HFD + C.c, 250 mg/kg), HFD + phytosomes (HFD + P, 50, 100, and 200 mg/kg, respectively). Values are expressed as mean ± standard deviation (ANOVA followed by Tukey's test, $p < 0.001$, n = 6). Different letters (a, b, c) indicate statistically significant differences between groups.

Unlike other enzymes, the levels of GPx activity exhibit a significant 2-fold increase in the HFD group compared to all other groups (p values ranging from 0.0001 to 0.027) (Figure 7B). Our results showed that, regardless of the phytosome concentration, all treatments had the same effect on the oxidative stress marker and antioxidant enzymes. However, phytosomes loaded with 50 mg/kg of *C. citrinus* extract demonstrated similar antioxidant potential to higher doses, suggesting that a lower dose might be more effective for achieving therapeutic benefits.

4. Discussion

This study utilized scanning electron microscopy (SEM), which provided valuable insights into the solid-state properties and surface morphology of complexes. In future studies, transmission electron microscopy (TEM) could be used to reveal surface topography along with detailed internal structure and crystallographic information sample. SEM analyses do not allow for definitive conclusions about the formation of phytosomes. Therefore, it is better to describe it as a phospholipid complex formulation with potential phytosomes characteristics. However, the phospholipid complex formulation prepared in this study was able to encapsulate 80.49 ± 0.07% of C. citrinus extract and had a size range of 129.98 ± 18.30 nm, which indicates that this small particle size is suitable for oral absorption. Additionally, the use of soybean phosphatidylcholine, which has a higher absorption rate, enables this formulation to be used orally administration.

Research findings confirm that *Callistemon citrinus* extract and its phytosomes formulation can help prevent weight gain and reduce oxidative stress in the brains of rats consuming a high-fat, high-fructose diet for 16 weeks. Previously, we reported that the anti-obesogenic, antioxidant, and anti-inflammatory properties of *C. citrinus* are due to its major bioactive compounds [20]. Ayala-Ruiz et al. [24] demonstrated that supplementation with 1,8-cineole, α-terpineol, limonene, and a combination of these terpenes decreases body weight gain in rats fed a high-fat–sucrose diet, and their behavior closely resembles that of the control group. Moreover, previous studies have suggested that the potential anti-obesogenic mechanism of *C. citrinus* may be linked to the inhibition of pancreatic lipase [20].

The increase in the intake of saturated and trans fats, along with excess sucrose or fructose, contributes to overweight or obesity and associated conditions, including type 2 diabetes, hyperlipidemia, hypertension, and hypercholesterolemia [37]. The accumulation of adipose tissue results in increased levels of free fatty acids in the blood and other tissues. High-fat, high-carbohydrate diets have been shown to alter the lipid profiles of organs such as the liver, heart, and skeletal muscle, resulting in higher levels of saturated fatty acids (SFAs) and a decrease in polyunsaturated fatty acids (PUFAs) [38–40]. The brain is the second tissue with a high content of lipids, but it has a lower antioxidant system; therefore, a rise in fatty acid levels from the diet results in an increase in ROS due to β-oxidation [41].

Kaplan and Greenwood [5] first reported the relationship between high-fat intake and cognitive impairment in rats fed soybean oil or lard. Conversely, Da Silva-Santi et al. [42] demonstrated that brain fatty acids are affected in mice fed high-fat or high-carbohydrate diets for 7, 14, 28, or 56 days. They observed that, initially, differences in fatty acid composition were evident; however, by the end, these differences had leveled out. Inflammatory markers were higher in the high-fat diet group compared to the high-carbohydrate diet, while markers of microglial infiltration were more prominent in the high-carbohydrate diet group. These diets have been linked to changes in the brain and are associated with the promotion of neuroinflammation [43]. Sighinolfi et al. [44] demonstrated that diets containing 60% fat (40% SFA) induce lasting changes in the lipid composition of specific brain regions, including the cortex, striatum, and hypothalamus—areas crucial for eating behavior and responses related to peripheral metabolism.

Although lipid composition is tightly regulated in the brain, this study aimed to determine whether phytosomes loaded with *Callistemon citrinus* affect the fatty acid profile in the brains of rats fed a diet rich in animal and vegetable fats, along with added fructose for 16 weeks. The primary fatty acids in the brain are arachidonic acid (ARA, 20:4*n*-6) and docosahexaenoic acid (DHA, 22:6*n*-3). Since the brain's de novo synthesis of polyunsaturated fatty acids (PUFAs) is limited, these fatty acids must be obtained from the diet or their precursors, α-linolenic acid (ALA, 18:3*n*-3) and linoleic acid (LA, 18:2*n*-6), which are metabolized into PUFAs in the liver. Conversely, eicosapentaenoic acid (EPA, 20:5*n*-3), ALA, and LA are generally found at lower concentrations in the brain compared to arachidonic acid (ARA) and docosahexaenoic acid (DHA).

Despite differences in treatment, this study found that the overall brain fatty acid profile across all groups showed minor changes in fatty acid concentrations. This aligns with the findings reported by Reichlmayr-Lais et al. [45], who examined the effects of various vegetable oil fats. The percentages of fatty acids (C16:0, C18:0, C20:4, and C22:6) observed in our study were similar to those previously reported in the brain [46]. The group that received a high-fat–fructose diet supplemented with *C. citrinus* extract showed values identical to those of the control group. In contrast, the other groups had slightly higher values than these two groups (Table 3, Figure 4). Additionally, the fatty aldehydes in the brain, such as C16:0 dimethyl acetaldehyde (DMA), C18:0 DMA, and C18:1 (*n*-7 or *n*-9), showed no significant differences between groups.

Although several studies report small changes in the brain lipid profile as a function of brain region, this study quantified the profile for the entire brain due to the limitation established by the University's Bioethical Committee Guidelines regarding the number of animals that can be used in these studies. Unfortunately, given the number of Wistar rats used in this study, it was not possible to determine the fatty acid profile by brain region.

Researchers have identified numerous natural products with antioxidant and anti-inflammatory properties; however, few can cross the blood–brain barrier when present in complex mixtures. This study used the in vitro parallel artificial membrane permeability assay to identify potential compounds in *Callistemon citrinus* that might cross the intestinal and blood–brain barriers. Mony et al. [47] described the ability of terpenoid compounds to cross the blood–brain barrier (BBB), attributing their restorative effects to damage caused by oxidative stress within it, thereby normalizing its function and promoting brain health. Jäger et al. [48] revealed that 1,8-cineole can enter the body through inhalation, with its highest concentration in blood serum occurring at 18 min after exposure. They also demonstrated that body fat influences the half-life of cineole, as it is eliminated more quickly by organisms with lower body fat. Our findings suggest that 1,8-cineole and α-terpineol can permeate both the intestinal and blood–brain barriers in vitro assays. This monoterpene has also shown effectiveness against inflammatory intestinal diseases, such as Crohn's disease and ulcerative colitis, by increasing the activity of the PPARγ receptor [49]. According to Satou et al. [50], other monoterpenes such as limonene, α-pinene, and linalool can also be found in brain tissue after inhalation. However, in this study, we were unable to detect limonene as a compound capable of permeating the brain through the PAMPA BBB assay, despite its presence in the *C. citrinus* extract. Dao et al. [51] confirmed that 1,8-cineole can cross the blood–brain barrier (BBB), even when administered orally. These findings suggest that terpenes from *C. citrinus* may help enhance the brain's antioxidative status. However, a future pharmacokinetic study of compounds in the plasma or brain following oral administration will be considered.

Callistemon citrinus derives its antioxidant properties mainly from various secondary metabolites, including 1,8-cineole, limonene, and α-terpineol [22], as well as phenolic compounds and flavonoids [23]. Ayala-Ruiz et al. [24] demonstrated that 1,8-cineole, limonene, and α-terpineol, both alone and in combination, exhibit strong antioxidant effects by reducing oxidative stress markers (MDA, HNE, AOPP) and pro-inflammatory cytokines (TNFα, IL-6, and leptin) in rat livers fed a high-fat–sucrose diet. In our study, we observed that oxidative stress in brain tissue was triggered in the group fed a high-fat–fructose diet, leading to an increase in lipid peroxidation products, specifically MDA and 4-HNE. Meanwhile, MDA is believed to be mutagenic, and 4-HNE is especially toxic due to its reactivity with thiol and amino acid groups [52]. As reported by Hou et al. [53], the concentration of MDA rises in the brains of rats consuming a high-fat diet. Similarly, Maciejczyk et al. [54] found significantly higher levels of MDA and 4-HNE in the cortex and hypothalamus of rats on a high-fat diet compared to the control group.

Another biomarker commonly used to measure oxidative stress is advanced protein oxidation products (AOPPs). These mainly consist of damaged protein aggregates, often containing dityrosine residues that contribute to crosslinking, disulfide bonds, and carbonyl groups. Initially, AOPP were only detected in plasma, formed by the reaction of plasma proteins (albumin) with chlorinated oxidants, which are products of myeloperoxidase activity [30]. However, AOPP levels are also reported in the cerebrospinal fluid, liver, kidney, and brain [55–58]. Furthermore, elevated levels of AOPP have been found in the brains of people with Alzheimer's disease [59]. In this study, a significant increase in AOPP levels was observed in the HFD group compared to the other groups. The groups fed an HFD and supplemented with the extract and phytosomes of *Callistemon citrinus* showed a reduction in oxidative stress biomarkers (MDA, 4-HNE, and AOPP), making them statistically comparable to the control group.

Reduced glutathione (GSH) is a tripeptide vital for the brain, playing a key role in detoxifying reactive oxygen species (ROS). The brain requires a substantial amount of oxygen, making it particularly susceptible to the formation of reactive species [60]. Additionally, it is the second most tissue rich in polyunsaturated fatty acids. Moreover, the antioxidant system in the brain is weaker than in other tissues, increasing its vulnerability to oxidative stress [61]. Alzoubi et al. [62] noted a decline in GSH levels in the cortex but found no differences in the hippocampus and cerebellum of rats fed a high-fat diet. Similarly, Cavaliere et al. [63] reported a significant decrease in reduced glutathione levels in the brains of mice on a high-fat diet. These findings support our results and highlight the effectiveness of *C. citrinus* extract and phytosomal formulation in reducing oxidative stress caused by a high-fat diet (HFD) in the brains of Wistar rats.

The primary enzymatic defense against oxidative stress involves three enzymes: SOD, which catalyzes the dismutation of superoxide anion radicals into hydrogen peroxide, followed by CAT and GPx, the enzymes that reduce hydrogen peroxide, thereby preventing the accumulation of harmful reactive species ROS. However, these enzymes are affected by HDF feeding in different ways. Liu et al. [64] reported a decrease in the expression of SOD and CAT in the brain of mice fed a high-fat diet for 6 months. Similarly, El Azab and Abdulmalek [65] showed significant reductions in the levels of both enzymes in the brains of both young and aged rats on a high-fat diet. Our study shows that oral supplementation with the extract and phytosomes of *C. citrinus* increases SOD and CAT activities in the brain, even in rats fed a high-fat diet (HFD), which contrasts with the findings in the HFD group. This result highlights that *C. citrinus*, whether extracted or in phytosome form, exhibits a significant antioxidant effect in the brain.

The third enzyme, GPx, catalyzes the conversion of peroxides and hydroperoxides into water or alcohols. GPx plays a crucial role in maintaining cellular redox balance and protecting neural cells from oxidative stress [66]. The activity of this enzyme decreases in obesity models; for example, Amri et al. [67] reported low GPx activity levels in the brains of Wistar rats fed a high-fat, high-fructose diet for 12 weeks. Similarly, Charradi et al. [68] observed reduced glutathione peroxidase (GPx) enzyme activity in the brains of rats fed a high-fat diet for 6 weeks. Conversely, Savaskan et al. [69] demonstrated an overexpression of GPx4 (cytosolic) in response to brain damage. An increase in GPx activity has been linked to higher levels of lipid peroxidation products [70]. In our study, elevated levels of MDA and 4-HEN were found only in the high-fat–fructose diet (HFD) group, which was associated with an increase in GPx. Conversely, all groups consuming an HFD supplemented with extracts and phytosomes of *C. citrinus* exhibited GPx activity similar to that of the control group. A previous study by Ortega-Pérez et al. [21] found that HFD increases GPx activity in the liver and heart, which is associated with decreased GSH levels and elevated levels of MDA, 4-HNE, and AOPP.

Paraoxonase is another enzyme used as a biomarker of oxidative stress; anti-inflammatory activity has been reported in models of Alzheimer's disease and hyperlipidemia, where reduced PON activity has been observed [71]. Shih et al. [72] reported that PON2 is essential for preventing obesity. The first reported PON activity involves protecting LDL and HDL against lipid peroxidation. Nguyen and Sok [73] reported that monoenoic acid inhibited its activity. In our study, the HFD group exhibited a high concentration of palmitoleic acid as well as elevated levels of 4-HNE, which is also known to inhibit PON activity. Meanwhile, supplementation with *C. citrinus* extract and phytosomes helps reduce the decline of palmitoleic acid and 4-HNE levels, maintaining PON activity similar to that of the control group.

Orlistat is a medication used to assist weight loss mainly by blocking pancreatic lipase, which prevents fats from being absorbed into the intestine. Zakaria et al. [74] demonstrated that orlistat can help prevent insulin resistance and oxidative stress, and it protects the liver from damage caused by increased fat intake during a high-fat diet. Othman et al. [75] reported that orlistat can reduce oxidative stress in the myocardial tissue of obese rats. Mahmoudi et al. [76] demonstrated that orlistat, when combined with grape seed extract, resulted in a reduction in brain lipoperoxidation. AL-Dalaeen et al. [77] showed that orlistat has beneficial anti-inflammatory effects on the brain.

In our study, orlistat was administered via oral gavage simultaneously with the induction of obesity, whereas in other studies, obesity was first established before treatment. The HFD plus orlistat group showed a reduction in weight gain but experienced loose stools. In contrast, the group fed HFD supplemented with *C. citrinus* extract and phytosomes reduced body weight without this side effect. The antioxidant activity of orlistat in this study was found to be similar, based on levels of MDA, 4-HNE, and AOPP, with an increase in SOD and GPx activity compared to the treatment groups with *C. citrinus*. However, orlistat had minimal effects on CAT, PON, and GSH. Our results support the evidence that a high-fat–fructose diet is closely linked to brain oxidative stress, characterized by increased levels of MDA, 4-HNE, and AOPP, as well as decreased levels of GSH. Meanwhile, the extracts and phytosomes of *C. citrinus* exhibited antioxidant properties, which lowered the biomarkers of oxidative stress (MDA, 4-HNE, and AOPP) and increased the enzymatic antioxidant activity of CAT, SOD, and PON1, while also raising the levels of GSH.

Most of the observed antioxidant activity is attributed to the major terpenes and phenols in *C. citrinus*, including 1,8-cineole, α-terpineol, gallic acid, p-coumaric acid, and ellagic acid [18,21]. 1,8-Cineole and α-terpineol have been demonstrated to exhibit antioxidant and neuroprotective effects in the brain [78,79]. Gallic acid provided a protective effect on the blood–brain barrier (BBB) [80]. Conversely, p-coumaric and ellagic acids demonstrated both antioxidant and anti-neuroinflammatory impact in the brain [81,82].

Although phytosomes improve the bioavailability, solubility, and stability of compounds in the extract, no significant differences were observed among the various phytosomes loaded with C. citrinus at doses 50, 100, and 200 mg/kg, or when compared to the extract at 200 mg/kg, in their ability to reduce oxidative stress in the brain. One possible explanation is that, according to the in vitro PAMPA, 1,8-cineole and α-terpineol are able to cross both the intestinal and blood–brain barriers. These compounds may help prevent oxidative stress in the brains of Wistar rats fed a high-fat–fructose diet. However, using a lower dose of phytosomes to achieve the same effect as a higher dose is a promising approach. Using male Wistar rats simplified the process because, in female rats, hormone fluctuations significantly affect some parameters. However, future studies should consider female rats in the tests. Another limitation of the study was the small number of animals, which prevented the examination of the fatty acid profile in different brain regions and the extension of the study period to collect samples at specific times. Future studies should consider dose-dependent tests, evaluate animal behavior, assess pro-inflammatory cytokines, such as TNF-α, IL-6, and IL-1β, along with leptin levels, as well as enzymes involved in the inflammatory process, including myeloperoxidase, cyclooxygenase-2, 5-lipoxygenase, and xanthine oxidase. Finally, analyzing the expression of antioxidative enzymes and testing the major compounds should be interesting to validate the possible mechanism of action of *Callistemon citrinus* extract and phytosomes act.

5. Conclusions

Our findings provide the first evidence that a high-fat–fructose diet causes oxidative stress in the brain, which can be prevented by administering phytosomes loaded with *Callistemon citrinus* extracts at a lower concentration than the pure extract. Moreover, our study shows that 1,8-cineole and α-terpineol, the main bioactive compounds identified in *C. citrinus*, are likely responsible for its antioxidant effects. Therefore, the phytosomes loaded with *C. citrinus* might be seen as a promising supplement to prevent oxidative stress in the brain. In future studies, it would be advisable to perform toxicology and pharmacokinetic studies of phytosomes to gather additional data that supports the use of this formulation in combating oxidative stress in the brain caused by obesity.

Supplementary Materials: The following supporting information can be downloaded at: https://www.mdpi.com/article/10.3390/biom15081129/s1, Figure S1. GC/MS chromatogram (arbitrary units) of *C. citrinus* leaf extract illustrating its compounds. Table S1. Compounds of *Callistemon citrinus* leaves extract found in GC/MS. Table S2. Composition of standard rat chow diet and high-fat-fructose diet.

Author Contributions: Investigation, O.R.M.-R. and L.G.O.-P.; visualization, O.R.M.-R. and P.R.-C.; formal analysis, A.J.G.-C., L.A.A.-R. and O.R.M.-R.; data curation, L.G.O.-P. and J.S.P.-S.; software, A.A.-M. and D.G.-H.; methodology, O.R.M.-R., A.J.G.-C. and L.G.O.-P.; writing—original draft, O.R.M.-R. and P.R.-C.; supervision, P.R.-C.; writing—review and editing, P.R.-C., D.G.-H. and A.A.-M. All authors have read and agreed to the published version of the manuscript.

Funding: This research did not receive a specific grant from public, commercial, or not-for-profit funding agencies.

Institutional Review Board Statement: All experimental procedures were approved and adhered to the laboratory animal care guidelines and use established by the Mexican Official Standard (NOM-062-ZOO-1999) and were approved by the Institutional Committee for the use of Animals of the UMSNH (approval date: 1 December 2023; protocol ID IIQB-CIBE-06-2023.

Informed Consent Statement: Not applicable.

Data Availability Statement: The raw data supporting the conclusions of this article will be made available by the authors on request.

Acknowledgments: The authors thank the Coordinación de la Investigación Científica de la Universidad Michoacana de San Nicolás de Hidalgo in México for the facilities provided for the study.

Conflicts of Interest: The authors declare no conflicts of interest.

References

1. GBD 2021 Adult BMI Collaborators. Global, regional, and national prevalence of adult overweight and obesity, 1990–2021, with forecast to 2050: A forecasting study for the Global Burden of Disease Study 2021. *Lancet* **2025**, *405*, 813–838. [CrossRef] [PubMed]
2. Pietrocola, F.; Bravo-San Pedro, J.M. Targeting autophagy to counteract obesity-associated oxidative stress. *Antioxidants* **2021**, *10*, 102. [CrossRef] [PubMed]
3. Jiang, S.; Liu, H.; Li, C. Dietary regulation of oxidative stress in chronic metabolic diseases. *Foods* **2021**, *10*, 1854. [CrossRef]
4. Maurizi, G.; Della, L.; Maurizi, A.; Poloni, A. Adipocytes properties and crosstalk with immune system in obesity-related inflammation. *J. Cell. Physiol.* **2018**, *233*, 88–97. [CrossRef]
5. Kaplan, R.J.; Greenwood, C.E. Dietary saturated fatty acids and brain function. *Neurochem. Res.* **1998**, *23*, 615–626. [CrossRef] [PubMed]

6. Lei, X.; Yang, Y. Vitexin and an HMG-Co A reductase inhibitor prevent the risks of atherosclerosis in high-fat atherogenic diet fed rats. *J. King Saud Univ. Sci.* **2020**, *32*, 2088–2095. [CrossRef]
7. Galaly, S.R.; Hozayen, W.G.; Amin, K.A.; Ramadan, S.M. Effects of Orlistat and herbal mixture extract on brain, testes functions and oxidative stress biomarkers in a rat model of high fat diet. *Beni Suef Univ. J. Basic Appl. Sci.* **2014**, *3*, 93–105. [CrossRef]
8. Berdún, R.; Obis, È.; Mota-Martorell, N.; Bassols, A.; Valent, D.; Serrano, J.C.; Jové, M. High-fat diet-induced obesity increases brain mitochondrial complex I and lipoxidation-derived protein damage. *Antioxidants* **2024**, *13*, 161. [CrossRef]
9. Pistell, P.J.; Morrison, C.D.; Gupta, S.; Knight, A.G.; Keller, J.N.; Ingram, D.K.; Bruce-Keller, A.J. Cognitive impairment following high fat diet consumption is associated with brain inflammation. *J. Neuroimmunol.* **2010**, *219*, 25–32. [CrossRef]
10. Chen, X.; Fu, G.; Zhao, Q.; Ke, Z.; Zhang, R. Selenoprotein GPX1 is a prognostic and chemotheraphy-related biomarker for brain lower grade glioma. *J. Trace Elem. Med. Biol.* **2022**, *74*, 127082. [CrossRef]
11. Ruck, L.; Wiegand, S.; Kühnen, P. Relevance and consequence of chronic inflammation for obesity development. *Mol. Cell. Pediatr.* **2023**, *10*, 16. [CrossRef]
12. Johnson, R.J.; Gomez-Pinilla, F.; Nagel, M.; Nakagawa, T.; Rodriguez-Iturbide, B.; Sanchez-Lozada, L.G.; Tolan, D.R.; Lanaspa, M.A. Cerebral fructose metabolism as a potential mechanism driving Alzheimer's disease. *Front. Aging Neurosci.* **2020**, *12*, 560865. [CrossRef]
13. Suleiman, J.B.; Nna, V.U.; Zakaria, Z.; Othman, Z.A.; Bakar, A.B.A.; Mohamed, M. Obesity-induced testicular oxidative stress, inflammation and apoptosis: Protective and therapeutic effects of orlistat. *Reprod. Toxicol.* **2020**, *95*, 113–122. [CrossRef]
14. Barani, M.; Sangiovanni, E.; Angarano, M.; Rajizadeh, M.A.; Mehrabani, M.; Piazza, S.; Gangadharappa, H.V.; Pardakhty, A.; Mehrbani, M.; Dell' Agli, M.; et al. Phytosomes as innovative delivery system for phytochemicals: A comprehensive review of literature. *Int. J. Nanomed.* **2021**, *16*, 6983–7022. [CrossRef]
15. Deleanu, M.; Toma, L.; Sanda, G.M.; Barbălată, T.; Niculescu, L.Ș.; Sima, A.V.; Deleanu, C.; Săcărescu, L.; Suciu, A.; Alexandru, G.; et al. Formulation of phytosomes with extracts of Ginger rhizomes and rosehips with improved bioavailability, antioxidant and anti-inflammatory effects in vivo. *Pharmaceutics* **2023**, *15*, 1066. [CrossRef] [PubMed]
16. Cock, I.E. Antimicrobial activity of Callistemon citrinus and Callistemon salignus methanolic extracts. *Pharmacogn. Commun.* **2012**, *28*, 50–57. [CrossRef]
17. Kumar, R.; Gupta, A.; Singh, A.K.; Bishayee, A.; Pandey, A.K. The antioxidant and antihyperglycemic activities of bottlebrush plant (*Callistemon lanceolatus*) stem extracts. *Medicines* **2020**, *7*, 11. [CrossRef]
18. López-Mejía, A.; Ortega-Pérez, L.G.; Magaña-Rodríguez, O.R.; Ayala-Ruiz, L.A.; Piñón-Simental, J.S.; Hernández, D.G.; Rios-Chavez, P. Protective effect of *Callistemon citrinus* on oxidative stress in rats with 1, 2-dimethylhydrazine-induced colon cancer. *Biomed. Pharmacother.* **2021**, *142*, 112070. [CrossRef]
19. Piñón-Simental, J.S.; Ayala-Ruiz, L.A.; Ortega-Pérez, L.G.; Magaña-Rodríguez, O.R.; Meléndez-Herrera, E.; Aguilera-Méndez, A.; Rios-Chavez, P. Use of *Callistemon citrinus* as a gastroprotective and anti-inflammatory agent on indomethacin-induced gastric ulcers in obese rats. *PeerJ* **2024**, *12*, e17062. [CrossRef]
20. Ortega-Pérez, L.G.; Piñón-Simental, J.S.; Magaña-Rodríguez, O.R.; Lopéz-Mejía, A.; Ayala-Ruiz, L.A.; García-Calderón, A.J.; Godínez-Hernández, D.; Rios-Chavez, P. Evaluation of the toxicology, anti-lipase, and antioxidant effects of *Callistemon citrinus* in rats fed with a high fat-fructose diet. *Pharm. Biol.* **2022**, *60*, 1384–1393. [CrossRef]
21. Ortega-Pérez, L.G.; Hernández-Soto, J.A.; Padilla-Avalos, O.; Ayala-Ruiz, L.A.; Magaña-Rodríguez, O.R.; Piñón-Simental, J.S.; Aguilera-Méndez, A.; Godínez-Hernández, D.; Rios-Chavez, P. Role of Callistemon citrinus Leaf Phytosomes Against Oxidative Stress and Inflammation in Rats Fed with a High-Fat-Fructose Diet. *Antioxidants* **2024**, *13*, 1263. [CrossRef]
22. Petronilho, S.; Rocha, S.; Ramírez-Chávez, E.; Molina-Torres, J.; Rios-Chavez, P. Assessment of the terpenic profile of *Callistemon citrinus* (Curtis) Skeels from Mexico. *Ind. Crops Prod.* **2013**, *46*, 369–379. [CrossRef]
23. Ortega-Pérez, L.G.; Ayala-Ruiz, L.A.; Magaña-Rodríguez, O.R.; Piñón-Simental, J.S.; Aguilera-Méndez, A.; Godínez-Hernández, D.; Rios-Chavez, P. Development and evaluation of phytosomes containing *Callistemon citrinus* leaf extract: A preclinical approach for the treatment of obesity in a rodent model. *Pharmaceutics* **2023**, *15*, 2178. [CrossRef]
24. Ayala-Ruiz, L.A.; Ortega-Pérez, L.G.; Piñón-Simental, J.S.; Magaña-Rodríguez, O.R.; Meléndez-Herrera, E.; Rios-Chavez, P. Role of the major terpenes of *Callistemon citrinus* against the oxidative stress during a hypercaloric diet in rats. *Biomed. Pharmacother.* **2022**, *153*, 113505. [CrossRef]
25. Nájera-Maldonado, J.M.; Salazar, R.; Alvarez-Fitz, P.; Acevedo-Quiroz, M.; Flores-Alfaro, E.; Hernández-Sotelo, D.; Espinoza-Rojo, M.; Ramírez, M. Phenolic compounds of therapeutic interest in neuroprotection. *J. Xenobiot.* **2024**, *14*, 227–246. [CrossRef]
26. Rule, D.C.; Melson, E.A.; Alexander, B.M.; Brown, T.E. Dietary fatty acid composition impacts the fatty acid profiles of different regions of the bovine brain. *Animals* **2022**, *12*, 2696. [CrossRef]
27. Ulmann, L.; Mimouni, V.; Roux, S.; Porsolt, R.; Poisson, J.P. Brain and hippocampus fatty acid composition in phospholipid classes of aged-relative cognitive deficit rats. *Prostaglandins Leukot. Essent. Fat. Acids* **2001**, *64*, 189–195. [CrossRef]

28. Especificaciones Técnicas para la Producción, Cuidado y uso de los Animales de Laboratorio. Diario Oficial de la Federación, México Distrito Federal, Mexico. Available online: https://www.gob.mx/cms/uploads/attachment/file/203498/NOM-062-ZOO-1999_220801.pdf (accessed on 22 August 2001).
29. Bélair, V.; Orstat, V.; Maheux, M.; Lafrance, C.P.; Brochu, M.; Lightburn, B.; Moss, R. Permeability of native and digestive polyphenols from apple, blueberry and cranberry extracts using PAMPA membrane permeability assays. *J. Food Compos. Anal.* **2021**, *101*, 103945. [CrossRef]
30. Witko-Sarsat, V.; Friedlander, M.; Capeillere-Blandin, C.; Nguyen-Khoa, T.; Nguyen, A.T.; Zingraff, J.; Jungers, P.; Descamps-Latscha, B. Advanced oxidation protein products as a novel marker of oxidative stress in uremia. *Kidney Int.* **1996**, *49*, 1304–1313. [CrossRef]
31. Johnston, J.W.; Horne, S.; Harding, K.; Benson, E.E. Evaluation of the 1-methyl-2-phenylindole colorimetric assay for aldehydic lipid peroxidation products in plants: Malondialdehyde and 4-hydroxynonenal. *Plant Physiol. Biochem.* **2007**, *45*, 108–112. [CrossRef] [PubMed]
32. Sedlak, J.; Lindsay, R.H. Estimation of total, protein-bound, and nonprotein sulfhydryl groups in tissue with Ellman's reagent. *Anal. Biochem.* **1968**, *25*, 192–205. [CrossRef]
33. Aebi, H. Catalase in vitro. *Methods. Enzimol.* **1984**, *105*, 121–126.
34. Prabhu, K.S.; Reddy, P.V.; Gumpricht, E.; Hildenbrandt, G.R.; Scholz, R.W.; Sordillo, L.M.; Reddy, C.C. Microsomal glutathione S-transferase A1-1 with glutathione peroxidase activity from sheep liver: Molecular cloning, expression and characterization. *Biochem. J.* **2001**, *360*, 345–354. [CrossRef]
35. Bouhalit, S.; Kechrid, Z. Assessment of the potential role of l-methionine on nickel sulfate induced renal injury and oxidative stress in rat. *Assessment* **2018**, *11*, 390. [CrossRef]
36. Dantoine, T.F.; Debord, J.; Charmes, J.P.; Merle, L.; Marquet, P.; Lachatre, G.; Leroux-Robert, C. Decrease of serum paraoxonase activity in chronic renal failure. *J. Am. Soc. Nephrol.* **1998**, *9*, 2082–2088. [CrossRef]
37. Liu, T.W.; Heden, T.D.; Matthew Morris, E.; Fritsche, K.L.; Vieira-Potter, V.J.; Thyfault, J.P. High-fat diet alters serum fatty acid profiles in obesity prone rats: Implications for in vitro studies. *Lipids* **2015**, *50*, 997–1008. [CrossRef]
38. Xie, Z.; Li, H.; Wang, K.; Lin, J.; Wang, Q.; Zhao, G.; Jia, W.; Zhang, Q. Analysis of transcriptome and metabolome profiles alterations in fatty liver induced by high-fat diet in rat. *Metabolism* **2010**, *59*, 554–560. [CrossRef]
39. Pakiet, A.; Jakubiak, A.; Mierzejewska, P.; Zwara, A.; Liakh, I.; Sledzinski, T.; Mika, A. The effect of a high-fat diet on the fatty acid composition in the hearts of mice. *Nutrients* **2020**, *12*, 824. [CrossRef]
40. Antunes, M.M.; Godoy, G.; de Almeida-Souza, C.B.; Da Rocha, B.A.; da Silva-Santi, L.G.; Masi, L.N.; Carbonera, F.; Visentainer, J.V.; Curi, R.; Bazotte, R.B. A high-carbohydrate diet induces greater inflammation than a high-fat diet in mouse skeletal muscle. *Braz. J. Med. Biol. Res.* **2020**, *53*, e9039. [CrossRef]
41. Schönfeld, P.; Reiser, G. How the brain fights fatty acids' toxicity. *Neurochem. Int.* **2021**, *148*, 105050. [CrossRef]
42. Da Silva-Santi, L.; Antunes, M.; Mori, M.A.; De Almeida-Souza, C.; Vergílio Visentainer, J.; Carbonera, F.; Crisma, A.R.; Masi, L.N.; Hirabara, S.M.; Curi, R.; et al. Brain fatty acid composition and inflammation in mice fed with high-carbohydrate diet or high-fat diet. *Nutrients* **2018**, *10*, 1277. [CrossRef] [PubMed]
43. Denver, P.; Gault, V.A.; McClean, P.L. Sustained high-fat diet modulates inflammation, insulin signalling and cognition in mice and a modified xenin peptide ameliorates neuropathology in a chronic high-fat model. *Diabetes Obes. Metab.* **2018**, *20*, 1166–1175. [CrossRef]
44. Sighinolfi, G.; Clark, S.; Blanc, L.; Cota, D.; Rhourri-Frih, B. Mass spectrometry imaging of mice brain lipid profile changes over time under high fat diet. *Sci. Rep.* **2021**, *11*, 19664. [CrossRef]
45. Reichlmayr-Lais, A.M.; Stangl, G.I.; Kirchgessner, M.; Eder, K. Fatty acid composition of brain and heart of rats fed various dietary oils. *Nutr. Res.* **1994**, *14*, 829–840. [CrossRef]
46. Angelis-Pereira, M.C.; Barcelos, M.F.P.; Pereira, J.D.A.R.; Pereira, R.C.; de Souza, R.V. Effect of different commercial fat sources on brain, liver and blood lipid profiles of rats in growth phase. *Acta Cir. Bras.* **2017**, *32*, 1013–1025. [CrossRef]
47. Mony, T.J.; Elahi, F.; Choi, J.W.; Park, S.J. Neuropharmacological Effects of Terpenoids on Preclinical Animal Models of Psychiatric Disorders: A Review. *Antioxidants* **2022**, *11*, 1834. [CrossRef]
48. Jäger, W.; Našel, B.; Našel, C.; Binder, R.; Stimpfl, T.; Vycudilik, W.; Buchbauer, G. Pharmacokinetic studies of the fragrance compound 1, 8-cineol in humans during inhalation. *Chem. Senses* **1996**, *21*, 477–480. [CrossRef]
49. Venkataraman, B.; Almarzooqi, S.; Raj, V.; Bhongade, B.A.; Patil, R.B.; Subramanian, V.S.; Attoub, S.; Rizvi, T.A.; Adrian, T.E.; Subramanya, S.B. Molecular docking identifies 1,8-cineole (eucalyptol) as a novel PPARγ agonist that alleviates colon inflammation. *Int. J. Mol. Sci.* **2023**, *24*, 6160. [CrossRef]
50. Satou, T.; Hayakawa, M.; Kasuya, H.; Masuo, Y.; Koike, K. Mouse brain concentrations of α-pinene, limonene, linalool, and 1,8-cineole following inhalation. *Flavour Fragr. J.* **2017**, *32*, 36–39. [CrossRef]

51. Dao, L.; Jiang, L.; Chen, Y.; Ayisa; Sa, C. Study on the opening effect of eucalyptol on the blood–brain barrier and its brain pharmacokinetics. *Biomed. Chromatogr.* **2023**, *37*, e5631. [CrossRef]
52. Žarković, N.; Gęgotek, A.; Łuczaj, W.; Jaganjac, M.; Šunjić, S.B.; Žarković, K.; Skrzydlewska, E. Overview of the Lipid Peroxidation Measurements in Patients by the Enzyme-Linked Immunosorbent Assay Specific for the 4-Hydroxynonenal-Protein Adducts (4-HNE-ELISA). *Front. Biosci. Landmark* **2024**, *29*, 153. [CrossRef]
53. Hou, J.; Jeon, B.; Baek, J.; Yun, Y.; Kim, D.; Chang, B.; Kim, S.; Kim, S. High fat diet-induced brain damaging effects through autophagy-mediated senescence, inflammation and apoptosis mitigated by ginsenoside F1-enhanced mixture. *J. Ginseng Res.* **2022**, *46*, 79–90. [CrossRef]
54. Maciejczyk, M.; Żebrowska, E.; Zalewska, A.; Chabowski, A. Redox balance, antioxidant defense, and oxidative damage in the hypothalamus and cerebral cortex of rats with high fat diet—Induced insulin resistance. *Oxidative Med. Cell. Longev.* **2018**, *2018*, 6940515. [CrossRef]
55. Djordjevic, G.; Ljubisavljevic, S.; Sretenovic, S.; Kocic, G.; Stojanovic, I.; Stojanovic, S. The cerebrospinal fluid values of advanced oxidation protein products and total thiol content in patients with amyotrophic lateral sclerosis. *Clin. Neurol. Neurosurg.* **2017**, *163*, 33–38. [CrossRef]
56. Zhao, Y.; Zhang, L.; Ouyang, X.; Jiang, Z.; Xie, Z.; Fan, L.; Zhu, D.; Li, L. Advanced oxidation protein products play critical roles in liver diseases. *Eur. J. Clin. Investig.* **2019**, *49*, e13098. [CrossRef]
57. Cao, W.; Hou, F.F.; Nie, J. AOPPs and the progression of kidney disease. *Kidney Int. Suppl.* **2014**, *4*, 102–106. [CrossRef]
58. Gürler, H.Ş.; Bilgici, B.; Akar, A.K.; Tomak, L.; Bedir, A. Increased DNA oxidation (8-OHdG) and protein oxidation (AOPP) by low level electromagnetic field (2.45 GHz) in rat brain and protective effect of garlic. *Int. J. Radiat. Biol.* **2014**, *90*, 892–896. [CrossRef] [PubMed]
59. Aksenov, M.Y.; Aksenova, M.V.; Butterfield, D.A.; Geddes, J.W.; Markesbery, W.R. Protein oxidation in the brain in Alzheimer's disease. *Neuroscience* **2001**, *103*, 373–383. [CrossRef] [PubMed]
60. Dwivedi, D.; Megha, K.; Mishra, R.; Mandal, P.K. Glutathione in brain: Overview of its conformations, functions, biochemical characteristics, quantitation and potential therapeutic role in brain disorders. *Neurochem. Res.* **2020**, *45*, 1461–1480. [CrossRef] [PubMed]
61. Aoyama, K. Glutathione in the brain. *Int. J. Mol. Sci.* **2021**, *22*, 5010. [CrossRef]
62. Alzoubi, K.H.; Hasan, Z.A.; Khabour, O.F.; Mayyas, F.A.; Al Yacoub, O.N.; Banihani, S.A.; Alomari, M.A.; Alrabadi, N.N. Vitamin E modifies high-fat diet-induced reduction of seizure threshold in rats: Role of oxidative stress. *Physiol. Behav.* **2019**, *206*, 200–205. [CrossRef]
63. Cavaliere, G.; Trinchese, G.; Penna, E.; Cimmino, F.; Pirozzi, C.; Lama, A.; Annunziata, C.; Catapano, A.; Raso, G.M.; Meli, R.; et al. High-fat diet induces neuroinflammation and mitochondrial impairment in mice cerebral cortex and synaptic fraction. *Front. Cell. Neurosci.* **2019**, *13*, 509. [CrossRef]
64. Liu, Y.; Zhang, Q.; Wang, L.; Wang, H.; Sun, T.; Xia, H.; Yang, Y.; Zhang, L. The α-lipoic acid improves high-fat diet-induced cerebral damage through inhibition of oxidative stress and inflammatory reaction. *Environ. Toxicol. Pharmacol.* **2017**, *56*, 219–224. [CrossRef]
65. El Azab, E.F.; Abdulmalek, S. Amelioration of age-related multiple neuronal impairments and inflammation in high-fat diet-fed rats: The prospective multitargets of geraniol. *Oxidative Med. Cell. Longev.* **2022**, *2022*, 4812993. [CrossRef]
66. Pei, J.; Pan, X.; Wei, G.; Hua, Y. Research progress of glutathione peroxidase family (GPX) in redoxidation. *Front. Pharmacol.* **2023**, *14*, 1147414. [CrossRef] [PubMed]
67. Amri, Z.; Ghorbel, A.; Turki, M.; Akrout, F.M.; Ayadi, F.; Elfeki, A.; Hammami, M. Effect of pomegranate extracts on brain antioxidant markers and cholinesterase activity in high fat-high fructose diet induced obesity in rat model. *BMC Complement. Altern. Med.* **2017**, *17*, 339. [CrossRef] [PubMed]
68. Charradi, K.; Elkahoui, S.; Karkouch, I.; Limam, F.; Hassine, F.B.; Aouani, E. Grape seed and skin extract prevents high-fat diet-induced brain lipotoxicity in rat. *Neurochem. Res.* **2012**, *37*, 2004–2013. [CrossRef]
69. Savaskan, N.E.; Borchert, A.; Bräuer, A.U.; Kuhn, H. Role for glutathione peroxidase-4 in brain development and neuronal apoptosis: Specific induction of enzyme expression in reactive astrocytes following brain injury. *Free Radic. Biol. Med.* **2007**, *43*, 191–201. [CrossRef]
70. Grim, J.M.; Hyndman, K.A.; Kriska, T.; Girotti, A.W.; Crockett, E.L. Relationship between oxidizable fatty acid content and level of antioxidant glutathione peroxidases in marine fish. *J. Exp. Biol.* **2011**, *214*, 3751–3759. [CrossRef]
71. Garrit, J.M.; Dao, K.; Costa, L.G.; Marsillach, J.; Furlong, C.E. Examining the role of paraoxonase 2 in the dopaminergic system of the mouse brain. *BMC Neurosci.* **2022**, *23*, 52. [CrossRef]
72. Shih, D.M.; Meng, Y.; Sallam, T.; Vergnes, L.; Shu, M.L.; Reue, K.; Tontonoz, P.; Fogelman, A.M.; Lusis, A.J.; Reddy, S.T. PON2 deficiency leads to increased susceptibility to diet-induced obesity. *Antioxidants* **2019**, *8*, 19. [CrossRef] [PubMed]

73. Nguyen, S.D.; Sok, D.-E. Preferential inhibition of paraoxonase activity of human paraoxonase 1 by negatively charged lipids. *J. Lipid Res.* **2004**, *45*, 2211–2220. [CrossRef]
74. Zakaria, Z.; Othman, Z.A.; Bagi Suleiman, J.; Jalil, N.A.C.; Ghazali, W.S.W.; Mohamed, M. Protective and therapeutic effects of orlistat on metabolic syndrome and oxidative stress in high-fat diet-induced metabolic dysfunction-associated fatty liver disease (MAFLD) in rats: Role on Nrf2 activation. *Vet. Sci.* **2021**, *8*, 274. [CrossRef] [PubMed]
75. Othman, Z.A.; Zakaria, Z.; Suleiman, J.B.; Mustaffa, K.M.F.; Jalil, N.A.C.; Wan Ghazali, W.S.; Zulkipli, N.N.; Mohamed, M. Orlistat mitigates oxidative stress-linked myocardial damage via NF-κβ and caspase-dependent activities in obese rats. *Int. J. Mol. Sci.* **2022**, *23*, 10266. [CrossRef]
76. Mahmoudi, M.; Charradi, K.; Limam, F.; Aouani, E. Grape seed and skin extract as an adjunct to xenical therapy reduced obesity, brain lipotoxicity and oxidative stress in high fat diet fed rats. *Obes. Res. Clin. Pract.* **2018**, *12*, 115–126. [CrossRef] [PubMed]
77. AL-Dalaeen, A.; Batarseh, N.; Abdelhadi, N.N.; Atawneh, S.; AbuKashef, R.; Al-Yasari, A.M.R.S. Protective and therapeutic effects of orlistat in combination with Elettaria cardamomum "Cardamom" extract on learning, memory, anxiety, and neuroinflammation in obese mice. *Medicina* **2025**, *61*, 263. [CrossRef]
78. Avola, R.; Furnari, A.G.; Graziano, A.C.E.; Russo, A.; Cardile, V. Management of the brain: Essential oils as promising neuroinflammation modulator in neurodegenerative diseases. *Antioxidants* **2024**, *13*, 178. [CrossRef]
79. Alipour, H.R.; Yaghmaei, P.; Ahmadian, S.; Ghobeh, M.; Ebrahim-Habibi, A. A study on alpha-terpineol in Alzheimer's disease with the use of rodent in vivo model, restraint stress effect and in vitro Amyloid beta fibrils. *Braz. J. Pharm. Sci.* **2022**, *58*, e19090. [CrossRef]
80. Qu, Y.; Wang, L.; Mao, Y. Gallic acid attenuates cerebral ischemia/re-perfusion-induced blood–brain barrier injury by modifying polarization of microglia. *J. Immunotoxicol.* **2022**, *19*, 17–26. [CrossRef]
81. Daroi, P.A.; Dhage, S.N.; Juvekar, A.R. p-Coumaric acid mitigates lipopolysaccharide induced brain damage via alleviating oxidative stress, inflammation and apoptosis. *J. Pharm. Pharmacol.* **2022**, *74*, 556–564. [CrossRef] [PubMed]
82. Liu, Y.L.; Huang, H.J.; Sheu, S.Y.; Liu, Y.C.; Lee, I.J.; Chiang, S.C.; Lin, A.M.Y. Oral ellagic acid attenuated LPS-in duced neuroinflammation in rat brain: MEK1 interaction and M2 microglial polarization. *Exp. Biol. Med.* **2023**, *248*, 656–664. [CrossRef] [PubMed]

biomolecules

MDPI

Review

Chinese Medicine-Derived Natural Compounds and Intestinal Regeneration: Mechanisms and Experimental Evidence

Fengbiao Guo [1] and Shaoyi Zhang [1,2,*]

1 Institute of Chinese Medical Sciences, State Key Laboratory of Mechanism and Quality of Chinese Medicine, University of Macau, Macau SAR 999078, China; fengbiaoguo@um.edu.mo

2 Department of Pharmaceutical Sciences, Faculty of Health Sciences, University of Macau, Macau SAR 999078, China

* Correspondence: shaoyizhang@um.edu.mo; Tel.: +853-8822-8785

Abstract: Intestinal regeneration is essential for maintaining epithelial integrity and repairing mucosal damage caused by inflammation, infections, or injuries. Traditional Chinese Medicine (TCM) has long utilized herbal remedies for gastrointestinal disorders, and accumulating evidence highlights that natural compounds derived from TCM possess significant regenerative potential. This review summarizes the multifaceted mechanisms by which these bioactive compounds promote intestinal healing. Key actions include the stimulation of intestinal stem cell (ISC) proliferation and differentiation, the modulation of inflammatory responses, the reinforcement of epithelial barrier integrity, the attenuation of oxidative stress, and the reshaping of the gut microbiota. Representative compounds such as Astragalus polysaccharides, berberine, curcumin, puerarin, and flavonoids like quercetin exhibit these effects through signaling pathways, including HIF-1, Wnt/β-catenin, NF-κB, Nrf2, and IL-22. Evidence from in vitro organoid models and in vivo studies in colitis, radiation injury, antibiotic-associated diarrhea, and intestinal dysmotility and diarrhea models demonstrates that these compounds enhance crypt villus regeneration, preserve tight junctions, and improve clinical outcomes. The holistic, multi-target actions of Chinese medicine-derived natural products make them promising candidates for therapeutic strategies aimed at intestinal repair. Further clinical validation and mechanistic studies are warranted to facilitate their integration into modern gastrointestinal medicine.

Keywords: intestinal stem cell; herbal medicine; regenerative pharmacology; microbiome metabolite axis

1. Introduction

The intestinal epithelium is one of the most rapidly renewing tissues in the human body, capable of complete self-renewal roughly every 3–5 days. This continuous regeneration is driven by intestinal stem cells (ISCs) residing at the base of crypts, which proliferate and differentiate to replenish various epithelial cell lineages [1]. Efficient intestinal regeneration is critical for maintaining gut homeostasis and for healing after an injury or inflammation. In conditions such as inflammatory bowel disease (IBD), an acute injury (e.g., radiation exposure), an infection, or other stressors, normal regenerative processes may be impaired, leading to mucosal damage, barrier dysfunction, and illness. Enhancing the regenerative capacity of the intestine—for instance, by protecting or stimulating ISCs, modulating the immune response, and improving the gut microenvironment—is therefore an important therapeutic goal in gastrointestinal medicine [2–4].

Traditional Chinese Medicine (TCM) has long employed herbal remedies and natural products for gastrointestinal ailments, and many of these remedies are now recognized to contain bioactive compounds with potent biological effects. In recent years, there has been growing scientific interest in Chinese medicine derived natural compounds as promoters of intestinal repair and regeneration. Many plant-derived molecules (e.g., alkaloids, flavonoids, polysaccharides, etc.) and herbal formulations have demonstrated the ability to enhance the healing of the intestinal mucosa in experimental models. These compounds often act via multiple converging mechanisms—such as stimulating intestinal stem cell activity, regulating inflammation, strengthening the epithelial barrier, reducing oxidative stress, and even reshaping the gut microbiota [5,6]. Such multi-targeted actions align well with the holistic approach of TCM and suggest that these natural compounds could be valuable adjuncts or alternatives to conventional therapies for disorders involving impaired intestinal regeneration (for example, IBD or intestinal radiation injury).

In this review, we provide a comprehensive overview of how Chinese medicine-derived natural compounds influence intestinal regeneration. We discuss a broad range of mechanisms by which these compounds exert their effects, including the modulation of intestinal stem cells, the regulation of inflammation, the improvement of the epithelial barrier function, the reduction in oxidative stress, and interactions with the gut microbiota. We highlight evidence from both in vitro studies (including intestinal organoid models) and in vivo experiments that illustrate these mechanisms. Through detailed examples from peer-reviewed studies, we aim to delineate the pharmacological actions of these natural compounds and their potential therapeutic implications for promoting intestinal mucosal healing.

2. Mechanisms of Action of Natural Compounds in Intestinal Regeneration

2.1. Modulation of Intestinal Stem Cells and Epithelial Lineage Differentiation

A fundamental aspect of intestinal regeneration is the activation and differentiation of ISCs to replace the damaged epithelium. Various natural compounds from Chinese medicinal herbs have demonstrated the ability to support ISCs' survival, proliferation, and lineage commitment. For example, Astragalus polysaccharide (APS) (Figure 1a)—a bioactive polysaccharide from *Astragalus membranaceus* (Huangqi)—has been shown to robustly promote ISC-driven regeneration in injured intestines. In a recent study of radiation-induced intestinal injury, an APS treatment improved the survival of irradiated mice and markedly enhanced crypt regeneration, evidenced by increased numbers of ISCs and crypt survival, along with an elevated expression of ISC markers (e.g., Lgr5) and tight junction proteins in the intestine. Intestinal organoid assays confirmed that APS directly promotes the regeneration of ISCs ex vivo. Mechanistically, APS was found to activate the hypoxia-inducible factor 1 (HIF-1) signaling pathway, which is known to be crucial for intestinal stem cell function and mucosal homeostasis. Notably, blocking HIF-1 signaling abrogated the radioprotective and pro-regenerative effects of APS, indicating that APS acts through HIF-1 to foster ISC-mediated repair. These findings suggest that enhancing HIF-1-dependent pathways in ISCs is one strategy by which TCM-derived compounds can accelerate intestinal regeneration [7].

Another example of ISC modulation comes from Dendrobium species (a genus of medicinal orchids). Polysaccharides (Figure 1a) isolated from *Dendrobium fimbriatum* have demonstrated pro-regenerative effects on the intestinal epithelium by engaging ISCs' activity via immune modulation. In an intestinal injury model, *Dendrobium fimbriatum* polysaccharide promoted the regeneration of ISCs and protected the mucosal integrity, an effect mediated by lamina propria lymphocytes (LPLs) secreting interleukin-22 (IL-22). IL-22 is a cytokine known to stimulate intestinal stem cells and progenitor cells to proliferate and differentiate, especially toward epithelial protective lineages. The co-culture of intestinal organoids with immune cells in this study provided an in vitro simulation of the gut immune microenvironment, and treatment with the *Dendrobium* polysaccharide led to the upregulation of IL-22 from LPLs, which in turn drove ISC-dependent epithelial repair. This highlights how herbal polysaccharides can indirectly activate intestinal stem cell regeneration by modulating immune cell signals in the niche (in this case, IL-22) [8,9].

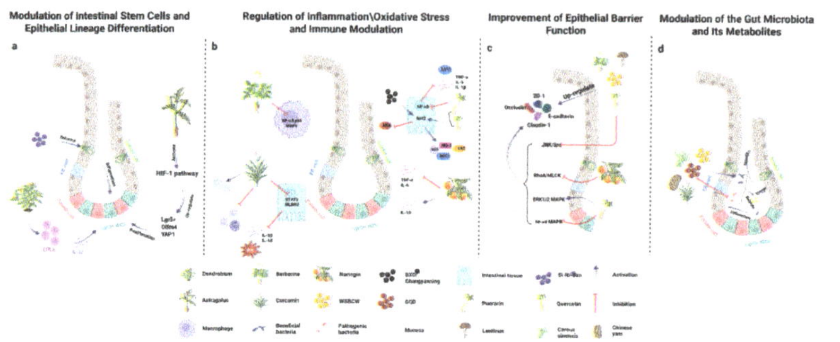

Figure 1. Mechanisms of action of Chinese medicine-derived natural compounds in intestinal regeneration. This schematic illustrates four key mechanisms by which natural compounds promote intestinal regeneration, integrating molecular, cellular, and microbial interactions to depict the multifaceted roles of natural compounds in intestinal repair and homeostasis. These include (**a**) the regulation of the intestinal stem cell and epithelial cell lineage differentiation; (**b**) the modulation of inflammation/oxidative stress and immune regulation; (**c**) the improvement of the epithelial barrier function; and (**d**) the modulation of the intestinal microbiota and its metabolites.

Traditional multi-herb formulations can also influence stem cell-driven regeneration by shifting the balance of the epithelial lineage differentiation. Si-Ni-San (Sinisan) (Figure 1a), a classic herbal medicine compound, was recently reported to alleviate colitis injuries by promoting secretory cell lineage commitment and mucin production in colonic crypts (i.e., enhancing goblet cell differentiation). By facilitating the restoration of mucus-secreting goblet cells, Sinisan helps rebuild the protective mucus layer and supports epithelial healing [10]. This suggests that certain formulas can bias ISCs towards specific lineages (secretory cells, in this case) that are beneficial for mucosal repair. Likewise, preserving the stem cell niche is important—Paneth cells, for instance, are niche cells that support ISCs, and some Chinese herbal components may help maintain such niche factors, indirectly aiding stem cell function.

Overall, compounds like APS and Dendrobium polysaccharides illustrate a key mechanism: enhancing the intrinsic regenerative capacity of the gut via ISCs. By upregulating pathways that govern ISC proliferation, like Wnt/β-catenin or HIF-1, or by providing pro-regenerative cytokine signals (IL-22) and maintaining niche support, Chinese medicine-derived molecules can accelerate the replacement of lost or damaged intestinal epithelial cells. This ISC modulation often translates into the faster restoration of normal crypt–villus structures and improved outcomes after injury.

2.2. Regulation of Inflammation and Immune Modulation

Chronic or excessive inflammation in the gut can impede regeneration by causing ongoing tissue damage and by creating a hostile environment for stem cells and progenitors.

Many natural products from TCM exert significant anti-inflammatory and immunomodulatory effects, which help heal the intestinal mucosa. A prominent example is berberine, an isoquinoline alkaloid from herbs like Coptis chinensis (Huanglian) [11]. Berberine has been extensively studied in models of colitis and is known to suppress key inflammatory signaling pathways, such as Nuclear Factor κB (NF-κB) and mitogen-activated protein kinases (MAPKs) [12]. In mouse colitis induced by dextran sulfate sodium (DSS), berberine (Figure 1b) treatment reduced colonic inflammation by downregulating pro-inflammatory cytokines and signaling—for instance, it inhibited the activation of NF-κB p65 and MAPK pathways in macrophages, leading to the decreased production of inflammatory mediators [13]. By dampening inflammation, berberine enabled improved crypt architecture and mucosal healing. Similarly, curcumin (from Curcuma longa turmeric, used in TCM formulations) is a well-known natural anti-inflammatory agent [14]. Curcumin can block signaling cascades like the Signal Transducer and Activator of Transcription 3 (STAT3) and the NOD-like receptor protein 3 (NLRP3) inflammasome that drive inflammation in IBD. In DSS colitis models, curcumin (Figure 1b) suppressed the phosphorylation and DNA-binding activity of STAT3 in colonic tissue, thereby reducing the downstream expression of inflammatory cytokines [15]. It also mechanistically inhibited NLRP3 inflammasome activation (by preventing upstream events like K^+ efflux and reactive oxygen species (ROS) accumulation), which led to the lower maturation of IL-1β/IL-18 and reduced inflammatory damage [16,17]. These actions create a more permissive environment for tissue regeneration.

Beyond single compounds, multi-component herbal formulas exhibit potent immunomodulatory effects. Puerarin, an isoflavonoid from Pueraria lobata (kudzu root, often included in formulas like Gegen-Qinlian Decoction), has demonstrated both anti-inflammatory and anti-apoptotic properties in colitis [18]. In a DSS-induced colitis mouse study, puerarin (Figure 1b) markedly reduced colon inflammation by inhibiting the NF-κB pathway—treated mice had significantly lower myeloperoxidase activity and reduced levels of TNF-α, IL-1β, and IL-6 in colonic tissue. Puerarin also activated nuclear factor erythroid 2-related factor 2 (Nrf2) in the colon, which upregulated antioxidant enzymes (HO-1, SOD, etc.) and further mitigated inflammation and oxidative injury [19]. The net effect was less tissue damage and enhanced barrier preservation (discussed below). Another compound, Naringin (a flavonoid from citrus fruits, sometimes utilized in TCM) (Figure 1b), showed immune-regulating effects in a sepsis-induced intestinal injury model. Naringin-treated septic mice had significantly lower levels of TNF-α and IL-6 and higher IL-10 (an anti-inflammatory cytokine) levels, indicating a shift toward an anti-inflammatory profile. This immunomodulation was accompanied by an improved intestinal mucosal structure and higher survival rates in the Naringin group [20]. These examples underscore that many TCM-derived compounds promote intestinal regeneration indirectly by curbing inflammation—they reduce the influx of inflammatory cells and cytokines that would otherwise perpetuate the mucosal injury, thereby giving the epithelium a chance to recover.

It is also notable that some natural compounds can modulate gut immune cell populations in more specific ways. Curcumin (Figure 1b), for instance, has been reported to influence T cell differentiation in colitis, restricting pro-inflammatory Th1/Th17 responses and increasing regulatory or anti-inflammatory cell types. By altering dendritic cell and T-cell interactions, curcumin helped restore the immune balance in the gut mucosa [21]. Likewise, cannabinoids from Cannabis sativa (occasionally classed as Chinese herbal compounds in modern contexts) have immunosuppressive effects that might protect the gut during inflammation (though research is still emerging) [22]. Overall, attenuating pathological inflammation is a key mechanism by which Chinese medicine-derived compounds create conditions conducive to mucosal regeneration.

2.3. Improvement in Epithelial Barrier Function

The intestine is only effectively healed if the epithelial barrier—formed by tightly joined enterocytes and mucus—is re-established. A leaky or impaired barrier not only fails to absorb nutrients properly but also allows bacterial toxins and antigens to drive further inflammation, hampering regeneration. Antibiotic exposure can also precipitate profound disruptions in the gut ecosystem and intestinal mucosal barrier, evident as increased paracellular permeability and the disruption of the tight junction architecture.

Many natural compounds have been shown to strengthen the intestinal barrier by upregulating junctional proteins and enhancing mucus production. The restoration of tight junction (TJ) integrity is a common outcome observed with TCM-derived treatments. For example, in the aforementioned study of the WangshiBaochi pill (WSBCW, a Chinese herbal medicine for gastrointestinal disorders) (Figure 1c), treated mice exhibited significantly increased expression of tight junction and adherens junction proteins, such as zonula occludens-1 (ZO-1), occludins, claudins, and E-cadherin, in the intestine. Correspondingly, the WSBCW improved intestinal permeability (measured by FITC-dextran assays) in healthy mice and prevented hyper-permeability in a diarrhea model, indicating a more intact barrier. Histologically, the WSBCW increased the villus length and crypt depth, signifying enhanced mucosal structural regeneration [23]. This example illustrates how an herbal formula can promote barrier repair at the molecular level (junctional proteins) and tissue level (villus architecture).

Another example is lentinan (Figure 1c), a β-glucan from Lentinus mushrooms, which has demonstrated both preventive and therapeutic efficacy in antibiotic-induced dysbiosis. Treated mice had higher levels of butyrate and propionate in the colon, correlating with an improved epithelial energy supply and water absorption. Consequently, lentinan-treated mice exhibited strengthening of the intestinal barrier—with the colonic ZO-1 and occludin expression being restored to normal levels—and a blunted inflammatory response relative to antibiotic-only mice. Mechanistic analyses showed that lentinan attenuated NF-κB pathway activation, leading to lower concentrations of TNF-α, IL-6, and IL-1β in intestinal tissues [24].

Quercetin is a natural compound (a flavonol present in many fruits and in TCM herbs like Sophora japonica) that has strong barrier-protective effects [25]. In intestinal epithelial cell cultures, a quercetin pretreatment markedly counteracted the barrier disruption induced by injurious stimuli. Mechanistically, quercetin (Figure 1c) was shown to increase the gene and protein expression of key tight junction proteins, including ZO-1, occludin, and claudin-1, thereby maintaining the tight junction integrity. For example, in IEC-6 cell monolayers injured by indomethacin, quercetin inhibited stress-activated JNK/Src signaling that would normally downregulate TJs, and as a result, the quercetin-treated cells preserved high levels of ZO-1/occludin and avoided barrier leakage [26]. In vivo, quercetin similarly protected the intestinal barrier in a rat model of acute necrotizing pancreatitis: it prevented the loss of TJs (ZO-1, claudin-1, and occludin) and lowered markers of permeability (DAO and endotoxin) while reducing inflammatory damage [27]. These data suggest that quercetin can directly stabilize the epithelial junctional complex under stress conditions, thus supporting barrier function during regeneration.

Several other TCM-derived compounds mirror these barrier-enhancing actions. Berberine (Figure 1c) in DSS colitis models has been shown to increase the colonic expression of ZO-1, ZO-2, occludin, and E-cadherin, correlating with reduced mucosal permeability [28]. Notably, berberine also modulated the expression of genes like Zeb1 and Slug that affect the epithelial–mesenchymal transition, thereby preserving epithelial phenotypes and junctions in inflamed tissues [29,30]. Puerarin (Figure 1c) protected the integrity of TJs in ethanol-injured Caco-2 cell monolayers by influencing signaling pathways (it activated ERK1/2

MAPK and inhibited NF-κB/MLCK, which led to TJ stabilization) [31]. In DSS colitis mice, puerarin treatment increased colonic ZO-1 and occludin levels and prevented the rise in the intestinal permeability, which is consistent with a more intact barrier. Naringin (Figure 1c) was also shown to promote tight junction protein expression (ZO-1 and claudin-1) in both in vivo and in vitro sepsis models—an effect tied to its ability to inhibit the RhoA/MLCK pathway and NF-κB activation that would otherwise disrupt junctions [32]. Furthermore, compounds that increase mucus secretion or goblet cell numbers (such as Sinisan's effect on mucin-producing cells) also improve the barrier by reinforcing the mucus layer. In summary, a critical mechanism of these natural products is reinforcing the physical epithelial barrier—by upregulating tight junctions and adherens junctions and restoring the mucus layer—which not only helps prevent further injury but also creates conditions for proper tissue regeneration and function.

2.4. Reduction in Oxidative Stress

Oxidative stress is a common feature of intestinal injury and inflammation, where excess ROS are generated by the immune cells and damaged epithelium. ROS and other free radicals can kill stem cells and progenitors, damage DNA, and oxidatively impair proteins and lipids, thus hindering regenerative processes. Many Chinese medicinal compounds have inherent antioxidant properties or can activate cellular antioxidant pathways, thereby mitigating oxidative stress in the gut environment and promoting healing.

A number of flavonoids in TCM, including quercetin and Icariin (from Epimedium species), are well-documented antioxidants. Quercetin (Figure 1b) not only scavenges ROS directly but also activates the Nrf2 pathway, a master regulator of antioxidant defenses in cells. In porcine enterocyte studies, quercetin prevented oxidative injury induced by the pro-oxidant diquat by activating Nrf2, which led to the upregulation of enzymes like glutathione peroxidase and superoxide dismutase. Through Nrf2, quercetin also modulated the intracellular redox state in intestinal cells and even appeared to tighten junctions in an Nrf2-dependent manner (linking antioxidant action with barrier enhancement) [33]. Puerarin has similarly been shown to elevate Nrf2 signaling and downstream antioxidant enzymes (HO-1, NQO1, catalase, etc.) in colitis mice, reducing lipid peroxidation and tissue oxidative damage [34]. By enhancing the antioxidant capacity, puerarin reduced mucosal inflammation and protected TJs, ultimately aiding in the faster recovery of the epithelium.

Herbal formulas can also reduce oxidative stress through immunomodulation. The Changyanning formula (Figure 1b), when tested in an intestinal organoid co-culture model, was found to control excessive ROS production by modulating the intestinal immune microenvironment. In that study, co-cultured immune cells produced high ROS in an untreated enteritis model; the addition of the Changyanning formula suppressed the ROS levels, likely by tempering the activation of reactive immune cells. The reduction in oxidative stress was associated with a therapeutic benefit in the enteritis model [35]. Another classical formula, Banxia Xiexin Decoction (BXD) (Figure 1b), was reported to exert anti-colitis effects by suppressing oxidative stress in the colorectum via Nrf2 activation. BXD-treated DSS colitis mice had higher colonic Nrf2 contents and lower malondialdehyde (an oxidative damage marker) levels, along with reduced inflammatory cytokines, suggesting that its amelioration of colitis was partly achieved through an antioxidant mechanism [36].

Oxidative stress and inflammation are often intertwined, and many TCM compounds address both phenomena simultaneously (e.g., curcumin's activation of the SIRT1/Nrf2 pathway and inhibition of TLR4 in models of necrotizing colitis, which reduced both oxidative damage and inflammation). By reducing ROS levels and boosting cellular antioxidant defenses, Chinese medicine-derived compounds protect intestinal stem cells and differentiated cells from oxidative injury, thereby preserving the regenerative potential of the tissue [37]. This mechanism is especially crucial in scenarios like radiation-induced intestinal damage, where oxidative stress is a major driver of tissue injury. The radioprotective effect of Astragalus polysaccharide in irradiated intestines, for instance, is partly attributed to its antioxidant and anti-apoptotic functions, as noted in other studies. In summary, combating oxidative stress is a vital component of how these natural products foster a hospitable milieu for intestinal regeneration.

2.5. The Modulation of the Gut Microbiota and Its Metabolites

The gut microbiota plays an integral role in intestinal homeostasis and regeneration. Commensal bacteria produce metabolites (such as short-chain fatty acids) that nourish epithelial cells and modulate inflammation, whereas dysbiosis (an imbalanced microbiota) can perpetuate the injury by encouraging pathogen overgrowth or aberrant immune activation. One reason is the overuse of antibiotics. Broad-spectrum antibiotics sharply reduce the diversity of commensal microbiota and deplete key SCFA-producing bacteria. In antibiotic-treated mice, colonic concentrations of acetate, propionate, and butyrate drop to roughly half of the normal levels [38]. Fascinatingly, many Chinese herbal compounds and formulas can reshape the gut microbiota composition, and this microbiota modulation is increasingly recognized as a mechanism contributing to their therapeutic effects on the intestine.

For example, the WSBCW (Figure 1d) herbal preparation mentioned earlier not only acted directly on the host (improving junctions and reducing cytokines) but also significantly altered the gut microbiota profile of mice. The WSBCW treatment increased the relative abundance of beneficial genera like Bifidobacterium (often associated with gut health) and Desulfovibrio while suppressing the overgrowth of potentially harmful bacteria like the Bacteroides fragilis group [23]. These changes in flora were correlated with improved intestinal function and reduced diarrhea in the treated animals. Similarly, berberine has well-documented microbiota effects: it exhibits broad antimicrobial activity that can reduce pathogens, but it tends to favor an increase in short-chain fatty acid-producing bacteria at sub-antimicrobial doses. One study found that the berberine treatment led to metabolic changes, as seen in healthy volunteers' stool samples, consistent with an increase in the butyrate production pathway [39]. In animal models, berberine has been shown to restore a dysbiotic microbiota toward a more balanced state in colitis—for instance, in DSS-treated cats, berberine increased levels of beneficial gut bacteria and improved the overall microbial diversity [40]. By promoting butyrate and other SCFA productions, berberine can indirectly support the intestinal epithelial energy supply and enhance mucosal repair.

Cereus sinensis polysaccharide (CSP-1) (Figure 1d)—a purified natural polysaccharide—helped reverse antibiotic-induced microbial imbalances. In a mouse model of antibiotic-associated diarrhea (AAD), CSP-1 enriched SCFA-producing genera (e.g., Phascolarctobacterium and Bifidobacterium); restored microbial diversity; and elevated cecal acetate, propionate, and butyrate to normal levels. In line with this, CSP-1-treated mice showed reduced intestinal fluid loss and improved weight gain compared to untreated AAD controls, as well as a significantly dampened secretion of inflammatory cytokines (TNF-α, IL-1β, and IL-2) back to baseline levels [41]. Beyond that, a polysaccharide from Chinese yam (Figure 1d) significantly alleviated AAD in mice by upregulating the abundance of probiotic bacteria, suppressing the overgrowth of opportunists, and increasing cecal SCFA levels [42].

Complex herbal formulas often have even more pronounced microbiota-modulating capabilities. A striking example is Gegen-Qinlian Decoction (GQD) (Figure 1d), which contains herbs like Pueraria (gegen) and Coptis (which has berberine). In a UC rat model, a modified GQD significantly raised the abundance of Akkermansia and Romboutsia—both genera associated with mucosal health—and decreased Escherichia–Shigella populations that include pathogenic bacteria [43]. These microbiota changes were accompanied by the increased production of short-chain fatty acids (acetate, propionate, and butyrate) in the colon and an improvement in colonic inflammation and histology. SCFAs like butyrate are known to fuel colonocytes and promote ISC differentiation into the functional epithelium; they also have anti-inflammatory properties [44]. Thus, GQD's ability to foster SCFA-producing commensals and reduce opportunistic bacteria likely created a biochemical environment favorable for regeneration (e.g., through butyrate-mediated enhancement of epithelial barrier and regulatory immune responses). In diarrheic piglet models, GQD similarly increased beneficial commensals (e.g., Lactobacillus and Ruminococcus) and SCFA levels while inhibiting pathogenic E. coli, leading to the restoration of the mucosal structure and function [45].

Curcumin (Figure 1d) has also been shown to selectively modulate the gut flora composition. In one study, curcumin supplementation in colitis mice helped restore a healthier microbiome, which was linked to reduced inflammation and even improvements in gut–brain-axis-related symptoms [46]. New formulations like nanoparticle-encapsulated curcumin appear to enhance this effect; a recent report noted that a nano-curcumin greatly alleviated experimental colitis by shifting the gut microbiota structure and influencing mucosal immune cell regulation [47]. This suggests that curcumin's benefits are partly microbiota-mediated, as microbiome changes can lead to secondary effects like altered bile acid profiles or immune modulation.

In summary, microbiota interactions are a crucial mechanism through which Chinese medicine-derived compounds exert regenerative effects. By pruning the overgrowth of pathogenic bacteria and enriching commensal strains that produce beneficial metabolites (SCFAs, etc.), these interventions reduce mucosal insults (e.g., less toxins and pathogen-associated inflammation) and increase factors that promote healing (e.g., butyrate to enhance barrier function and ISC differentiation). The result is a more balanced microbial ecosystem that supports the intestinal epithelium's recovery. This multi-hit approach—simultaneously targeting host pathways and microbial communities—underscores the holistic nature of TCM therapeutics in intestinal health.

3. Experimental Evidence from In Vitro and In Vivo Studies

3.1. In Vitro Models: Intestinal Organoids and Cell Cultures

Advanced in vitro models have been instrumental in elucidating how Chinese medicinal compounds act on intestinal cells and stem cells. Intestinal organoids (Figure 2a)—three-dimensional mini-gut structures grown from ISCs—provide a particularly powerful platform to study regeneration in a controlled setting. Researchers have utilized organoid cultures to directly observe the pro-regenerative actions of TCM-derived compounds. For example, the effect of Dendrobium polysaccharides on ISC regeneration was demonstrated using a co-culture of mouse intestinal organoids with immune cells. In this system, organoids (containing stem cells and epithelium) normally suffered damage when exposed to injurious stimuli, but treatment with the Dendrobium fimbriatum polysaccharide, in the presence of LPL immune cells, led to significantly improved organoid growth and recovery [9]. The IL-22 secretion from immune cells (stimulated by the polysaccharide) was identified as the key factor driving organoid (and ISC) regeneration. This organoid co-culture approach closely mimics the in vivo intestinal microenvironment and clearly showed that a TCM compound can activate intestinal stem cell-mediated repair via immune modulation.

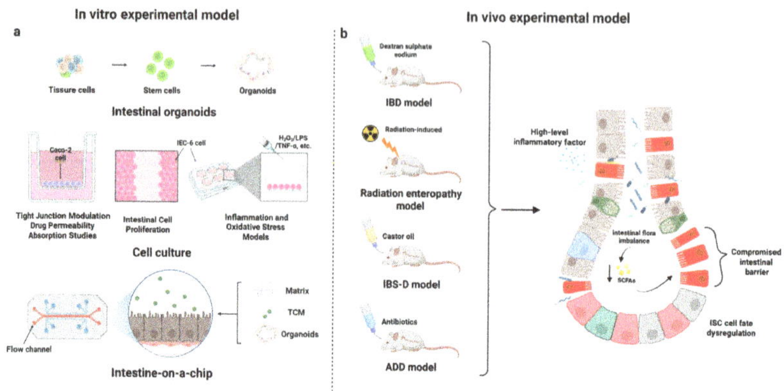

Figure 2. In vitro and in vivo intestinal experimental models for natural compounds derived from Chinese medicine. This infographic provides an overview of experimental models used to study intestinal regeneration and disease mechanisms, integrating in vitro and in vivo approaches. In vitro models utilize (**a**) intestinal organoids, intestinal epithelial mimic cells (Caco-2 and HT29), and intestinal organ chips to conduct drug permeability and absorption studies and assess intestinal barrier integrity, cell proliferation, inflammation (e.g., TNF-α), and oxidative stress. In vivo models focus on chemical and physical induction, including (**b**) inflammatory bowel disease (IBD) induced by dextran sulfate sodium (DSS), intestinal damage induced by ionizing radiation (e.g., γ rays and X rays), diarrhea-predominant irritable bowel syndrome (IBS-D) induced by castor oil gavage, and antibiotic-associated diarrhea (AAD) induced by antibiotics. These models comprehensively simulate intestinal inflammation, microbial imbalances, epithelial barrier loss, and intestinal stem cell dysregulation.

In another study, organoids were used to evaluate the Astragalus polysaccharide (APS). Ding et al. established intestinal organoid cultures from mice and found that APSs markedly enhanced organoid-forming efficiency and growth after radiation damage. Essentially, APS-treated organoids had higher survival and regeneration of crypt-like domains, aligning with the compound's stimulation of ISC proliferation. These ex vivo organoid results reinforced the in vivo findings that the APS promotes ISC-driven regeneration through HIF-1 signaling [7]. Organoid assays thus confirmed that the APS has a direct effect on epithelial stem cells' capacity to regrow tissue.

Aside from organoids, traditional cell culture models have provided insights into barrier-related mechanisms. Caco-2 and IEC-6 cell monolayers (Figure 2a) (which model intestinal epithelial layers) have been treated with compounds like quercetin, puerarin, and berberine to test their ability to prevent barrier disruption. Quercetin's ability to preserve tight junctions under chemical stress was demonstrated in IEC-6 cells, as described earlier [25]. Puerarin prevented ethanol-induced tight junction dissociation in Caco-2 cells by modulating NF-κB and MAPK signaling, maintaining high levels of ZO-1 and occludin in the cell junctions [30]. Berberine protected Caco-2/HT29 co-culture barriers from inflammatory (TNF-α) injuries, helping maintain E-cadherin and other junctions dose-dependently [48]. These cell culture experiments allow the precise dissection of pathways—for instance, using siRNA knockdowns or pathway inhibitors, researchers have verified that blocking Nrf2 or inhibiting certain kinases will negate the barrier-protective effects of these compounds. Such findings strongly indicate that the compounds act on specific epithelial cell signaling cascades (e.g., quercetin on JNK/Src and Nrf2, puerarin on NF-κB/MLCK) to exert their pro-regenerative, barrier-stabilizing outcomes.

Another interesting in vitro system is organ-on-a-chip technology (Figure 2a), though its use in TCM research is still in its early stages. To date, few organ-on-chip studies have been performed for TCM and the gut; one report examined the potential nephrotoxicity of a plant compound (kaempferol) on a kidney chip and found no damage at relevant doses, hinting at the value of such models for safety [49]. Gut-on-chip models may be applied to study dynamic interactions (peristalsis and flow) with herbal compounds. Organoids are currently the primary in vitro tool demonstrating the intestinal regenerative effects of Chinese herbal compounds, complemented by simpler cell cultures for mechanistic probing.

In summary, in vitro experiments (particularly organoids) have provided compelling evidence that Chinese medicine-derived compounds can directly act on intestinal epithelial systems to promote regeneration. They allow the observation of stem cell proliferation, crypt-like domain formation, differentiation, and barrier integrity in response to these natural products, under controlled conditions free from systemic influences. These studies validate that these compounds have intrinsic efficacy on gut tissues, thus substantiating their observed potential in whole-animal models.

3.2. In Vivo Studies: Animal Models of Intestinal Injury and Disease

A wealth of in vivo evidence from animal studies supports the regenerative effects of Chinese medicinal compounds on the intestine. These studies span various models—from chemically induced colitis and diarrhea models to radiation injury and sepsis—and consistently show improved outcomes in animals treated with natural compounds or herbal formulas.

One of the most common models, DSS-induced colitis (Figure 2b) in mice or rats, simulates ulcerative colitis with epithelial erosion and ulceration. Multiple compounds have been tested in this model. Berberine-treated colitis mice exhibit significantly milder disease, showing reduced weight loss and diarrhea, longer colon lengths, and lower histopathology scores compared to untreated controls. At the cellular level, berberine reduces neutrophil infiltration and preserves the crypt architecture; molecularly, it inhibits colonic NF-κB activation and upregulates antioxidant enzymes (consistent with its anti-inflammatory/antioxidant actions noted earlier) [35]. Curcumin has even progressed to clinical trials after showing promising animal data—in DSS colitis mice, it ameliorated inflammation (via STAT3/NLRP3 suppression) [15], and a randomized controlled trial involving ulcerative colitis patients showed that oral curcumin (1500 mg daily) improved clinical remission rates and quality of life while lowering inflammatory markers (C-reactive protein, etc.) [50]. This translation from animals to humans highlights the therapeutic potential of such compounds for enhancing mucosal healing in IBD.

Classic TCM formulas have also been validated in vivo. BXD was found to significantly attenuate colitis in DSS-treated mice, improving body weight recovery and reducing colon damage. Its mechanism involved blocking pro-inflammatory NF-κB signaling and boosting Nrf2-driven antioxidant responses in colonic tissue [35]. As a result, BXD-treated mice had less crypt loss and better preservation of the mucosal barrier than the controls. Si-Ni-San (Sinisan), mentioned earlier for promoting goblet cell regeneration, has also shown efficacy in a rat colitis model by restoring mucosal goblet cell numbers and mucin levels, leading to an improved stool form and reduced inflammatory ulceration [10]. GQD, besides modulating the microbiota as described, ameliorated the pathology in UC mice—reducing diarrhea and bleeding—through downregulating hyperactive signaling pathways like PI3K/Akt and EGFR that contribute to epithelial injury [51].

In models of intestinal dysmotility and diarrhea, such as IBS-like (Figure 2b) or infective diarrhea models, TCM compounds also enhance gut structure recovery. The herbal pill WSBCW improved a castor oil-induced diarrhea model in mice, not only by firming up stools but by increasing the villus height and crypt depth in the small intestine, indicating enhanced epithelial regeneration post-insult. It also upregulated tight junctions and reduced plasma cytokines in these mice, demonstrating both barrier and immune improvements. These changes translated into better gastrointestinal functional parameters: WSBCW-treated animals experienced faster restoration of normal gut transit and permeability than the controls [23]. Another formula, Sishen Wan, was shown to protect the intestinal barrier in a diarrhea-predominant IBS model by reducing endoplasmic reticulum stress in enterocytes, thereby improving tight junction integrity and relieving symptoms [52].

Another noteworthy example is the AAD model (Figure 2b). A growing body of research indicates that natural compounds—including those derived from Traditional Chinese Medicine—can be used to prevent or ameliorate AAD. These natural compounds restore a healthy microbiota, strengthen the intestinal barrier, and modulate immune responses. Lizhong-Tang—composed of ginseng, Atractylodes, ginger, and licorice—demonstrated potent anti-AAD effects in a lincomycin-induced mouse model. Lizhong-Tang-treated mice had markedly lower diarrhea scores and maintained their body weight better than untreated animals. Microbiome analyses (T-RFLP and 16S sequencing) confirmed that Lizhong-Tang corrected antibiotic-induced shifts in the gut microbiota composition, helping to ameliorate dysbiosis to achieve a healthier profile [53]. Berberine exhibits broad antimicrobial activity in the gut and favorably reshapes the microbiota composition. In the context of antibiotic-associated C. difficile infections, berberine and the flavone baicalein (from Scutellaria baicalensis) have been shown to inhibit the growth and virulence of C. difficile. These compounds also downregulated the expression of toxin synthesis genes in C. difficile, thereby attenuating the pathogen's cytotoxic effects. By curbing opportunistic pathogens and their toxins, such phytochemicals can prevent the exacerbation of antibiotic-associated colitis and contribute to preserving the epithelial barrier [54].

Severe injury models provide further evidence. In radiation-induced intestinal syndrome (Figure 2b), where massive ISC loss leads to lethal GI tract failure, the APS from Astragalus dramatically improved outcomes: APS-treated mice had a higher survival rate after high-dose irradiation, the preservation of crypts, and the rapid re-epithelialization of the intestine. Mice in the APS group showed robust ISC recovery and proliferating cell nuclear antigen (PCNA) staining in regenerating crypts compared to the irradiated controls, aligning with APS's HIF-1-mediated stem cell protection [7]. This suggests that some TCM compounds might serve as radioprotectants to enhance post-injury regeneration when conventional options (like cytokine therapies) are limited. In polymicrobial sepsis (which can cause mucosal degeneration), Naringin significantly improved survival in a cecal puncture model by preserving the gut barrier and modulating cytokines. Naringin-treated septic mice had notably less intestinal damage and lower bacterial translocation, likely because their tight junctions were maintained and the inflammatory injury was controlled [20].

Across these diverse models, a common theme emerges: animals receiving Chinese medicine-derived compounds or formulas consistently show faster and more complete regeneration of the intestinal lining after injury. This is evidenced by metrics like the increased villus length, deeper crypts with abundant goblet and absorptive cells, lower injury scores, and restored barrier function tests (e.g., permeability assays). The mechanistic hallmarks—reduced inflammatory infiltrates, higher ISC counts ($Lgr5^+$ cells), greater tight junction protein staining, and lower oxidative damage—corroborate the modes of action discussed in the previous section.

It is worth noting that, in some cases, these compounds are moving toward clinical evaluation. Curcumin's positive trial in ulcerative colitis and accumulating clinical observations that TCM formulations (e.g., herbal combinations like Qing-dai or Plantain extract) can induce mucosal healing in IBD contribute to an optimistic belief that the preclinical findings are translatable. However, clinical evidence is still relatively sparse, and further trials are needed for most compounds. It is crucial to address factors such as the bioavailability (some compounds like curcumin are poorly absorbed unless formulated specifically) and quality control of herbal extracts.

In summary, in vivo studies strongly support the efficacy of Chinese medicine-derived natural compounds in promoting intestinal regeneration in various contexts of disease and injury. They validate that these compounds' multifaceted mechanisms—anti-inflammatory, barrier-enhancing, antioxidant, microbiome-modulating, and stem cell-activating effects—indeed culminate in the tangible healing of the gut lining in living organisms. These findings lay a strong foundation for clinical trials and eventually integrating such natural therapies into the management of gastrointestinal disorders where regeneration is needed.

4. Translational Challenges of Natural Compounds in Intestinal Therapy: Safety, Standardization, and Bioavailability

Natural compounds such as curcumin, berberine, ginsenosides, plant polysaccharides, and quercetin show therapeutic promise for intestinal repair, yet each one faces specific translational hurdles. Table 1 presents the key pathways, evidence basis, and clinical progress of major natural compounds. Curcumin is generally safe, but rare hepatotoxicity and potential drug interactions (e.g., with anticoagulants) warrant caution. Its extreme variability in turmeric preparations complicates standardization, and its poor oral bioavailability—due to rapid metabolism—requires enhancers like piperine or nanoformulations [55]. Berberine has a favorable safety profile but is contraindicated in pregnancy and may interact with drugs via CYP3A4 or P-glycoprotein inhibition. Its defined structure allows for easier standardization, but poor systemic absorption and high first-pass metabolism demand high doses or advanced delivery systems [56]. Ginsenosides, though typically safe, may affect sleep and interfere with warfarin or immunosuppressants. Variability in species, processing, and compound ratios hinders standardization. Most ginsenosides have low absorption, relying on the gut microbiota to generate bioactive metabolites like Compound K [57]. Polysaccharides (e.g., from Astragalus or mushrooms) are safe and act locally via prebiotic effects and immune modulation, but their large size precludes systemic absorption. Their complex, heterogeneous structures pose major standardization challenges; batch-to-batch consistency is difficult without advanced analytical techniques [58,59]. Quercetin, a widely studied flavonoid, is safe at moderate doses but can affect cytochrome enzymes and drug levels. While its purified form simplifies standardization, its bioavailability is limited by its poor solubility and metabolism [60].

Table 1. A summary of the major natural compounds reviewed, including their proposed molecular targets, evidence base, and status in clinical testing.

Compound	Key Pathways/Actions	Evidence Level	Clinical Trial Status
Curcumin (turmeric)	Inhibits NF-κB, TLR4/AP-1, and pro-inflammatory cytokines; activates Nrf2 antioxidant response; modulates gut microbiome (promotes SCFA-producing flora).	Extensive preclinical evidence (in vitro and animal models of colitis); multiple clinical trials in IBD and other GI disorders.	Completed: Several RCTs in ulcerative colitis showed symptom improvement and remission maintenance (approved for use as adjunct in IBD in some settings; further trials ongoing [61]).
Quercetin (flavonoid)	Antioxidant (scavenges ROS and upregulates HO-1 via Nrf2); suppresses NF-κB and MAPK signaling (reduces IL-8 and TNF-α); stabilizes mast cells; helps restore tight junction proteins in gut epithelium.	Preclinical evidence from studies in vitro (cell protection) and in rodents (DSS colitis models); some human data (dietary intake correlational studies and small pilot trials).	Completed: A small RCT in ulcerative colitis reported reduced disease activity and inflammation with quercetin supplementation [62]. Ongoing: Further clinical studies in IBD and IBS are in early phases (exploratory).

Table 1. *Cont.*

Compound	Key Pathways/Actions	Evidence Level	Clinical Trial Status
Berberine (Coptis alkaloid)	Activates AMPK; inhibits NF-κB/STAT3 and inflammasome pathways (IL-1β, IL-6, and TNF-α); modulates gut microbiota (antibacterial against harmful flora; promotes short-chain fatty acid producers); may enhance Wnt/β-catenin signaling for mucosal repair.	Strong preclinical evidence (multiple animal colitis studies showing improved mucosal healing); numerous clinical trials for metabolic and GI outcomes (diabetes, H. pylori, and diarrhea).	Completed: RCTs in ulcerative colitis and chronic diarrhea/IBS have shown benefit (e.g., berberine improved histological scores in UC [63]). Ongoing: Trials in IBD (phase II in China) and colon polyp prevention are in progress.
Ginsenosides (Panax ginseng)	Multi-target immunomodulation. Attenuate NF-κB and MAPK signaling (inflammatory cytokines IL-1β and IL-17); some ginsenosides (Rg1 and Rb1) activate Wnt/β-catenin and EGFR pathways supporting epithelial regeneration. Antioxidative effects (reduce ROS, upregulate SOD, etc.); promote balance of Th17/Treg in gut.	Preclinical evidence only: in vitro studies on immune cells; several mouse colitis models with isolated ginsenosides (Rb1, Rg1, Rk3, etc.) showing reduced inflammation and faster mucosal healing [64,65].	Not yet tested clinically for intestinal injury. (Ginseng extracts have been clinically studied for fatigue, metabolism, etc., but no clinical trials to date have focused on intestinal regeneration/IBD. Currently used as a supplement in humans and is not an established therapy for gut disease.)
Astragalus Polysaccharide (APS)	Enhances intestinal stem cell (ISC) activation via HIF-1α signaling; promotes crypt cell survival and proliferation; immunomodulatory—reduces TNF-α, IL-1β, and IFN-γ levels in injured gut; regulates gut microbiota and increases SCFA production (butyrate), leading to improved tight junction integrity.	Preclinical evidence: robust evidence in animal models of intestinal injury (radiation enteropathy and DSS colitis [7]); organoid culture studies confirming ISC regeneration effects.	Not yet tested clinically. (Experimental stage—no human trials so far. Astragalus-based TCM remedies are used empirically for GI health, but APS as an isolated compound has not yet undergone clinical trials.)
Lentinan (shiitake β-glucan)	Stimulates immune cells (macrophages and dendritic cells) via Dectin-1 and other pattern recognition receptors, inducing cytokines (IL-2 and TNF-α) that aid in tissue repair; may enhance IgA secretion and mucosal immunity; antioxidant and anti-apoptotic effects on epithelial cells reported. Indirectly supports gut barrier by fostering a protective immune microenvironment.	Preclinical evidence: in vitro immunological studies; animal models of infection and cancer (shows gut protection as side observation). In GI injury models, data are sparse, but related mushroom polysaccharides show decreased inflammation and oxidative damage.	Completed (other indications): Used clinically as an injected cancer adjuvant (extensive experience in humans for oncology). Not formally trialed for IBD or mucosal healing. (Only anecdotal or compassionate use in GI disorders; no RCTs in intestinal regeneration.)
CSP-1 (sulfated polysaccharide fraction)	Anti-inflammatory and prebiotic: inhibits TLR4/MyD88/NF-κB signaling cascade, leading to reduced NF-κB activation; downregulates IL-23/IL-17 axis (fewer Th17 pro-inflammatory cells); helps restore tight junction proteins and barrier function (likely via its sulfate groups, binding gut mucins and modulating microbiota).	Preclinical evidence: early-stage studies in cell culture and rodent antibiotic-associated diarrhea models. Shown to stabilize gut barrier and lower inflammatory mediators in these models [59].	Not yet tested clinically. (Research compound only—no human studies. Needs further animal validation before any clinical trials.)

Abbreviations: NF-κB—nuclear factor κB; TLR4—toll-like receptor 4; AP-1—activator protein 1; Nrf2—nuclear factor erythroid 2-related factor 2; ROS—reactive oxygen species; AMPK—AMP-activated protein kinase; STAT3—signal transducer and activator of transcription 3; Wnt—Wnt/β-catenin pathway; HIF-1—hypoxia-inducible factor 1; ISC—intestinal stem cell; SCFA—short-chain fatty acid; Th17—T helper 17 cells; DSS—dextran sulfate sodium; IBD—inflammatory bowel disease; and RCT—randomized controlled trial.

In summary, while these compounds have demonstrated local and systemic benefits in preclinical models, their clinical translation is constrained by an inconsistent quality, poor pharmacokinetics, and potential interactions. Overcoming these challenges—through better drug delivery systems, rigorous clinical trials with standardized preparations, and a deeper understanding of herb–drug interactions—will be key to integrating natural compounds into mainstream therapies for intestinal regeneration.

5. Conclusions

Natural compounds derived from Traditional Chinese Medicine exhibit remarkable capacity to promote intestinal regeneration through multiple converging mechanisms. As detailed above, these compounds can stimulate intestinal stem cells, helping to replenish the epithelium after injury, and they can favor the differentiation of protective cell lineages (such as goblet and Paneth cells) that support mucosal healing. At the same time, they exert anti-inflammatory effects, dampening cytokine storms and immune overactivation that impede healing in conditions like IBD. In addition to reducing inflammation, many of these agents strengthen the gut's barrier function—by upregulating tight junction proteins and restoring the integrity of the mucosal layer—which prevents further damage and sets the stage for proper tissue recovery. Additionally, their antioxidant actions protect vulnerable stem and progenitor cells from oxidative injury, ensuring the regenerative machinery of the gut can function optimally even in stress conditions. Finally, by modulating the gut microbiota imbalance caused by antibiotics and other factors, Chinese medicine-derived compounds help re-establish a symbiotic microbial environment that produces beneficial metabolites (such as SCFAs) and curbs pathogenic insults. This holistic, multi-target approach is a signature feature of TCM therapies and is ideally suited to a complex process like intestinal regeneration, which involves an interplay between epithelial cells, immune cells, and microbiota.

The experimental evidence supporting these benefits is extensive. From controlled in vitro studies with intestinal organoids to a variety of animal models of intestinal injury, TCM compounds have consistently demonstrated enhanced mucosal healing—with outcomes such as the faster regeneration of villi/crypts, improved histological scores, reduced leakiness, and higher survival in severe injury models. Notably, some compounds like curcumin have translated this promise into early clinical success, indicating that preclinical mechanisms manifest in human disease to some extent. Herbal formulas, which combine multiple active ingredients, often show synergistic effects on multiple pathways (for example, simultaneously reducing NF-κB-driven inflammation and increasing Nrf2-driven antioxidant responses). This multi-component synergy might be crucial for tackling complex diseases like IBD where no single target is sufficient.

Although these mechanisms are promising and some natural compounds have made clinical progress, the clinical translation of natural compounds still faces significant challenges, including safety concerns, standardization issues, and poor bioavailability. For example, the hepatotoxic risk and drug interactions of curcumin, the absorption limitation of berberine, and the structural complexity of polysaccharides all require targeted solutions such as nanoformulations or advanced delivery systems. Standardization and rigorous clinical trials of herbal extracts are essential to bridge the gap between preclinical efficacy and clinical applicability.

These findings encourage further scientific and clinical exploration of Chinese medicine-derived compounds as therapeutic agents for intestinal repair. There is a need for well-designed clinical trials to evaluate efficacy and safety in patients (for example, in ulcerative colitis remission or in mitigating the gastrointestinal side effects of radiation/chemotherapy). Issues of standardization, dosage, and bioavailability must be addressed, given that natural extracts can vary in composition. It is also important to pinpoint biomarkers of response—for instance, identifying which subset of patients (perhaps defined by the microbiome profile or genetic markers) might benefit most from a given herbal treatment. Modern techniques, including metabolomics and advanced imaging, could help in understanding exactly how these compounds interact with the gut ecosystem in humans.

In conclusion, Chinese medicine-derived natural compounds represent a promising and mechanistically rich source of therapies for enhancing intestinal regeneration. They embody a multi-modal strategy: protecting and activating stem cells, tempering the immune response, fortifying the barrier, quenching oxidative stress, and cultivating a healing-friendly microbiome. Such a comprehensive approach is highly relevant for diseases characterized by epithelial injury and defective healing. By bridging traditional knowledge with modern biomedical research, we can potentially develop novel treatments that safely and effectively promote the restoration of the gut integrity and function in patients suffering from inflammatory and injurious gastrointestinal conditions.

Author Contributions: Original draft preparation and review and editing, F.G.; supervision, project administration, and funding acquisition, S.Z. All authors have read and agreed to the published version of the manuscript.

Funding: This work is supported by the National Natural Science Foundation of China (Grant No. 82100677), the University of Macau (SRG2025-00024-ICMS), Startup Funding from the State Key Laboratory of Mechanism and Quality of Chinese Medicine (University of Macau), and Macau Science and Technology Development Fund, Macau SAR (005/2023/SKL).

Institutional Review Board Statement: Not applicable.

Informed Consent Statement: Not applicable.

Data Availability Statement: Not applicable.

Conflicts of Interest: The authors declare no conflicts of interest.

Abbreviations

The following abbreviations are used in this manuscript:

TCM	Traditional Chinese Medicine
ICS	Intestinal Stem Cell
IBD	Inflammatory Bowel Disease
APS	Astragalus Polysaccharide
LPLs	Lamina Propria Lymphocytes
IL-22	Interleukin-22
NF-κB	Nuclear Factor κB
NLRP3	NOD-Like Receptor Protein 3
MAPKs	Mitogen-Activated Protein Kinases
DSS	Dextran Sulfate Sodium
Nrf2	Nuclear Factor Erythroid 2-Related Factor 2
WSBCW	WangshiBaochi Pill
TJ	Tight Junction
ZO-1	Zonula Occludens-1
ROS	Reactive Oxygen Species
BXD	Banxia Xiexin Decoction
CSP-1	Cereus Sinensis Polysaccharide
AAD	Antibiotic-Associated Diarrhea
SCFA	Short-Chain Fatty Acid
GQD	Gegen-Qinlian Decoction
PCNA	Proliferating Cell Nuclear Antigen

References

1. Barker, N. Adult intestinal stem cells: Critical drivers of epithelial homeostasis and regeneration. *Nat. Rev. Mol. Cell Biol.* **2013**, *15*, 19–33. [CrossRef]
2. Rees, W.D.; Tandun, R.; Yau, E.; Zachos, N.C.; Steiner, T.S. Regenerative Intestinal Stem Cells Induced by Acute and Chronic Injury: The Saving Grace of the Epithelium? *Front. Cell Dev. Biol.* **2020**, *8*, 583919. [CrossRef]
3. Bialkowska, A.B.; Yang, V.W.; Giarrizzo, M.; LaComb, J.F.; Orzechowska-Licari, E.J. Intestinal Epithelial Regeneration in Response to Ionizing Irradiation. *J. Vis. Exp.* **2022**, e64028. [CrossRef]
4. Kwon, S.J.; Khan, M.S.; Kim, S.G. Intestinal Inflammation and Regeneration–Interdigitating Processes Controlled by Dietary Lipids in Inflammatory Bowel Disease. *Int. J. Mol. Sci.* **2024**, *25*, 1311. [CrossRef] [PubMed]

5. Yang, Y.; Xiao, G.; Cheng, P.; Zeng, J.; Liu, Y. Protective Application of Chinese Herbal Compounds and Formulae in Intestinal Inflammation in Humans and Animals. *Molecules* **2023**, *28*, 6811. [CrossRef]
6. Peng, J.; Li, H.; Olaolu, O.A.; Ibrahim, S.; Ibrahim, S.; Wang, S. Natural Products: A Dependable Source of Therapeutic Alternatives for Inflammatory Bowel Disease through Regulation of Tight Junctions. *Molecules* **2023**, *28*, 6293. [CrossRef] [PubMed]
7. Ding, Q.; Zu, X.; Chen, W.; Xin, J.; Xu, X.; Lv, Y.; Wei, X.; Wang, J.; Wei, Y.; Li, Z.; et al. Astragalus polysaccharide promotes the regeneration of intestinal stem cells through HIF-1 signalling pathway. *J. Cell. Mol. Med.* **2023**, *28*, e18058. [CrossRef]
8. Yang, J.; Jiang, Y.; Li, M.; Wu, K.; Wei, S.; Zhao, Y.; Shen, J.; Du, F.; Chen, Y.; Deng, S.; et al. Organoid, organ-on-a-chip and traditional Chinese medicine. *Chin. Med.* **2025**, *20*, 1–19. [CrossRef]
9. Wang, Y.-J.; Wang, H.-Y.; Li, Q.-M.; Zha, X.-Q.; Luo, J.-P. *Dendrobium fimbriatum* polysaccharide ameliorates DSS-induced intestinal mucosal injury by IL-22-regulated intestinal stem cell regeneration. *Int. J. Biol. Macromol.* **2023**, *230*, 123199. [CrossRef]
10. Cai, Y.-J.; Lan, J.-H.; Li, S.; Feng, Y.-N.; Li, F.-H.; Guo, M.-Y.; Liu, R.-P. Sinisan, a compound Chinese herbal medicine, alleviates acute colitis by facilitating colonic secretory cell lineage commitment and mucin production. *J. Integr. Med.* **2025**, *23*, 429–444. [CrossRef]
11. Deng, C.; Zhang, H.; Li, Y.; Cheng, X.; Liu, Y.; Huang, S.; Cheng, J.; Chen, H.; Shao, P.; Jiang, B.; et al. Exosomes derived from mesenchymal stem cells containing berberine for ulcerative colitis therapy. *J. Colloid. Interface Sci.* **2024**, *671*, 354–373. [CrossRef]
12. Wang, K.; Yin, J.; Chen, J.; Ma, J.; Si, H.; Xia, D. Inhibition of inflammation by berberine: Molecular mechanism and network pharmacology analysis. *Phytomedicine* **2024**, *128*, 155258. [CrossRef]
13. Gangwar, T.; Poonia, N.; Subudhi, R.N.; Arora, V. Therapeutic potential and underlying mechanisms of phytoconstituents: Emphasizing on resveratol, curcumin, quercetin, berberine, and hesperidin in ulcerative colitis. *Naunyn-Schmiedeberg's Arch. Pharmacol.* **2025**, *398*, 6579–6596. [CrossRef]
14. Lin, Y.; Liu, H.; Bu, L.; Chen, C.; Ye, X. Review of the Effects and Mechanism of Curcumin in the Treatment of Inflammatory Bowel Disease. *Front. Pharmacol.* **2022**, *13*, 908077. [CrossRef]
15. Liu, L.; Liu, Y.L.; Liu, G.X.; Chen, X.; Yang, K.; Yang, Y.X.; Xie, Q.; Gan, H.K.; Huang, X.L.; Gan, H.T. Curcumin ameliorates dextran sulfate sodium-induced experimental colitis by blocking STAT3 signaling pathway. *Int. Immunopharmacol.* **2013**, *17*, 314–320. [CrossRef]
16. Jena, A.B.; Dash, U.C.; Duttaroy, A.K. An in silico investigation on the interactions of curcumin and epigallocatechin-3-gallate with NLRP3 Inflammasome complex. *Biomed. Pharmacother.* **2022**, *156*, 113890. [CrossRef] [PubMed]
17. Peng, Y.; Ao, M.; Dong, B.; Jiang, Y.; Yu, L.; Chen, Z.; Hu, C.; Xu, R. Anti-Inflammatory Effects of Curcumin in the Inflammatory Diseases: Status, Limitations and Countermeasures. *Drug Des. Dev. Ther.* **2021**, *15*, 4503–4525. [CrossRef] [PubMed]
18. Zou, Y.; Ding, W.; Wu, Y.; Chen, T.; Ruan, Z. Puerarin alleviates inflammation and pathological damage in colitis mice by regulating metabolism and gut microbiota. *Front. Microbiol.* **2023**, *14*, 1279029. [CrossRef] [PubMed]
19. Jeon, Y.-D.; Lee, J.-H.; Lee, Y.-M.; Kim, D.-K. Puerarin inhibits inflammation and oxidative stress in dextran sulfate sodium-induced colitis mice model. *Biomed. Pharmacother.* **2020**, *124*, 109847. [CrossRef]
20. Li, Z.-L.; Yang, B.-C.; Gao, M.; Xiao, X.-F.; Zhao, S.-P.; Liu, Z.-L. Naringin improves sepsis-induced intestinal injury by modulating macrophage polarization via PPARγ/miR-21 axis. *Mol. Ther. Nucleic Acids* **2021**, *25*, 502–514. [CrossRef]
21. Rahimi, K.; Hassanzadeh, K.; Khanbabaei, H.; Haftcheshmeh, S.M.; Ahmadi, A.; Izadpanah, E.; Mohammadi, A.; Sahebkar, A. Curcumin: A Dietary Phytochemical for Targeting the Phenotype and Function of Dendritic Cells. *Curr. Med. Chem.* **2021**, *28*, 1549–1564. [CrossRef] [PubMed]
22. Varsha, K.K.; Nagarkatti, M.; Nagarkatti, P. Role of Gut Microbiota in Cannabinoid-Mediated Suppression of Inflammation. *Adv. Drug Alcohol. Res.* **2022**, *2*, 10550. [CrossRef] [PubMed]
23. Yin, S.; Sun, C.; Ji, Y.; Abdolmaleky, H.; Zhou, J.-R. Herbal medicine WangShiBaoChiWan improves gastrointestinal health in mice via modulation of intestinal tight junctions and gut microbiota and inhibition of inflammation. *Biomed. Pharmacother.* **2021**, *138*, 111426. [CrossRef] [PubMed]
24. Ji, X.; Su, L.; Zhang, P.; Yue, Q.; Zhao, C.; Sun, X.; Li, K.; Liu, X.; Zhang, S.; Zhao, L. Lentinan improves intestinal inflammation and gut dysbiosis in antibiotics-induced mice. *Sci. Rep.* **2022**, *12*, 1–12. [CrossRef]
25. Cherian, I.V.; Vijukumar, A.; Islam, M.M.; Janvi; Vikal, A. Assessing the therapeutic potential of quercetin, a widely spread flavonoid, in the prevention and management of chronic and degenerative diseases through a modern Chinese medicine perspective. *Pharmacol. Res. Mod. Chin. Med.* **2025**, *15*, 100630. [CrossRef]
26. Fan, J.; Li, B.-R.; Zhang, Q.; Zhao, X.-H.; Wang, L. Pretreatment of IEC-6 cells with quercetin and myricetin resists the indomethacin-induced barrier dysfunction via attenuating the calcium-mediated JNK/Src activation. *Food Chem. Toxicol.* **2021**, *147*, 111896. [CrossRef]

27. Junyuan, Z.; Hui, X.; Chunlan, H.; Junjie, F.; Qixiang, M.; Yingying, L.; Lihong, L.; Xingpeng, W.; Yue, Z. Quercetin protects against intestinal barrier disruption and inflammation in acute necrotizing pancreatitis through TLR4/MyD88/p38 MAPK and ERS inhibition. *Pancreatology* **2018**, *18*, 742–752. [CrossRef]
28. Du, M.; Liu, X.; Ji, X.; Wang, Y.; Liu, X.; Zhao, C.; Jin, E.; Gu, Y.; Wang, H.; Zhang, F. Berberine alleviates enterotoxigenic *Escherichia coli*-induced intestinal mucosal barrier function damage in a piglet model by modulation of the intestinal microbiome. *Front. Nutr.* **2025**, *11*, 1494348. [CrossRef]
29. Han, M.; Guo, Y.; Tang, S.; Li, D.; Wan, J.; Zhu, C.; Zuraini, Z.; Liang, J.; Gao, T.; Zhou, Z.; et al. Effects of berberine hydrochloride on antioxidant response and gut microflora in the *Charybdis japonica* infected with *Aeromonas hydrophila*. *BMC Microbiol.* **2024**, *24*, 1–14. [CrossRef]
30. Obchoei, S.; Detarya, M.; Boonnate, P.; Saranaruk, P.; Vaeteewoottacharn, K.; Mahalapbutr, P.; Okada, S.; Wongkham, S. Low Dose Berberine Suppresses Cholangiocarcinoma Cell Progression as a Multi-Kinase Inhibitor. *Asian Pac. J. Cancer Prev.* **2022**, *23*, 3379–3386. [CrossRef]
31. Che, S.-Y.; Yuan, J.-W.; Zhang, L.; Ruan, Z.; Sun, X.-M.; Lu, H. Puerarin prevents epithelial tight junction dysfunction induced by ethanol in Caco-2 cell model. *J. Funct. Foods* **2020**, *73*, 104079. [CrossRef]
32. Li, Z.; Gao, M.; Yang, B.; Zhang, H.; Wang, K.; Liu, Z.; Xiao, X.; Yang, M. Naringin attenuates MLC phosphorylation and NF-κB activation to protect sepsis-induced intestinal injury via RhoA/ROCK pathway. *Biomed. Pharmacother.* **2018**, *103*, 50–58. [CrossRef]
33. Jia, H.; Zhang, Y.; Si, X.; Jin, Y.; Jiang, D.; Dai, Z.; Wu, Z. Quercetin Alleviates Oxidative Damage by Activating Nuclear Factor Erythroid 2-Related Factor 2 Signaling in Porcine Enterocytes. *Nutrients* **2021**, *13*, 375. [CrossRef] [PubMed]
34. Zhou, J.; Lv, J.; Chen, X.; Li, T.; Shen, J.; Wang, Z.; Dai, C.; Hao, Z. Puerarin as a Phytochemical Modulator of Gastrointestinal Homeostasis in Livestock: Molecular Mechanisms and Translational Applications. *Antioxidants* **2025**, *14*, 756. [CrossRef] [PubMed]
35. Yu, W.; Zhang, Y.; Kang, C.; Zheng, Y.; Liu, X.; Liang, Z.; Yan, J. The pharmacological evidence of the chang-yan-ning formula in the treatment of colitis. *Front. Pharmacol.* **2022**, *13*, 1029088. [CrossRef] [PubMed]
36. Chen, G.; Yang, Y.; Liu, M.; Teng, Z.; Ye, J.; Xu, Y.; Cai, X.; Cheng, X.; Yang, J.; Hu, C.; et al. Banxia xiexin decoction protects against dextran sulfate sodium-induced chronic ulcerative colitis in mice. *J. Ethnopharmacol.* **2015**, *166*, 149–156. [CrossRef]
37. Yin, Y.; Wu, X.; Peng, B.; Zou, H.; Li, S.; Wang, J.; Cao, J. Curcumin improves necrotising microscopic colitis and cell pyroptosis by activating SIRT1/NRF2 and inhibiting the TLR4 signalling pathway in newborn rats. *Innate Immun.* **2020**, *26*, 609–617. [CrossRef]
38. Theriot, C.M.; Koenigsknecht, M.J.; Carlson, P.E., Jr.; Hatton, G.E.; Nelson, A.M.; Li, B.; Huffnagle, G.B.; Li, J.Z.; Young, V.B. Antibiotic-induced shifts in the mouse gut microbiome and metabolome increase susceptibility to Clostridium difficile infection. *Nat. Commun.* **2014**, *5*, 3114. [CrossRef]
39. Duan, H.; Hu, J.; Deng, Y.; Zou, J.; Ding, W.; Peng, Q.; Duan, R.; Sun, J.; Zhu, J. Berberine Mediates the Production of Butyrate to Ameliorate Cerebral Ischemia via the Gut Microbiota in Mice. *Nutrients* **2023**, *16*, 9. [CrossRef]
40. Li, X.; Xu, S.; Zhang, Y.; Li, K.; Gao, X.-J.; Guo, M.-Y.; Fu, Y.; Liu, J. Berberine Depresses Inflammation and Adjusts Smooth Muscle to Ameliorate Ulcerative Colitis of Cats by Regulating Gut Microbiota. *Microbiol. Spectr.* **2022**, *10*, e0320722. [CrossRef]
41. Cui, M.; Wang, Y.; Elango, J.; Wu, J.; Liu, K.; Jin, Y. Cereus sinensis Polysaccharide Alleviates Antibiotic-Associated Diarrhea Based on Modulating the Gut Microbiota in C57BL/6 Mice. *Front. Nutr.* **2021**, *8*, 751992. [CrossRef]
42. Zhang, C.; Shu, Y.; Li, Y.; Wang, F.; Gan, J.; Wang, Y.; Feng, X.; Guo, M. Chinese yam (Dioscorea) polysaccharide ameliorates ulcerative colitis in mice via modulating disorders of intestinal microecology and metabolism. *Int. J. Biol. Macromol.* **2025**, *315*, 144110. [CrossRef]
43. Li, Q.; Cui, Y.; Xu, B.; Wang, Y.; Lv, F.; Li, Z.; Li, H.; Chen, X.; Peng, X.; Chen, Y.; et al. Main active components of Jiawei Gegen Qinlian decoction protects against ulcerative colitis under different dietary environments in a gut microbiota-dependent manner. *Pharmacol. Res.* **2021**, *170*, 105694. [CrossRef]
44. Liu, C.-S.; Liang, X.; Wei, X.-H.; Jin, Z.; Chen, F.-L.; Tang, Q.-F.; Tan, X.-M. Gegen Qinlian Decoction Treats Diarrhea in Piglets by Modulating Gut Microbiota and Short-Chain Fatty Acids. *Front. Microbiol.* **2019**, *10*, 825. [CrossRef] [PubMed]
45. Luo, Z.; Liu, C.; Hu, Y.; Xia, T.; Zhang, B.; Chen, F.; Tan, X.; Zheng, Z. Gegen Qinlian decoction restores the intestinal barrier in bacterial diarrhea piglets by promoting Lactobacillus growth and inhibiting the TLR2/MyD88/NF-κB pathway. *Biomed. Pharmacother.* **2022**, *155*, 113719. [CrossRef] [PubMed]
46. Zhang, F.; Zhou, Y.; Chen, H.; Jiang, H.; Zhou, F.; Lv, B.; Xu, M.; Dominguez Perles, R. Curcumin Alleviates DSS-Induced Anxiety-Like Behaviors via the Microbial-Brain-Gut Axis. *Oxidative Med. Cell. Longev.* **2022**, *2022*, 6244757. [CrossRef] [PubMed]
47. Tian, W.; Li, Y.; Zhang, M.; Xiao, H.; Peng, Y.; Song, M.; Cao, Y.; Xiao, J. Engineering mucus-penetrating and enzyme-responsive nanostructured carriers for precision targeting of curcumin's pharmacokinetics and colitis-alleviating pathways. *Nano Today* **2025**, *61*, 102602. [CrossRef]

48. Jing, W.; Safarpour, Y.; Zhang, T.; Guo, P.; Chen, G.; Wu, X.; Fu, Q.; Wang, Y. Berberine Upregulates P-Glycoprotein in Human Caco-2 Cells and in an Experimental Model of Colitis in the Rat via Activation of Nrf2-Dependent Mechanisms. *J. Pharmacol. Exp. Ther.* **2018**, *366*, 332–340. [CrossRef]
49. Li, X.; Tian, T. Phytochemical Characterization of *Mentha spicata* L. Under Differential Dried-Conditions and Associated Nephrotoxicity Screening of Main Compound With Organ-on-a-Chip. *Front. Pharmacol.* **2018**, *9*, 1067. [CrossRef]
50. Sadeghi, N.; Mansoori, A.; Shayesteh, A.; Hashemi, S.J. The effect of curcumin supplementation on clinical outcomes and inflammatory markers in patients with ulcerative colitis. *Phytother. Res.* **2019**, *34*, 1123–1133. [CrossRef]
51. Liu, X.; Fan, Y.; Du, L.; Mei, Z.; Fu, Y. In Silico and In Vivo Studies on the Mechanisms of Chinese Medicine Formula (Gegen Qinlian Decoction) in the Treatment of Ulcerative Colitis. *Front. Pharmacol.* **2021**, *12*, 665102. [CrossRef]
52. Zhao, Y.; Zhan, J.; Sun, C.; Zhu, S.; Zhai, Y.; Dai, Y.; Wang, X.; Gao, X. Sishen Wan enhances intestinal barrier function via regulating endoplasmic reticulum stress to improve mice with diarrheal irritable bowel syndrome. *Phytomedicine* **2024**, *129*, 155541. [CrossRef]
53. Ahn, H.-R.; Park, D.H.; Shin, M.-S.; Nguyen, Q.N.; Park, J.Y.; Kim, D.-W.; Kang, K.S.; Lee, H.L. The Ameliorating Effect of Lizhong-Tang on Antibiotic-Associated Imbalance in the Gut Microbiota in Mouse. *Appl. Sci.* **2022**, *12*, 6943. [CrossRef]
54. Yang, X.; Zheng, D.; Yong, J.; Li, Y.; Sun, Y.; Zhao, F.; Tang, D.; Xie, Y.; Bi, D. Baicalein and Berberine Inhibit the Growth and Virulence of Clostridioides difficile. *Pathogens* **2025**, *14*, 662. [CrossRef]
55. Hassanzadeh, K.; Buccarello, L.; Dragotto, J.; Mohammadi, A.; Corbo, M.; Feligioni, M. Obstacles against the Marketing of Curcumin as a Drug. *Int. J. Mol. Sci.* **2020**, *21*, 6619. [CrossRef] [PubMed]
56. Li, Z.; Wang, Y.; Xu, Q.; Ma, J.; Li, X.; Yan, J.; Tian, Y.; Wen, Y.; Chen, T. Berberine and health outcomes: An umbrella review. *Phytother. Res.* **2023**, *37*, 2051–2066. [CrossRef] [PubMed]
57. Coon, J.T.; Ernst, E. Panax ginseng: A systematic review of adverse effects and drug interactions. *Drug Saf.* **2002**, *25*, 323–344. [CrossRef]
58. Liang, H.; Tao, S.; Wang, Y.; Zhao, J.; Yan, C.; Wu, Y.; Liu, N.; Qin, Y. Astragalus polysaccharide: Implication for intestinal barrier, anti-inflammation, and animal production. *Front. Nutr.* **2024**, *11*, 1364739. [CrossRef]
59. Jiang, Y.; Liao, Y.; Liu, Z.; Zhou, M.; Wang, H.; Qi, H.; Sun, S.; Xi, S.; Tang, Y. The effects of Cordyceps polysaccharides on ischemic brain injury in rats via intervening with IL-23/IL-17 axis and the intestinal barrier. *Int. J. Biol. Macromol.* **2024**, *283*, 137526. [CrossRef]
60. Andres, S.; Pevny, S.; Ziegenhagen, R.; Bakhiya, N.; Schafer, B.; Hirsch-Ernst, K.I.; Lampen, A. Safety Aspects of the Use of Quercetin as a Dietary Supplement. *Mol. Nutr. Food Res.* **2018**, *62*, 1700447. [CrossRef]
61. Peterson, C.T.; Vaughn, A.R.; Sharma, V.; Chopra, D.; Mills, P.J.; Peterson, S.N.; Sivamani, R.K. Effects of Turmeric and Curcumin Dietary Supplementation on Human Gut Microbiota: A Double-Blind, Randomized, Placebo-Controlled Pilot Study. *J. Evid. Based Integr. Med.* **2018**, *23*, 2515690 × 18790725. [CrossRef]
62. Wang, T.; Lu, S.Y.; Dan, L.; Sun, Y.; Fu, T.; Tian, L.; Chen, J. Higher Dietary Quercetin Intake Is Associated with Lower Risk of Adverse Outcomes among Individuals with Inflammatory Bowel Disease in a Prospective Cohort Study. *J. Nutr.* **2024**, *154*, 1861–1868. [CrossRef]
63. Xiong, X.; Cheng, Z.; Wu, F.; Hu, M.; Liu, Z.; Dong, R.; Chen, G. Berberine in the treatment of ulcerative colitis: A possible pathway through Tuft cells. *Biomed. Pharmacother.* **2021**, *134*, 111129. [CrossRef]
64. Kang, Z.; Zhonga, Y.; Wu, T.; Huang, J.; Zhao, H.; Liu, D. Ginsenoside from ginseng: A promising treatment for inflammatory bowel disease. *Pharmacol. Rep.* **2021**, *73*, 700–711. [CrossRef]
65. Yu, Z.L.; Gao, R.Y.; Lv, C.; Geng, X.L.; Ren, Y.J.; Zhang, J.; Ren, J.Y.; Wang, H.; Ai, F.B.; Wang, Z.Y.; et al. Notoginsenoside R1 promotes Lgr5(+) stem cell and epithelium renovation in colitis mice via activating Wnt/beta-Catenin signaling. *Acta Pharmacol. Sin.* **2024**, *45*, 1451–1465. [CrossRef]

biomolecules

MDPI

Article

Compound K Promotes Megakaryocytic Differentiation by NLRP3 Inflammasome Activation

Seonhwa Hwang [1,2], Min-Seo Park [1,2], Anthony Junhoe Koo [3], Eunsoo Yoo [4], Seh-Hyon Song [1], Hye-Kyung Kim [1,2], Min-Hi Park [1,2,*] and Jae-Seon Kang [1,2,*]

1 College of Pharmacy, Kyungsung University, 309 Suyeong-ro, Busan 48434, Republic of Korea; reseon17@naver.com (S.H.); minseo829@naver.com (M.-S.P.); shsong@ks.ac.kr (S.-H.S.); fiona30@ks.ac.kr (H.-K.K.)
2 Brain Busan 21 Plus Research Project Group, Kyungsung University, Busan 48434, Republic of Korea
3 College of Engineering, North Carolina A&T State University, Greensboro, NC 27411, USA; junhoekoo0727@gmail.com
4 Chemical, Biological, and Bioengineering Department, North Carolina A&T State University, Greensboro, NC 27411, USA; eyoo@ncat.edu
* Correspondence: parkmh@ks.ac.kr (M.-H.P.); jskang8551002@ks.ac.kr (J.-S.K.)

Abstract: Platelets are essential blood components that maintain hemostasis, prevent excessive bleeding, and facilitate wound healing. Reduced platelet counts are implicated in various diseases, including leukemia, hepatitis, cancer, and Alzheimer's disease. Enhancing megakaryocytic differentiation is a promising strategy to increase platelet production. Compound K (CK), a major bioactive metabolite of ginsenosides from *Panax ginseng*, has demonstrated anti-cancer and neuroprotective properties. In this study, we investigated the effects of CK on megakaryocytic differentiation and apoptosis in chronic myeloid leukemia (CML) cell lines K562 and Meg-01. CK treatment significantly upregulated the mRNA expression of key megakaryocytic differentiation markers, including CD61, CD41, and CD42a, and promoted the formation of large, multinucleated cells in K562 cells. Additionally, flow cytometry analysis revealed that CK at 5 μM induced apoptosis, a critical process in thrombocytopoiesis, in both K562 and Meg-01 cells. RT^2 Profiler PCR array analysis further identified a marked increase in the expression of genes associated with the activation of the NLRP3 inflammasome in CK-treated K562 and Meg-01 cells. This study is the first to demonstrate that CK promotes megakaryocytic differentiation and apoptosis through the activation of the ERK/EGR1 and NLRP3 inflammasome pathways. These findings suggest that CK may enhance platelet production, indicating its potential as a therapeutic candidate for platelet-related disorders and other associated diseases.

Keywords: compound K; megakaryocytic differentiation; leukemia cell; NLRP3 inflammasome; platelets

1. Introduction

Platelets, also known as thrombocytes, are indispensable blood cells involved in a range of physiological processes, including hemostasis, wound healing, inflammation, tumor metastasis, and host defense [1–3]. These small, disk-shaped, anucleate cells originate from megakaryocytes (MKs), which are primarily located in the lungs and bone marrow [4,5]. MKs are derived from hematopoietic stem cells (HSCs) and undergo a complex sequence of regulated processes—such as DNA endoreduplication, cytoplasmic maturation, and cellular expansion—to produce and release platelets into circulation [6–9]. Although MKs are relatively rare in the bone marrow, they play a critical role in hematopoiesis and the regulation of hemostasis, ensuring the continuous supply of platelets necessary for maintaining key physiological functions [6–8].

Megakaryopoiesis, the process by which MKs develop from HSCs, is primarily regulated by thrombopoietin (TPO), a key hormone that promotes megakaryocyte (MK) colony formation and stimulates the proliferation and maturation of MK progenitors [10,11]. TPO binds to c-Mpl receptors on MKs and platelets, triggering a critical signal transduction cascade essential for MK development. Additionally, cytokines such as interleukin (IL)-1β, IL-3, and IL-6 contribute to the regulation of MK proliferation and maturation, often acting synergistically with TPO to enhance these processes [12]. As MKs progress through maturation, they undergo profound structural changes, accumulating internal membrane systems, granules, and organelles in preparation for platelet biogenesis. MK differentiation is marked by a shift from mitotic division to polyploidization via endomitosis, a process that increases both cell size and ploidy levels, ultimately enabling a single MK to produce thousands of platelets [7,13].

Platelets play a role beyond their traditional function in hemostasis, being actively involved in various physiological and pathological processes such as inflammation, immune responses, and tumor metastasis [14–16]. Decreased platelet counts have been linked to a range of diseases, including leukemia, hepatitis, cancer, and Alzheimer's disease (AD) [17,18]. Moreover, platelet counts naturally decline with age, a factor that may contribute to the heightened platelet reactivity observed in older populations [19–21]. Recent investigations have begun to elucidate the relationship between platelet function and AD, particularly focusing on the influence of platelets on cerebral blood flow and cognitive decline [22]. These findings underscore the multifunctional nature of platelets and their critical role in maintaining systemic health across different physiological and pathological contexts.

Compound K (CK), a primary metabolite of ginsenosides from *Panax ginseng*, exhibits superior bioavailability compared to its precursor [23–29]. CK has been shown to possess diverse pharmacological properties, including anti-diabetic, anti-carcinogenic, anti-inflammatory, anti-aging, and hepatoprotective effects, positioning it as a promising candidate for therapeutic applications [30–33]. In this study, we explored the effects of CK on apoptosis, differentiation, and the associated molecular mechanisms in chronic myeloid leukemia (CML) K562 and Meg-01 cell lines. Our results demonstrate that CK effectively inhibits cell growth and induces megakaryocytic differentiation in both cell lines. Additionally, we propose that the activation of the NLRP3 inflammasome pathway is a key mechanism underlying CK-mediated megakaryocytic differentiation, suggesting its potential as a therapeutic agent for enhancing platelet production.

2. Materials and Methods

2.1. Materials

Study materials included ginsenoside compound K (CK; ChemFaces, Wuhan, China), RPMI 1640 medium (Gibco, Grand Island, NY, USA), FBS (Gibco), penicillin–streptomycin (Sigma-Aldrich, St. Louis, MO, USA), 2′,7′-Dichlorofluorescein Di-acetate (H2DCFDA) (Invitrogen, Carlsbad, CA, USA), dihydroergotamine 123 (DHR 123) (Thermo Fisher Scientific, Waltham, MA, USA), diethylenetriaminepentaacetic acid (Sigma-Aldrich), D-Penicillamine (Sigma-Aldrich), dimethyl sulfoxide (DMSO) (Sigma-Aldrich), Ez-Cytox (DoGenBio, Seoul, Republic of Korea), Hoechst 33258, pentahydrate (Invitrogen), FITC Annexin V/Dead Cell Apoptosis Kit (Invitrogen), Total RNA Extraction Kit (SJ bio science, Incheon, Republic of Korea), Compact cDNA Synthesis Kit (SJ bio science), and SYBR Green Master Mix (SJ bio science).

2.2. Cell Culture and Treatment

The K562 cells were obtained from Korea Cell Line Bank (Seoul, Republic of Korea). The Meg-01 cells were kindly provided by Dr. Hyung Soo Kim (Department of Pharmaceutical Science and Technology, Kyungsung University, Busan, Republic of Korea). Both cell lines were cultured in RPMI 1640 medium supplemented with 10% fetal bovine serum (FBS) and 1% penicillin–streptomycin. Cultures were maintained at 37 °C in a humidified incubator with 5% CO_2. For subculturing, cells and medium from a 100 × 10 mm culture dish were collected and centrifuged at 3000 rpm for 10 min. The supernatant was discarded, and the cell pellet was resuspended in 1 mL of fresh medium. The resuspended cells were then transferred to a new 100 × 10 mm culture dish containing 20 mL of medium. CK was dissolved in dimethyl sulfoxide (DMSO) to a concentration of 100 mM and stored at −20 °C until use. For drug treatment, CK and cells were diluted to twice the desired final concentration in a culture medium and then combined at a 1:1 volume ratio (*v*/*v*) to ensure simultaneous seeding and treatment. This approach was chosen to minimize the risk of cell damage due to high chemical concentrations, particularly in the suspension cultures.

2.3. Measurement of Cell Viability

Cell viability was assessed using the EZ-Cytox Kit, which employs a water-soluble tetrazolium salt (WST) for the MTT assay. Cells were seeded at a density of 2×10^5 cells/well in a 96-well plate and simultaneously treated with CK for 48 and 72 h to determine cell viability and apoptosis. The cells were incubated at 37 °C in a humidified incubator with 5% CO_2 throughout the treatment. Two hours before the completion of the treatment, 10 μL of EZ-Cytox reagent was added to each well containing the culture medium. Following the 2 h incubation with EZ-Cytox, absorbance was measured at 450 nm using an Epoch Microplate Spectrophotometer (BioTek, Winooski, VA, USA). Cell viability was calculated as a percentage relative to the vehicle control group.

2.4. Quantitative Real-Time PCR (q-PCR)

To evaluate the effect of CK on gene expression, K562 and Meg-01 cells were seeded at a density of 2.0×10^6 cells per well in a 12-well plate and treated with either vehicle control or 3 and 5 μM of CK for 72 h. Total RNA was extracted using the Total RNA Extraction Kit according to the manufacturer's instructions. The concentration of RNA was quantified, and cDNA synthesis was performed using a Compact cDNA Synthesis Kit. Quantitative real-time PCR (qPCR) was conducted in a total reaction volume of 20 μL, comprising 2 μL of cDNA, 10 μL of SYBR Green Master Mix, 6 μL of UltraPure™ DNase/RNase-Free Distilled Water (Invitrogen), and 1 μL each of forward and reverse primers. The thermal cycling conditions were as follows: an initial step at 50 °C for 2 min, followed by 10 min at 95 °C, then 40 cycles of denaturation at 95 °C for 15 s and annealing at 60 °C for 1 min. qPCR reactions were carried out using the QuantStudio 1 system (Applied Biosystems, Thermo Fisher Scientific) and analyzed with QuantStudio™ Design & Analysis Software v1.5.1 (Thermo Fisher Scientific). All gene expression levels were normalized to 18S ribosomal RNA using the comparative Ct method ($2^{-\Delta\Delta Ct}$). The primer sequences used in this study are provided in Table 1.

Table 1. Primer sequences used for q-PCR analysis.

Gene	Primer Sequence (5′→3′)		Scale
	Forward Primer	Reverse Primer	
18S	5′-CTCTAGATAACCTCGGGCCG-3′	5′-GTCGGGAGTGGGTAATTTGC-3′	25 nmole
CD41	5′-GATGAGACCCGAAATGTAGGC-3′	5′-GTCTTTTCTAGGACGTTCCAGTG-3′	25 nmole
CD42a	5′-ACCCTCGATGTGACGCAGA-3′	5′-CCAGAGGCGCAGATAGGTG-3′	25 nmole
CD61	5′-GTGACCTGAAGGAGAATCTGC-3′	5′-CCGGAGTGCAATCCTCTGG-3′	25 nmole
IL-1ß	5′-ATGATGGCTTATTACAGTGGCAA-3′	5′-GTCGGAGATTCGTAGCTGGA-3′	25 nmole
IL-6	5′-ACTCACCTCTTCAGAACGAATTG-3′	5′-CCATCTTTGGAAGGTTCAGGTTG-3′	25 nmole

2.5. Enzyme-Linked Immunosorbent Assay (ELISA)

The concentration of IL-1β was quantified using a quantitative sandwich enzyme immunoassay technique with the Human IL-1β ELISA Kit (Abclonal, Woburn, MA, USA). K562 and Meg-01 cells were cultured in a 48-well plate and divided into three groups: non-treated cells (vehicle control) and cells treated with 3 μM or 5 μM of CK. The cells were incubated at 37 °C for 72 h. After incubation, the culture samples were centrifuged at 3000 rpm for 10 min, and the supernatants were collected for analysis. Next, the samples and standards were added to the wells of microplates pre-coated with antibodies specific to human IL-1β. The plates were incubated at 37 °C for 2 h, followed by three washes. Subsequent incubation and washing steps for the enzyme conjugate were performed in accordance with the manufacturer's protocol. After completing the assay, 50 μL of stop solution was added to each well, and absorbance was measured using a microplate reader. IL-1β concentrations were calculated based on the standard curve generated during the experiment.

2.6. Western Blot Analysis

To investigate cell differentiation and the mechanisms of apoptosis, Western blot analysis was conducted using specific antibodies against CD61, NLRP3, and caspase-1. K562 and Meg-01 cells were seeded at a density of 6.0×10^5 cells per well in 6-well plates, treated with vehicle control or 3 μM and 5 μM of CK, and incubated for 72 h. Following incubation, cells were lysed with 200 μL of ProEXTM CETi Lysis Buffer containing Protease and Phosphatase Inhibitors and incubated on ice for 30 min. Lysates were centrifuged at 13,000 rpm for 10 min, and the supernatant was collected. Protein concentrations were determined using the Protein Assay Dye Reagent Concentrate. Cell lysates were prepared with 35 μg of protein in H_2O and 5X SDS-PAGE Sample Buffer (TransLab). Equal amounts of protein were separated on 6–10% SDS-PAGE gels and transferred to polyvinylidene difluoride (PVDF) membranes (Millipore, Carrigtwohill, Ireland) by electrophoresis. Membranes were blocked using a Protein-Free Block Solution and subsequently incubated overnight at 4 °C with specific primary antibodies (1:1000 dilution) on a shaker. The following day, membranes were washed three times with TBS-T and incubated with IRDye 680RD Goat anti-Rabbit secondary antibody (1:10,000 dilution) (LI-COR, Lincoln, NE, USA) for 1 h and 30 min at room temperature on a shaker. After washing the membranes three times with TBS-T, protein bands were detected using the Odyssey Imaging System (Odyssey-XF, LI-COR). Protein band intensities were quantified using Empiria Studio software (version 2.3, LI-COR), with β-actin serving as the loading control. Original Western blot images can be found in Supplementary File S1.

2.7. Analysis of Cell Morphology

Hoechst 33258 staining was employed to observe morphological changes in K562 and Meg-01 cells. Cells were seeded at a density of 2.0×10^5 cells/mL in 35 × 10 mm culture dishes, treated with vehicle control or 5 μM of CK, and incubated for 72 h. Following incubation, cells were stained with 5 μg/mL Hoechst 33258 and incubated for 20 min in the dark at 37 °C. After staining, cells were washed with PBS. Fluorescence microscopy was conducted using an Axio Observer microscope (Zeiss, Oberkochen, Germany) equipped with an LD Plan-NEOFLUAR 40X/0.6 Corr Ph2 objective. Image acquisition and analysis were performed using ZEN 3.6 (blue edition) microscopy software.

2.8. Flow Cytometric Analysis

The FITC Annexin V/Dead Cell Apoptosis Kit was employed to measure apoptosis in K562 and Meg-01 cells, following the manufacturer's instructions. Cells were seeded in 35 × 10 mm culture dishes and treated with vehicle control or CK at concentrations of 3 μM and 5 μM for 72 h. After treatment, cells were collected at a density of 1×10^6 cells/mL and washed with ice-cold PBS. The cells were then resuspended in a buffer containing propidium iodide (PI), Annexin V, and FITC and incubated at room temperature for 15 min. Following incubation, the cells were treated with 1X annexin-binding buffer. Apoptotic cells were subsequently analyzed using CytoFLEX flow cytometry (Beckman Coulter, Brea, CA, USA). For K562 cells, up to 30,000 events per sample were recorded, while up to 10,000 events per sample were detected during flow cytometric analysis of Meg-01 cells.

2.9. RT^2 Profiler PCR Arrays

A PCR array was conducted to identify apoptosis-related pathways. RNA samples were prepared following the same protocol as described in the qPCR section above. For the assay, vehicle control (Con) and 5 μM CK-treated samples were used, excluding the 3 μM CK condition. After quantification of the extracted mRNA, cDNA was synthesized using the RT^2 First Strand Kit. PCR reactions were assembled by adding RT^2 SYBR Green Master Mix, cDNA, and UltraPure™ DNase/RNase-Free Distilled Water to each well, following the manufacturer's instructions. Quantitative PCR (qPCR) arrays were performed using the RT^2 Profiler PCR arrays, which included 84 apoptosis-related genes, 5 housekeeping genes, and 3 control genes. The reverse transcription reactions were carried out using the QuantStudio 1 system (Applied Biosystems) under the same cycling conditions described previously. The resulting threshold cycle (Ct) value was uploaded to the web-based GeneGlobe Data Analysis Centre, and $2^{-\Delta\Delta Ct}$ was used to calculate relative gene expression. The list of genes used in the experiment is provided in Table S1.

2.10. Statistical Analysis

Statistical analysis of the experimental data was performed using GraphPad Prism version 5.0 (GraphPad Software, Inc., San Diego, CA, USA). A one-way analysis of variance (ANOVA) followed by Bonferroni and Dunnett's multiple comparison tests was used to assess differences between groups. A p-value of less than 0.05 ($p < 0.05$) was considered statistically significant. Data are presented as the mean ± standard error of the mean (SEM).

3. Results

3.1. Measurement of Cell Viability of K562 Cells and Meg-01 Cells

K562 and Meg-01 cell lines are well-established models of human CML [34,35]. To determine the optimal experimental concentrations of CK, we assessed its effects on the viability of these CML cell lines (Figure 1). Cells were treated with the vehicle control or CK at concentrations of 3 μM and 5 μM for 48 h and 72 h, and cell viability was measured using the MTT assay. As shown in Figure 1B, CK treatment significantly decreased the viability of K562 cells. After 48 h of treatment, cell viability decreased by 18.89% compared to the control for 5 μM CK. Following 72 h of treatment, cell viability further declined by 31.11% compared to the control (*** $p < 0.001$). Similarly, in Meg-01 cells (Figure 1C), CK treatment at 3 μM and 5 μM for 48 h reduced viability by 6.50% and 18.17% compared to the control, respectively. After 72 h of treatment, cell viability decreased further by 7.28% and 32.25% compared to the control, respectively (*** $p < 0.001$). Also, as shown in Figure 1C, the viability of Meg-01 cells treated with CK (3 and 5 μM) for 48 h decreased by 6.50% and 18.17% compared to the vehicle control, respectively, while the viability of cells treated for 72 h decreased by 7.28% and 32.25% compared to the control, respectively (*** $p < 0.001$). These findings demonstrate that CK significantly inhibits the viability of both K562 and Meg-01 cells in a dose- and time-dependent manner, with 5 μM CK exhibiting a pronounced inhibitory effect on cell growth.

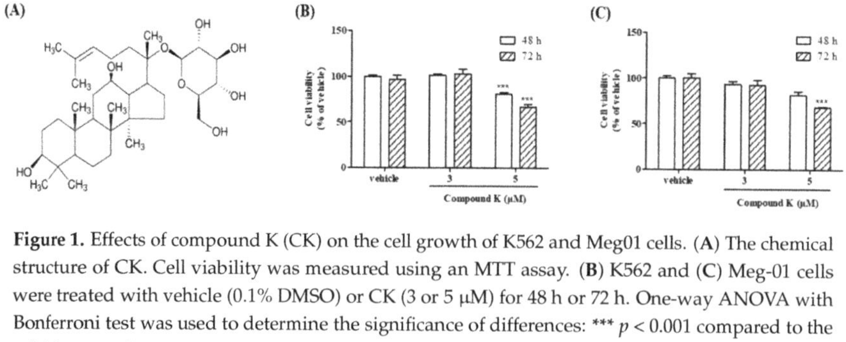

Figure 1. Effects of compound K (CK) on the cell growth of K562 and Meg01 cells. (**A**) The chemical structure of CK. Cell viability was measured using an MTT assay. (**B**) K562 and (**C**) Meg-01 cells were treated with vehicle (0.1% DMSO) or CK (3 or 5 μM) for 48 h or 72 h. One-way ANOVA with Bonferroni test was used to determine the significance of differences: *** $p < 0.001$ compared to the vehicle-treated group.

3.2. CK Promoted Megakaryocytic Differentiation of K562 and Meg-01 Cells

K562 and Meg-01 cells are widely used models for investigating hematopoietic differentiation, as they can be induced to differentiate into various hematopoietic lineages with appropriate agents [34,36]. In this study, we examined whether CK could induce megakaryocytic differentiation in K562 and Meg-01 cells. As shown in Figure 2, CK treatment (5 μM) significantly upregulated the mRNA expression of megakaryocytic differentiation markers, including CD41, CD42a, and CD61, in K562 cells after 72 h of treatment (** $p < 0.01$ and *** $p < 0.001$) (Figure 2A). Similarly, in Meg-01 cells, CK treatment at both 3 μM and 5 μM for 72 h significantly increased the mRNA expression of CD41, CD42a, and CD61 (* $p < 0.05$, ** $p < 0.01$, and *** $p < 0.001$) (Figure 2B). Consistent with the gene expression results, in K562 cells treated with 5 μM CK for 72 h, protein expression of CD61 and CD42a was significantly increased to 52.63% and 79.93%, respectively, compared to the vehicle-treated group (* $p < 0.05$). Meg-01 cells were also significantly increased at 61.33% and 96.79% compared to the vehicle-treated group (* $p < 0.05$, ** $p < 0.01$, and *** $p < 0.001$) (Table 2 and Figure 2C,D). These results indicate that CK markedly promotes the expression of CD61, a crucial megakaryocytic marker, at both the gene and protein levels in K562 and Meg-01 cells.

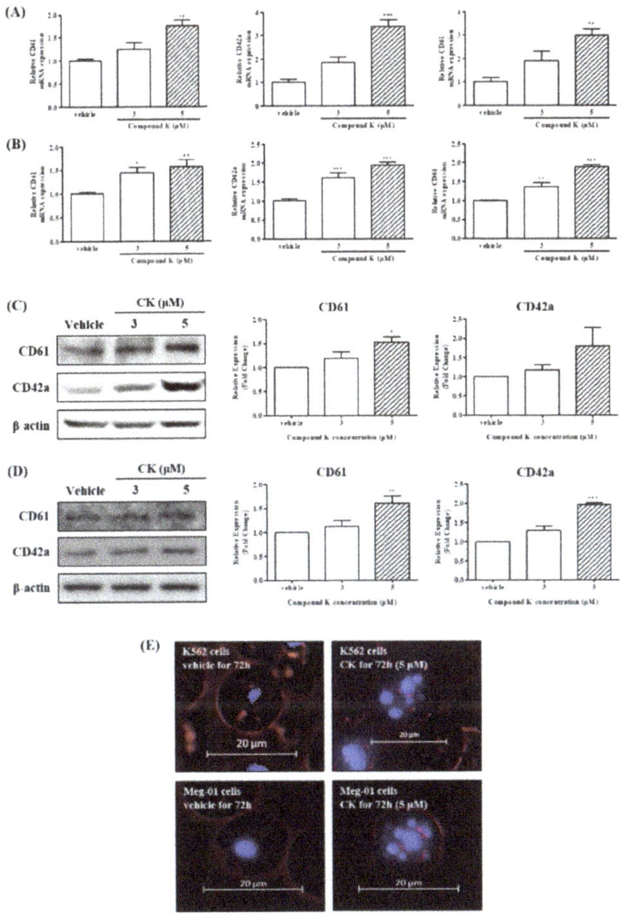

Figure 2. Effects of CK on the expression of megakaryocytic differentiation genes in K562 and Meg-01 cells. K562 and Meg-01 cells were treated with vehicle or CK (3 or 5 μM) for 72 h. The messenger

RNA (mRNA) expression levels in (**A**) K562 and (**B**) Meg-01 cells. CK induced the secretion of differentiation markers. The protein expression of CD61, CD42a, and β-actin was measured by Western blot analysis (n = 3). Quantification results of (**C**) K562 and (**D**) Meg-01 cells and a representative blot are shown. One-way ANOVA with Bonferroni test was used to determine the significance of differences: * $p < 0.05$, ** $p < 0.01$, and *** $p < 0.001$ compared to vehicle-treated cells. (**E**) Morphological changes of the cells and multi-lobulation of the nucleus were observed under microscope. K562 and Meg-01 cells were treated with vehicle (0.1% DMSO) or 5 μM CK for 72 h. Representative histograms are shown. Blue fluorescence indicates nuclear staining by Hoechst. Red indicates a cell membrane that has been unstained. The scale bar is 20 μm.

Table 2. Summary of apoptosis and megakaryocyte differentiation gene expression in CML.

Compared to the Vehicle-Treated Group		K562		Meg-01	
		3 μM	5 μM	3 μM	5 μM
The protein expression level of	CD61 (%)	18.58%	52.63%	12.02%	61.33%
	CD42a (%)	17.54%	79.93%	29.33%	96.79%
Cells with apoptosis (%)		17.71%	26.59%	13.01%	24.62%

3.3. CK Promoted Multi-Lobulated Nuclei in K562 and Meg-01 Cells

MKs are known to undergo characteristic morphological changes during differentiation and maturation, including the formation of multi-lobulated nuclei. To assess whether CK induces such morphological changes, we examined nuclear multi-lobulation in K562 and Meg-01 cells using Hoechst 33258 staining and fluorescence microscopy. As shown in Figure 2E, treatment with 0.1% DMSO (vehicle control) had no significant effect, whereas treatment with CK (5 μM) notably promoted the formation of multi-lobulated nuclei in both K562 and Meg-01 cells. These findings suggest that CK enhances megakaryocytic differentiation in CML K562 and Meg-01 cells by promoting nuclear multi-lobulation, a hallmark of MK maturation.

3.4. Expression of Egr-1 Induced by CK Could Promote Megakaryocytic Differentiation of CML Cells

Early growth response 1 (Egr-1) is a multifunctional transcription factor that regulates various cellular processes, including cell growth, differentiation, and apoptosis, in response to cytokine signaling and growth factors. To investigate whether the Egr-1 gene is involved in CK-induced megakaryocytic differentiation, we assessed its protein levels following CK treatment. As shown in Figure 3, treatment with CK (5 μM) for 72 h significantly increased Egr-1 protein levels in both K562 and Meg-01 cells (* $p < 0.05$). The induction of Egr-1 is known to be mediated by extracellular signal-regulated kinase (ERK). To further confirm CK-induced Egr-1 activation, we measured ERK protein levels under the same experimental conditions. As shown in Figure 3D, CK (5 μM) treatment significantly increased ERK levels in Meg-01 cells (* $p < 0.05$), consistent with the induction of Egr-1. These results suggest that CK-induced Egr-1 expression may be mediated through ERK activation in CML cells, promoting megakaryocytic differentiation.

3.5. Effects of CK on mRNA Expression of Megakaryocytic Differentiation-Related Cytokines in K562 and Meg-01 Cells

The proliferation and maturation of MK precursors are indirectly regulated by various cytokines, including interleukin (IL)-1β, IL-3, and IL-6. These cytokines stimulate TPO, which in turn promotes the differentiation and maturation of MK precursors, ultimately leading to platelet production [37]. In this study, we examined the effect of CK on the expression of IL-1β and IL-6, two key cytokines known to stimulate TPO, in K562 and Meg-01 cells. As shown in Figure 3E,F, treatment with 5 μM CK significantly increased the mRNA expression of both IL-1β and IL-6 in K562 and Meg-01 cells (* $p < 0.05$, ** $p < 0.01$, and *** $p < 0.001$). These findings suggest that CK may enhance TPO production by upregulating specific cytokines, thereby promoting the differentiation and maturation of MK precursors in these cell lines.

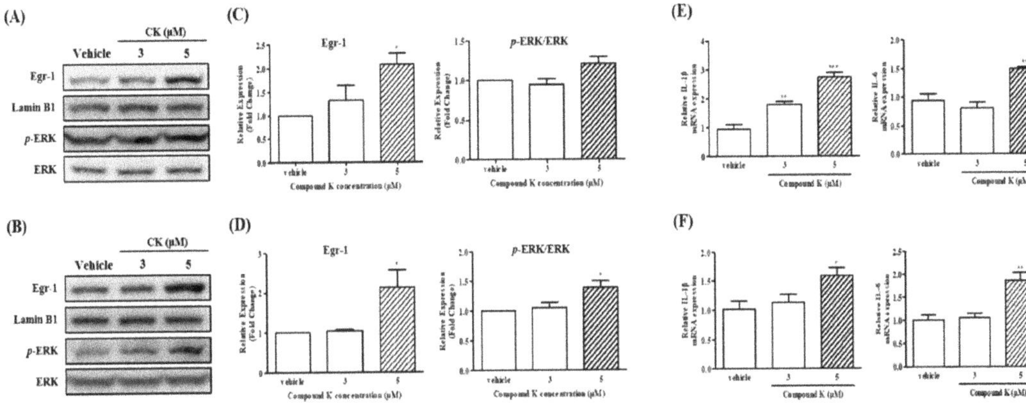

Figure 3. Involvement of CK in the megakaryocytic differentiation-related signaling pathway-dependent gene expression activation in CML cells. Cells were treated with vehicle or CK (3 or 5 μM) for 72 h. Western blot analysis measured the levels of Egr-1, Lamin B1, phosphor-ERK (p-ERK), and ERK protein. A representative blot of (**A**) K562 and (**B**) Meg-01 cells is shown. The experiments were performed in triplicate, and the data represent the mean ± SD of independent experiments of (**C**) K562 and (**D**) Meg-01 cells. The mRNA expression levels in (**E**) K562 and (**F**) Meg-01 cells. One-way ANOVA with Bonferroni test was used to determine the significance of differences: * $p < 0.05$, ** $p < 0.01$, and *** $p < 0.001$ compared to vehicle-treated cells.

3.6. Effects of CK on Apoptosis of K562 and Meg-01 Cells

It is widely accepted that apoptosis is essential for platelet biogenesis [38–41]. Recently, it was observed that MKs and platelets depend on apoptosis for their development and viability [42]. We investigated the effect of CK on apoptosis in K562 and Meg-01 cells. Cells were treated with the vehicle control or CK at concentrations of 3 μM and 5 μM for 72 h, and apoptosis was assessed via flow cytometry. As depicted in Table 2 and Figure 4, CK treatment significantly increased apoptosis in both cell lines. In K562 cells, apoptosis increased to 17.71% with 3 μM CK and 26.59% with 5 μM CK. Similarly, in Meg-01 cells, apoptosis levels rose to 13.01% with 3 μM CK and 24.62% with 5 μM CK. Furthermore, the proportion of early apoptotic and late apoptotic cells was significantly elevated in the 5 μM CK-treated group compared to the vehicle-treated control (** $p < 0.01$ and *** $p < 0.001$). Apoptosis shares several features with the platelet assembly process in MKs, including cytoskeletal reorganization. These findings suggest that CK induces apoptosis, an essential component of platelet biogenesis, in K562 and Meg-01 cells.

3.7. CK Activates NLRP3 Inflammasome in K562 and Meg-01 Cells

Apoptosis can be triggered through various pathways, including the NLRP3 inflammasome, endoplasmic reticulum (ER) stress, and mitochondrial dysfunction. To elucidate the molecular mechanisms underlying CK-induced apoptosis, we conducted pathway-focused gene expression profiling using the RT^2 Profiler PCR Array System, which targets 84 genes associated with apoptosis in K562 and Meg-01 cells. Differentially expressed genes were identified based on a fold change greater than 2.5 ($p < 0.001$). K562 and Meg-01 cells treated with 5 μM CK exhibited distinct gene expression profiles compared to vehicle-treated controls. As shown in Table S2 and Figure 5B, twelve genes were significantly upregulated (>2.5-fold) in K562 cells following CK treatment, including CASP14, CIDEB, PYCARD, TNFRSF9, and TRAF3, all of which are involved in the positive regulation of apoptosis. Additionally, anti-apoptotic genes such as BCL2, BIRC3, CD40LG, and CIDEA were upregulated. In Meg-01 cells, 16 genes were upregulated by more than 2.5-fold after CK exposure, including AKT1, CASP4, CASP6, CRADD, LTA, TRAF2, BAG3, BCL2L2, BFAR, BIRC2, IGF1R, and XIAP (Table S3 and Figure 5F). When comparing the upregulated apoptosis-related genes in K562 and Meg-01 cells, the positive regulation of the apoptosis pathway showed the highest number of upregulated genes, with five genes in K562 cells and six genes in Meg-01 cells (Figure 5C,G). Notably, four of these genes—CASP4, CASP6, CASP14, and PYCARD—are inflammasome-related (Figure 5D,H). These findings underscore the involvement of the inflammasome pathway, particularly the NLRP3 inflammasome, in CK-induced apoptosis, warranting further investigation.

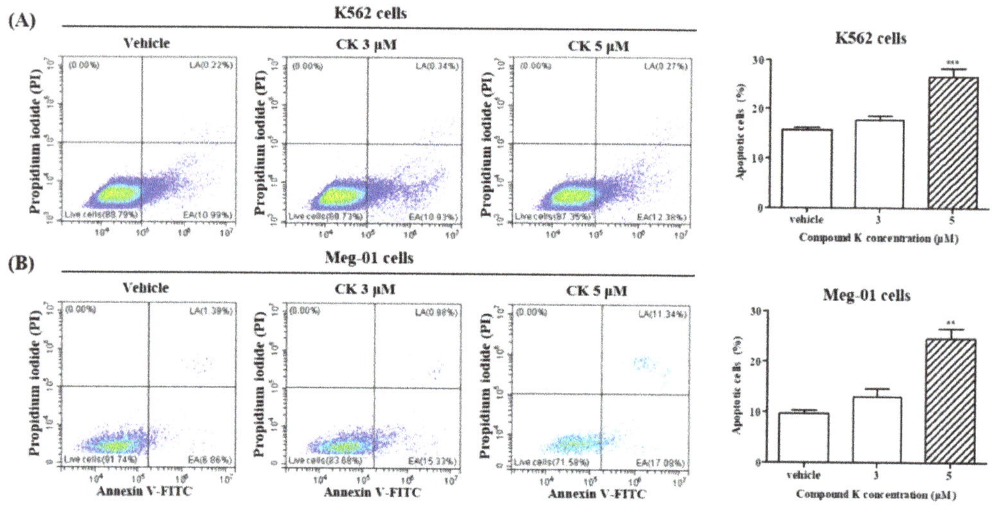

Figure 4. Effect of CK on apoptosis of K562 cells and Meg-01 cells. The population of early apoptotic (EA) and late apoptotic (LA) cells was detected using an Annexin V-FITC and propidium iodide (PI) staining kit, according to the manufacturer's instructions. A representative histogram is shown for (**A**) K562 and (**B**) Meg-01 cells. The apoptotic cell populations in EA and LA phases were quantified. The experiments were replicated three times, and the data represent the mean ± SEM of independent experiments. One-way ANOVA with Bonferroni test was used to determine the significance of differences: ** $p < 0.01$ and *** $p < 0.001$ compared to vehicle-treated cells.

3.8. NLRP3 Inflammasome Activation Induced by CK Could Promote Apoptosis of K562 and Meg-01 Cells

To further investigate the involvement of the inflammasome pathway in CK-induced apoptosis, we conducted Western blot analysis to assess the protein expression of key components of the NLRP3 inflammasome complex, including NLRP3 and caspase-1, in K562 and Meg-01 cells treated with 3 μM and 5 μM CK. As shown in Figure 6A,B, treatment with 5 μM CK led to significant increases in the protein expression of both NLRP3 and caspase-1 in K562 and Meg-01 cells (** $p < 0.01$ and *** $p < 0.001$). These findings suggest that CK may activate the NLRP3 inflammasome complex, potentially promoting platelet production by inducing apoptosis. Activation of the inflammasome is known to trigger the production of pro-inflammatory cytokines, such as IL-1β, and initiate apoptotic pathways [43]. To further confirm NLRP3 inflammasome activation, we measured IL-1β levels by ELISA under the same experimental conditions as previous studies. The results demonstrate a significant increase in IL-1β levels in the 5 μM CK-treated group compared to the vehicle group (*** $p < 0.001$).

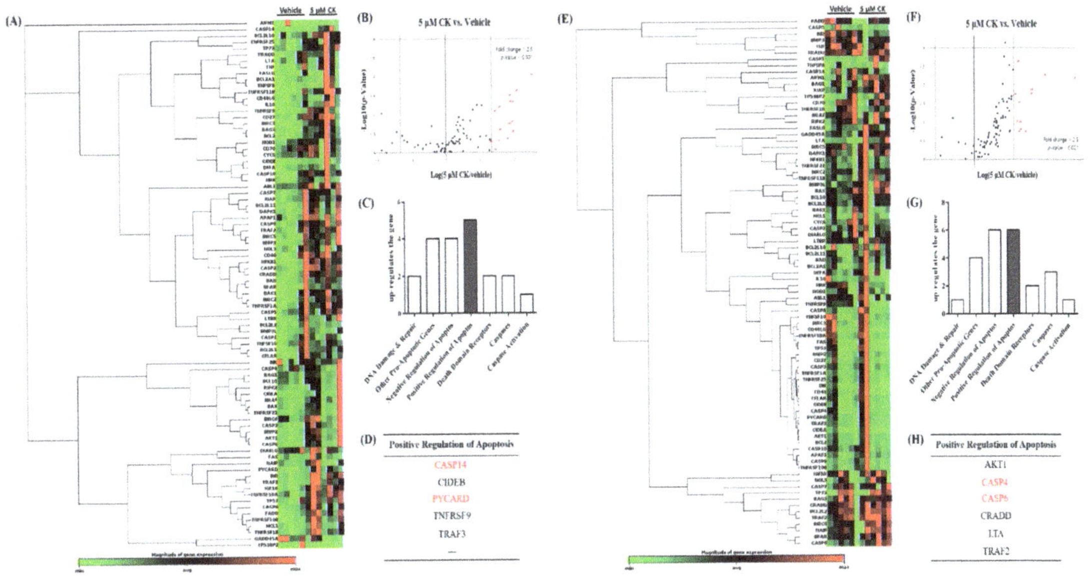

Figure 5. Summary of the RT[2] Profiler PCR array analysis of the apoptosis pathway. The hierarchical clustering of gene signatures was determined using PCR arrays in (**A**) K562 cells and (**E**) Meg-01 cells. (**B**,**F**) The volcano plots identify significant changes in gene expression between groups. Cluster analysis revealed that CK upregulates the genes in (**C**) K562 and (**G**) Meg-01 cells in the positive regulation signaling pathway. (**D**,**H**) List of positive regulation-related genes regulated by CK. There is a distinct increase in the inflammasome expression among the positive regulation-related genes. Red indicates upregulation of inflammasome genes (at least 2.5-fold).

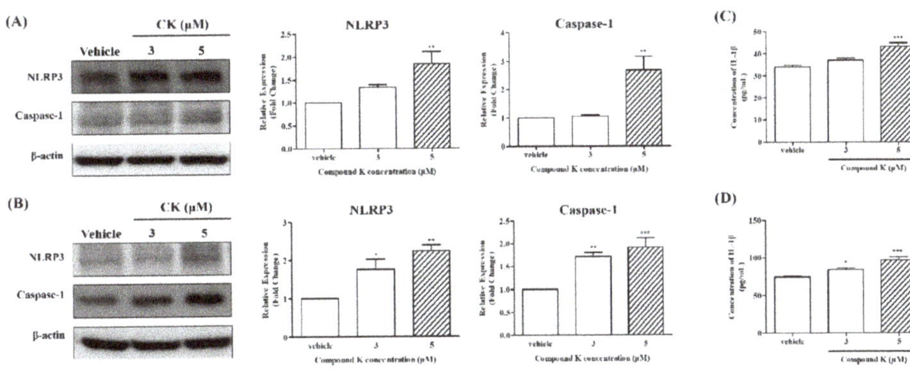

Figure 6. Effects of CK on the expression of the complexes of NLRP3 inflammatory markers in K562 and Meg-01 cells. Human chronic myeloid leukemia (CML) cells were treated with vehicle or CK (3 or 5 μM) for 72 h. The protein expression of NLRP3, caspase-1, and actin was measured by Western blot analysis (n = 3). Quantification results of (**A**) K562 and (**B**) Meg-01 cells and a representative blot is shown. Also shown are the ELISA results for inflammatory cytokine IL-1ß in the vehicle or CK (3 or 5 μM) groups. IL-1β concentration was significantly increased by CK treatment in (**C**) K562 and (**D**) Meg-01 cells. The mean ± SD of three independent experiments is also shown. One-way ANOVA with Bonferroni test was used to determine the significance of differences: * $p < 0.05$, ** $p < 0.01$, and *** $p < 0.001$ compared to vehicle-treated cells.

4. Discussion

Platelets, produced by MKs derived from bone marrow cells, are essential for maintaining hemostasis by facilitating blood clotting at injury sites [1,44]. Dysfunction in platelet production, characterized by a reduction in platelet count, can result from various underlying factors and is associated with several diseases, including leukemia, hepatitis, and AD [17,18]. Enhancing megakaryocytic differentiation is a critical process for augmenting the production of healthy platelets [13]. Recent studies have focused on strategies to improve megakaryocytic differentiation, which could lead to increased platelet production and offer potential therapeutic approaches for platelet-related disorders [45].

While platelets are crucial for hemostasis, they can be detrimental in certain conditions, such as cancer, where they promote metastasis [46]. In such cases, reducing platelet activation or count is the desired therapeutic approach. However, our study focuses on the potential therapeutic effects of CK in conditions where enhancing platelet production is beneficial. For instance, in thrombocytopenia or other platelet-related disorders characterized by abnormally low platelet counts, increasing platelet production is essential [47]. Additionally, recent research has suggested that reduced platelet counts may be associated with neurodegenerative diseases, including Alzheimer's disease [48].

Megakaryocytic differentiation is a complex, multi-step process characterized by the expression of MK-specific markers and the generation of multi-lobulated, polyploid cells through endomitosis. In this study, we explored the effects of CK on the induction of megakaryocytic differentiation and its potential underlying molecular mechanisms in K562 and Meg-01 cells. Our findings demonstrate that CK significantly promotes the expression of megakaryocytic differentiation markers (Figure 2) and facilitates multi-lobulated nucleus formation during megakaryocytic differentiation (Figure 2E), suggesting its capacity to enhance megakaryocyte maturation.

Thrombopoietin (TPO) is the primary hormone responsible for driving MK development [6,9]. TPO stimulates HSCs, thereby promoting the growth of MK colonies and facilitating the proliferation and differentiation of MK progenitors [11]. Pro-inflammatory cytokines such as IL-1β and IL-6, regulated by EGR1, have been identified as key enhancers of TPO production [12]. Notably, previous studies have demonstrated that EGR1 activation is linked to ERK signaling [49]. In line with these findings, our study shows that CK significantly stimulates the production of megakaryocytic differentiation-associated cytokines, particularly IL-1β and IL-6, via ERK-induced EGR1 activation in K562 and Meg-01 cells (Figure 3). Importantly, this study is the first to demonstrate that CK promotes megakaryocytic differentiation and elucidates the underlying molecular mechanisms, specifically through the activation of the ERK/EGR1 signaling pathway.

Recent studies have suggested that apoptosis is the final physiological fate of mature MKs [28,50–54]. In alignment with the induction of megakaryocytic differentiation, our findings indicate that CK-induced megakaryocytic differentiation is accompanied by an increase in apoptosis in K562 and Meg-01 cells (Figure 4). The increased apoptosis observed during CK-induced differentiation suggests that CK may facilitate the maturation and development of MKs by activating the apoptotic pathway, thereby promoting the production of functional platelets. The summary of apoptosis and megakaryocyte differentiation gene expression is included in Table 2.

Using RT^2 Profiler PCR array analysis, we observed a significant upregulation of genes associated with the NLRP3 inflammasome, indicating that CK may induce apoptosis through the activation of this inflammasome pathway (Figure 5). These findings are consistent with previous studies that have suggested a link between CK and the activation of the NLRP3 inflammasome. The NLRP3 inflammasome has been implicated in various conditions, including atherosclerosis, neurodegenerative diseases, and autoimmune disorders [43,55,56]. Interestingly, CK has also been reported to suppress NLRP3 inflammasome activation, which may have important therapeutic implications in mitigating inflammatory responses associated with these diseases [56]. While many studies emphasize the suppressive effects of CK on NLRP3 inflammasome activation, it is important to note that

NLRP3 inflammasome activation can also induce megakaryocytic differentiation. These seemingly contrasting findings may be attributable to differences in cell types or specific experimental conditions, warranting further investigation into the precise role of the NLRP3 inflammasome in CK-induced megakaryocytic differentiation.

Moreover, our results demonstrate that CK-induced megakaryocytic differentiation is mediated through the activation of the NLRP3 inflammasome, a key molecular complex involved in regulating inflammatory responses (Figure 6). Notably, CK-induced activation of the NLRP3 inflammasome also triggered apoptosis, independent of reactive oxygen species (ROS) production (Figure A1). These findings suggest that CK may enhance platelet production and improve platelet function, positioning it as a promising candidate for the therapeutic management of platelet-related disorders and other associated diseases. CK is a major bioactive component of Panax ginseng, belonging to the dammarane-type saponins. As a metabolite of protopanaxadiol (PPD), CK exhibits superior bioavailability compared to other ginsenosides and has demonstrated various pharmacological properties, including anti-diabetic, anti-cancer, and anti-aging effects [23,24,29–33]. Traditionally, differentiation-inducing substances have been associated with the induction of apoptosis via ROS generation [57]. Interestingly, our findings revealed that CK (5 μM) inhibited ROS levels without inducing cell death. These observations suggest that CK differs mechanistically from other differentiation-inducing agents, indicating a potential novel pathway for its effects. Further studies are warranted to elucidate the precise mechanisms by which CK influences differentiation and cell survival.

In conclusion, this study demonstrates that CK induces the upregulation of genes associated with the NLRP3 inflammasome, suggesting its activation (Figure 7). These findings are consistent with previous research and enhance our understanding of CK's role in modulating inflammatory responses. The potential therapeutic implications of CK-mediated NLRP3 inflammasome activation merit further investigation, particularly in inflammation-related diseases, to fully elucidate its broader clinical applications. While our study focuses on the potential effects of CK in promoting megakaryocytic differentiation and apoptosis, it is important to note that limited research has been conducted on the potential side effects or toxicity of CK in humans or other cell types. Further studies are needed to evaluate the long-term safety and possible adverse effects of CK and ensure its suitability for therapeutic agents.

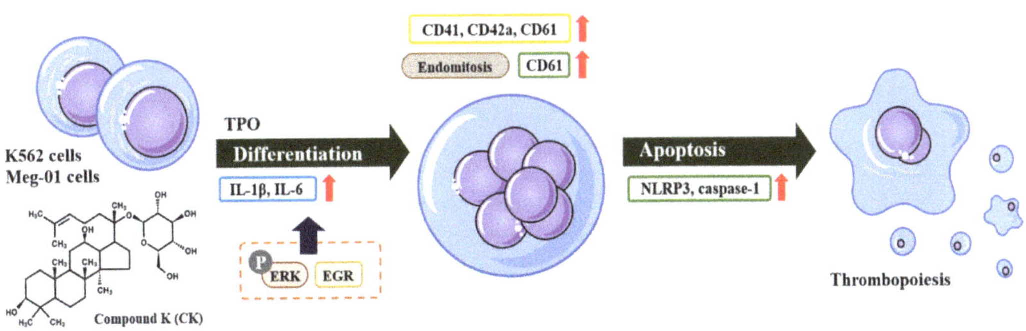

Figure 7. Summary of the experiments on the effect of CK on megakaryocytic differentiation in CML K562 and Meg-01 cells. The observation of specific expression markers in megakaryocytic differentiation and cell morphology confirmed CK's facilitation of differentiation. Additionally, CK reduces cell viability and promotes megakaryocytic differentiation through the activation of the NLRP3 inflammasome in human CML K562 and Meg-01 cells.

Supplementary Materials: The following supporting information can be downloaded at: https://www.mdpi.com/article/10.3390/biom14101257/s1, Table S1: Genes used for RT2 Profiler PCR array analysis; Table S2: Summarizes the effects of CK on apoptosis-related gene expression in K562 cells; Table S3: Summarizes the effects of CK on apoptosis-related gene expression in Meg-01 cells. Original Western blot images can be found in Supplementary File S1.

Author Contributions: Conceptualization, M.-H.P. and J.-S.K.; methodology, S.H. and M.-S.P.; software, S.H.; validation, M.-S.P.; formal analysis, S.H. and M.-S.P.; investigation, S.H. and M.-S.P.; resources, M.-H.P. and J.-S.K.; data curation, M.-H.P., S.-H.S. and H.-K.K.; writing—original draft preparation, S.H., M.-S.P. and M.-H.P.; writing—review and editing, A.J.K. and E.Y.; supervision, M.-H.P. and J.-S.K.; project administration, M.-H.P. and J.-S.K.; funding acquisition, M.-H.P. and J.-S.K. All authors have read and agreed to the published version of the manuscript.

Funding: This research was supported by a grant from the Korea Health Technology R&D Project through the Korea Health Industry Development Institute (KHIDI), funded by the Ministry of Health & Welfare, Republic of Korea (HR20C0026), and a National Research Foundation of Korea (NRF) grant funded by the Korea government (NRF-2021R1I1A3061377).

Data Availability Statement: Data are available upon request.

Conflicts of Interest: The authors declare no conflict of interest.

Appendix A

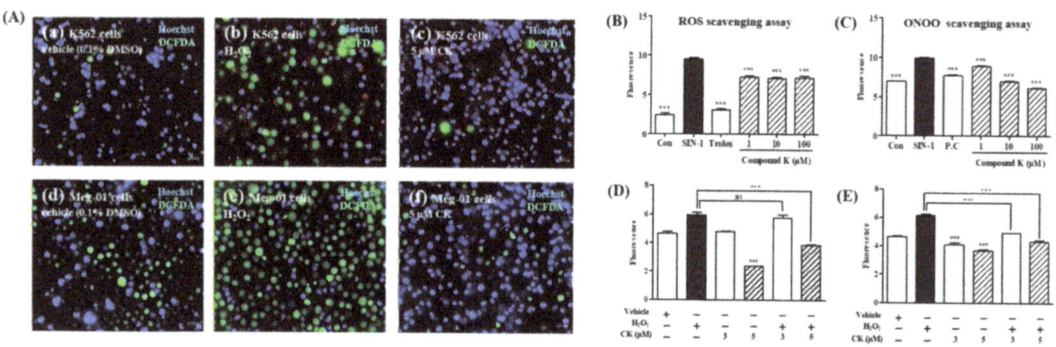

Figure A1. Effects of CK on reactive oxygen species (ROS) levels. (**A**) Intracellular ROS detection was performed using fluorescence microscopy. Intracellular ROS was measured by DCFDA staining after (**a**,**d**) vehicle, (**b**,**e**) H_2O_2, or (**c**,**f**) 5 μM CK treatment. Representative fluorescence images of (**a**–**c**) K562 and (**d**–**f**) Meg-01 cells. The intensity of the green fluorescence indicates ROS concentration in the cells. (**B**) ROS scavenging assay after CK exposure. The effect of CK on ROS production was evaluated. ROS levels were quantitatively determined using the H_2DCFDA ($n = 4$). Trolox was used as the positive control (P.C). (**C**) The effect of CK on peroxynitrite ($ONOO^-$) production was evaluated ($n = 4$). Penicillamine was used as the P.C. The scavenging activity of CK on ROS and $ONOO^-$ induced by 10 μM SIN-1 was estimated in a cell-free in vitro culture. One-way ANOVA with Bonferroni test was used to determine the significance of differences: *** $p < 0.001$ compared to SIN-1-treated cells. The scavenging activity of CK on ROS induced by H_2O_2 was estimated in (**D**) K562 and (**E**) Meg-01 cells. One-way ANOVA with Bonferroni test was used to determine the significance of differences: *** $p < 0.001$ compared to H_2O_2-treated cells. ### $p < 0.001$ compared to vehicle-treated cells. "ns" represents no significant differences compared to H_2O_2-treated cells.

References

1. Leslie, M. Cell biology. Beyond clotting: The powers of platelets. *Science* **2010**, *328*, 562–564. [CrossRef] [PubMed]
2. Broos, K.; Feys, H.B.; De Meyer, S.F.; Vanhoorelbeke, K.; Deckmyn, H. Platelets at work in primary hemostasis. *Blood Rev.* **2011**, *25*, 155–167. [CrossRef] [PubMed]
3. Jenne, C.; Urrutia, R.; Kubes, P. Platelets: Bridging hemostasis, inflammation, and immunity. *Int. J. Lab. Hematol.* **2013**, *35*, 254–261. [CrossRef] [PubMed]
4. Bain, B.J. Structure and function of red and white blood cells and platelets. *Medicine* **2021**, *49*, 183–188. [CrossRef]
5. Italiano, J., Jr.; Shivdasani, R. Megakaryocytes and beyond: The birth of platelets. *J. Thromb. Haemost.* **2003**, *1*, 1174–1182. [CrossRef]

6. Kacena, M.A.; Gundberg, C.M.; Horowitz, M.C. A reciprocal regulatory interaction between megakaryocytes, bone cells, and hematopoietic stem cells. *Bone* **2006**, *39*, 978–984. [CrossRef]
7. Guo, T.; Wang, X.; Qu, Y.; Yin, Y.; Jing, T.; Zhang, Q. Megakaryopoiesis and platelet production: Insight into hematopoietic stem cell proliferation and differentiation. *Stem Cell Investig.* **2015**, *2*, 3.
8. Levine, R. Isolation and characterization of normal human megakaryocytes. *Br. J. Haematol.* **1980**, *45*, 487–497. [CrossRef]
9. Patel, S.R.; Hartwig, J.H.; Italiano, J.E. The biogenesis of platelets from megakaryocyte proplatelets. *J. Clin. Investig.* **2005**, *115*, 3348–3354. [CrossRef]
10. Kaushansky, K. The molecular mechanisms that control thrombopoiesis. *J. Clin. Investig.* **2005**, *115*, 3339–3347. [CrossRef]
11. Machlus, K.R.; Italiano, J.E., Jr. The incredible journey: From megakaryocyte development to platelet formation. *J. Cell Biol.* **2013**, *201*, 785–796. [CrossRef] [PubMed]
12. Kaushansky, K.; Drachman, J.G. The molecular and cellular biology of thrombopoietin: The primary regulator of platelet production. *Oncogene* **2002**, *21*, 3359–3367. [CrossRef] [PubMed]
13. Deutsch, V.R.; Tomer, A. Megakaryocyte development and platelet production. *Br. J. Haematol.* **2006**, *134*, 453–466. [CrossRef] [PubMed]
14. Ruggeri, Z.M. Platelets in atherothrombosis. *Nat. Med.* **2002**, *8*, 1227–1234. [CrossRef] [PubMed]
15. Gay, L.J.; Felding-Habermann, B. Contribution of platelets to tumour metastasis. *Nat. Rev. Cancer* **2011**, *11*, 123–134. [CrossRef]
16. Engelmann, B.; Massberg, S. Thrombosis as an intravascular effector of innate immunity. *Nat. Rev. Immunol.* **2013**, *13*, 34–45. [CrossRef]
17. Qian, X.; Wen-jun, L. Platelet changes in acute leukemia. *Cell Biochem. Biophys.* **2013**, *67*, 1473–1479. [CrossRef]
18. Olariu, M.; Olariu, C.; Olteanu, D. Thrombocytopenia in chronic hepatitis C. *J. Gastrointestin. Liver Dis.* **2010**, *19*, 381–385.
19. Biino, G.; Santimone, I.; Minelli, C.; Sorice, R.; Frongia, B.; Traglia, M.; Ulivi, S.; Di Castelnuovo, A.; Gögele, M.; Nutile, T. Age-and sex-related variations in platelet count in Italy: A proposal of reference ranges based on 40987 subjects' data. *PLoS ONE* **2013**, *8*, e54289. [CrossRef]
20. Yang, J.; Zhou, X.; Fan, X.; Xiao, M.; Yang, D.; Liang, B.; Dai, M.; Shan, L.; Lu, J.; Lin, Z.; et al. mTORC1 promotes aging-related venous thrombosis in mice via elevation of platelet volume and activation. *Blood* **2016**, *128*, 615–624. [CrossRef]
21. Reilly, I.A.; FitzGerald, G.A. Eicosenoid biosynthesis and platelet function with advancing age. *Thromb. Res.* **1986**, *41*, 545–554. [CrossRef] [PubMed]
22. Leiter, O.; Walker, T.L. Platelets: The missing link between the blood and brain? *Progress in neurobiology* **2019**, *183*, 101695. [CrossRef] [PubMed]
23. Akao, T.; Kida, H.; Kanaoka, M.; Hattori, M.; Kobashi, K. Drug metabolism: Intestinal bacterial hydrolysis is required for the appearance of compound K in rat plasma after oral administration of ginsenoside Rb1 from Panax ginseng. *J. Pharm. Pharmacol.* **1998**, *50*, 1155–1160. [CrossRef] [PubMed]
24. Bae, E.-A.; Shin, J.-e.; Kim, D.-H. Metabolism of ginsenoside Re by human intestinal microflora and its estrogenic effect. *Biol. Pharm. Bull.* **2005**, *28*, 1903–1908. [CrossRef] [PubMed]
25. Ratan, Z.A.; Haidere, M.F.; Hong, Y.H.; Park, S.H.; Lee, J.-O.; Lee, J.; Cho, J.Y. Pharmacological potential of ginseng and its major component ginsenosides. *J. Ginseng Res.* **2021**, *45*, 199–210. [CrossRef]
26. Qi, L.-W.; Wang, C.-Z.; Yuan, C.-S. Isolation and analysis of ginseng: Advances and challenges. *Nat. Prod. Rep.* **2011**, *28*, 467–495. [CrossRef]
27. Li, T.; Mazza, G.; Cottrell, A.; Gao, L. Ginsenosides in roots and leaves of American ginseng. *J. Agric. Food Chem.* **1996**, *44*, 717–720. [CrossRef]
28. Lee, I.-S.; Kang, K.S.; Kim, S.-Y. Panax ginseng pharmacopuncture: Current status of the research and future challenges. *Biomolecules* **2019**, *10*, 33. [CrossRef]
29. Akao, T.; Kanaoka, M.; Kobashi, K. Appearance of compound K, a major metabolite of ginsenoside Rb1 by intestinal bacteria, in rat plasma after oral administration: Measurement of compound K by enzyme immunoassay. *Biol. Pharm. Bull.* **1998**, *21*, 245–249. [CrossRef]
30. Yoon, S.H.; Han, E.J.; Sung, J.H.; Chung, S.H. Anti-diabetic effects of compound K versus metformin versus compound K-metformin combination therapy in diabetic db/db mice. *Biol. Pharm. Bull.* **2007**, *30*, 2196–2200. [CrossRef]
31. Lee, H.U.; Bae, E.A.; Han, M.J.; Kim, N.J.; Kim, D.H. Hepatoprotective effect of ginsenoside Rb1 and compound K on tert-butyl hydroperoxide-induced liver injury. *Liver Int.* **2005**, *25*, 1069–1073. [CrossRef] [PubMed]
32. Park, J.-S.; Shin, J.A.; Jung, J.-S.; Hyun, J.-W.; Van Le, T.K.; Kim, D.-H.; Park, E.-M.; Kim, H.-S. Anti-inflammatory mechanism of compound K in activated microglia and its neuroprotective effect on experimental stroke in mice. *J. Pharmacol. Exp. Ther.* **2012**, *341*, 59–67. [CrossRef] [PubMed]
33. Kim, S.; Kang, B.Y.; Cho, S.Y.; Sung, D.S.; Chang, H.K.; Yeom, M.H.; Kim, D.H.; Sim, Y.C.; Lee, Y.S. Compound K induces expression of hyaluronan synthase 2 gene in transformed human keratinocytes and increases hyaluronan in hairless mouse skin. *Biochem. Biophys. Res. Commun.* **2004**, *316*, 348–355. [CrossRef] [PubMed]
34. Alitalo, R. Induced differentiation of K562 leukemia cells: A model for studies of gene expression in early megakaryoblasts. *Leuk. Res.* **1990**, *14*, 501–514. [CrossRef]
35. Ogura, M.; Morishima, Y.; Ohno, R.; Kato, Y.; Hirabayashi, N.; Nagura, H.; Saito, H. Establishment of a novel human megakaryoblastic leukemia cell line, MEG-01, with positive Philadelphia chromosome. *Blood* **1985**, *66*, 1384–1392. [CrossRef]

36. Takeuchi, K.; Satoh, M.; Kuno, H.; Yoshida, T.; Kondo, H.; Takeuchi, M. Platelet-like particle formation in the human megakaryoblastic leukaemia cell lines, MEG-01 and MEG-01s. *Br. J. Haematol.* **1998**, *100*, 436–444. [CrossRef]
37. Gordge, M.P. Megakaryocyte apoptosis: Sorting out the signals. *Br. J. Pharmacol.* **2005**, *145*, 271. [CrossRef]
38. Zauli, G.; Vitale, M.; Falcieri, E.; Gibellini, D.; Bassini, A.; Celeghini, C.; Columbaro, M.; Capitani, S. In vitro senescence and apoptotic cell death of human megakaryocytes. *Blood J. Am. Soc. Hematol.* **1997**, *90*, 2234–2243.
39. Morison, I.M.; Cramer Bordé, E.M.; Cheesman, E.J.; Cheong, P.L.; Holyoake, A.J.; Fichelson, S.; Weeks, R.J.; Lo, A.; Davies, S.M.; Wilbanks, S.M. A mutation of human cytochrome c enhances the intrinsic apoptotic pathway but causes only thrombocytopenia. *Nat. Genet.* **2008**, *40*, 387–389. [CrossRef]
40. Houwerzijl, E.J.; Blom, N.R.; van der Want, J.J.; Esselink, M.T.; Koornstra, J.J.; Smit, J.W.; Louwes, H.; Vellenga, E.; de Wolf, J.T.M. Ultrastructural study shows morphologic features of apoptosis and para-apoptosis in megakaryocytes from patients with idiopathic thrombocytopenic purpura. *Blood* **2004**, *103*, 500–506. [CrossRef]
41. Kaluzhny, Y.; Yu, G.; Sun, S.; Toselli, P.A.; Nieswandt, B.; Jackson, C.W.; Ravid, K. BclxL overexpression in megakaryocytes leads to impaired platelet fragmentation. *Blood J. Am. Soc. Hematol.* **2002**, *100*, 1670–1678. [CrossRef] [PubMed]
42. McArthur, K.; Chappaz, S.; Kile, B.T. Apoptosis in megakaryocytes and platelets: The life and death of a lineage. *Blood J. Am. Soc. Hematol.* **2018**, *131*, 605–610. [CrossRef] [PubMed]
43. Hoseini, Z.; Sepahvand, F.; Rashidi, B.; Sahebkar, A.; Masoudifar, A.; Mirzaei, H. NLRP3 inflammasome: Its regulation and involvement in atherosclerosis. *J. Cell. Physiol.* **2018**, *233*, 2116–2132. [CrossRef] [PubMed]
44. Shin, B.-K.; Kwon, S.W.; Park, J.H. Chemical diversity of ginseng saponins from Panax ginseng. *J. Ginseng Res.* **2015**, *39*, 287–298. [CrossRef]
45. De Thé, H. Differentiation therapy revisited. *Nat. Rev. Cancer* **2018**, *18*, 117–127. [CrossRef]
46. Bailey, S.E.; Ukoumunne, O.C.; Shephard, E.A.; Hamilton, W. Clinical relevance of thrombocytosis in primary care: A prospective cohort study of cancer incidence using English electronic medical records and cancer registry data. *Br. J. Gen. Pract.* **2017**, *67*, e405–e413. [CrossRef]
47. Izak, M.; Bussel, J.B. Management of thrombocytopenia. *F1000prime Rep.* **2014**, *6*, 45. [CrossRef]
48. Frith, J.; Watson, S.; Maggs, P.H.B.; Newton, J.L. Cognitive symptoms are common in immune thrombocytopenia and associate with autonomic symptom burden. *Eur. J. Haematol.* **2012**, *88*, 224–228. [CrossRef]
49. Shan, J.; Dudenhausen, E.; Kilberg, M.S. Induction of early growth response gene 1 (EGR1) by endoplasmic reticulum stress is mediated by the extracellular regulated kinase (ERK) arm of the MAPK pathways. *Biochim. Biophys. Acta (BBA)-Mol. Cell Res.* **2019**, *1866*, 371–381. [CrossRef]
50. Shibata, S.; Tanaka, O.; Ando, T.; Sado, M.; Tsushima, S.; Ohsawa, T. Chemical studies on oriental plant drugs. XIV. Protopanaxadiol, a genuine sapogenin of ginseng saponins. *Chem. Pharm. Bull.* **1966**, *14*, 595–600. [CrossRef]
51. Zhou, P.; Xie, W.; He, S.; Sun, Y.; Meng, X.; Sun, G.; Sun, X. Ginsenoside Rb1 as an anti-diabetic agent and its underlying mechanism analysis. *Cells* **2019**, *8*, 204. [CrossRef] [PubMed]
52. Yang, Y.; Lee, J.; Rhee, M.H.; Yu, T.; Baek, K.-S.; Sung, N.Y.; Kim, Y.; Yoon, K.; Kim, J.H.; Kwak, Y.-S. Molecular mechanism of protopanaxadiol saponin fraction-mediated anti-inflammatory actions. *J. Ginseng Res.* **2015**, *39*, 61–68. [CrossRef] [PubMed]
53. Qiu, X.-M.; Bai, X.; Jiang, H.-F.; He, P.; Wang, J.-H. 20-(s)-ginsenoside Rg3 induces apoptotic cell death in human leukemic U937 and HL-60 cells through PI3K/Akt pathways. *Anti-Cancer Drugs* **2014**, *25*, 1072–1080. [CrossRef] [PubMed]
54. Jang, H.-J.; Han, I.-H.; Kim, Y.-J.; Yamabe, N.; Lee, D.; Hwang, G.S.; Oh, M.; Choi, K.-C.; Kim, S.-N.; Ham, J. Anticarcinogenic effects of products of heat-processed ginsenoside Re, a major constituent of ginseng berry, on human gastric cancer cells. *J. Agric. Food Chem.* **2014**, *62*, 2830–2836. [CrossRef] [PubMed]
55. Grebe, A.; Hoss, F.; Latz, E. NLRP3 inflammasome and the IL-1 pathway in atherosclerosis. *Circ. Res.* **2018**, *122*, 1722–1740. [CrossRef] [PubMed]
56. Karasawa, T.; Takahashi, M. Role of NLRP3 inflammasomes in atherosclerosis. *J. Atheroscler. Thromb.* **2017**, *24*, 443–451. [CrossRef]
57. Gutti, U.; Komati, J.K.; Kotipalli, A.; Saladi, R.G.V.; Gutti, R.K. Justicia adhatoda induces megakaryocyte differentiation through mitochondrial ROS generation. *Phytomedicine* **2018**, *43*, 135–139. [CrossRef]

biomolecules

MDPI

Review

Maresins

Fernanda Berrocal-Navarrete [1,2], Paz Marín-Sanhueza [1,3], Ramón Norambuena-González [1,2], Matías Quiñones San Martín [1,4], Francisca Herrera-Vielma [1,4], Daniel R. González [1] and Jessica Zúñiga-Hernández [1,*]

1 Laboratory of Pharmacology and Physiology, Faculty of Health Sciences, University of Talca, Av. Lircay s/n, Talca 3460000, Chile; fberrocal18@alumnos.utalca.cl (F.B.-N.); pmarin19@alumnos.utalca.cl (P.M.-S.); ramon.norambuena@utalca.cl (R.N.-G.); mquinones@utalca.cl (M.Q.S.M.); francisca.ingbiotec@gmail.com (F.H.-V.); dagonzalez@utalca.cl (D.R.G.)
2 Doctorate in Biomedical Sciences Program, Faculty of Health Sciences, University of Talca, Av. Lircay s/n, Talca 3460000, Chile
3 Institute of Biological Sciences, School of Biochemistry, University of Talca, Av. Lircay s/n, Talca 3460000, Chile
4 Doctorate in Science Program, Mention in Research and Development of Bioactive Products, Institute of Chemistry of Natural Resources, University of Talca, Av. Lircay s/n, Talca 3460000, Chile
* Correspondence: jezuniga@utalca.cl

Abstract

Polyunsaturated fatty acids (PUFAs), particularly omega-3 derivatives such as docosahexaenoic acid (DHA), are precursors of specialized pro-resolving mediators (SPMs) that actively orchestrate the resolution of inflammation. Among these, maresins (MaRs) have gathered increasing attention due to their potent immunomodulatory and tissue-regenerative properties. This review provides a comprehensive synthesis of the current knowledge on the biosynthesis, structural diversity, and biological functions of MaRs, with a focus on MaR1. We discuss the enzymatic pathways involved in the generation of MaR1, MaR2, MaRs conjugates in tissue regeneration (MCTRs), and maresin-like lipid mediators (MaR-Ls), highlighting their roles in modulating inflammatory responses, promoting phagocytosis, and restoring tissue homeostasis. Preclinical evidence from in vitro and in vivo models demonstrates that MaRs exert protective effects in a wide range of pathological contexts, including neuroinflammation, liver injury, cardiovascular dysfunction, pulmonary diseases, and metabolic disorders. Although their therapeutic promise is well-supported, key gaps remain in the understanding of MaRs biosynthesis, receptor specificity, and translational applicability. This review emphasizes the importance of advancing mechanistic and clinical research to fully harness MaRs as part of next-generation therapeutics in inflammation-driven diseases.

Keywords: omega-3; nutrition; pro-resolution; PUFAs

1. Introduction

The health benefits of polyunsaturated fatty acid (PUFAs) consumption have been recognized since 1971, when Bang, Dyerberg and Nielsen published an article about the lower levels of plasma lipids found in Eskimo populations (from the northern west coast of Greenland) compared to other populations (Danes and Eskimos living in Denmark) with similar fat intake. The difference was that Eskimos consumed mostly fish, seals and whales, in which the fat content is rich in long-chain PUFAs, particularly C18–C22 fatty acids, most of them omega-3 fatty acids [1]. These seminal findings laid the foundation for the current understanding of the nutritional and therapeutic relevance of omega-3 fatty acids, and to a lesser extent omega-6, fatty acids, stimulating extensive research into the biological roles and health benefits of PUFAs consumption.

Among PUFAs, which include omega-3 (n-3) and omega-6 (n-6) fatty acids, only linoleic acid (LA, C18:2n-6) and α-linolenic acid (ALA, C18:3n-3) are strictly essential, since mammals lack the Δ12 and Δ15 desaturases required for their synthesis [2]. From these, long-chain fatty acids precursors, such as arachidonic acid (AA; C20:4n-6), an omega-6 derivative, and eicosapentaenoic acid (EPA; C20:5n-3) and docosahexaenoic acid (DHA; C22:6n-3), both omega-3, can be synthesized through elongation and desaturation steps. However, this conversion is relatively slow, making dietary intake of LA and ALA necessary to maintain adequate levels needed for several processes [3], such as brain and retinal development [4,5], prevention of cardiovascular diseases and hepatoprotection [6–8]. Once incorporated into the organism, these fatty acids undergo elongation and desaturation reactions to form their respective derivatives such as AA for omega-6, and EPA and DHA for omega-3 (Figure 1) [2]. There is competition between omega-6 and omega-3 for the enzymes responsible for desaturation, in a way that an increase in omega-6 generates a decrease in the amount of omega-3 [9]. The modern diet is characterized by an unbalanced ratio of omega-6 to omega-3 fatty acids, especially LA, that can increase the production of proinflammatory eicosanoids such as prostaglandins, thromboxanes and leukotrienes derived from AA [9]. This metabolic imbalance contributes to a chronic low-grade inflammatory state, which is closely associated with metabolic disorders such as obesity and type 2 diabetes mellitus (T2DM) [10,11]. In recent years, the scientific focus has shifted from the preventive role of PUFAs toward understanding their active involvement in the resolution of inflammation through the generation of bioactive lipid mediators. Omega-3 PUFAs have been shown to actively promote the resolution of inflammation through the generation of specialized pro-resolving mediators (SPMs) derived from these fatty acids, including lipoxins (LXs), resolvins (Rvs), protectins (PDs), and maresins (MaRs) [12,13]. Among these, MaRs have emerged as promising bioactive molecules with significant therapeutic potential in pathological conditions such as asthma, hepatic injury, and cardiovascular diseases [14–16]. Therefore, this review focuses on the current research, biosynthetic pathways, and biological functions of MaRs, given their growing importance as novel therapeutic targets in inflammation (Figure 1).

https://doi.org/10.3390/biom16010139

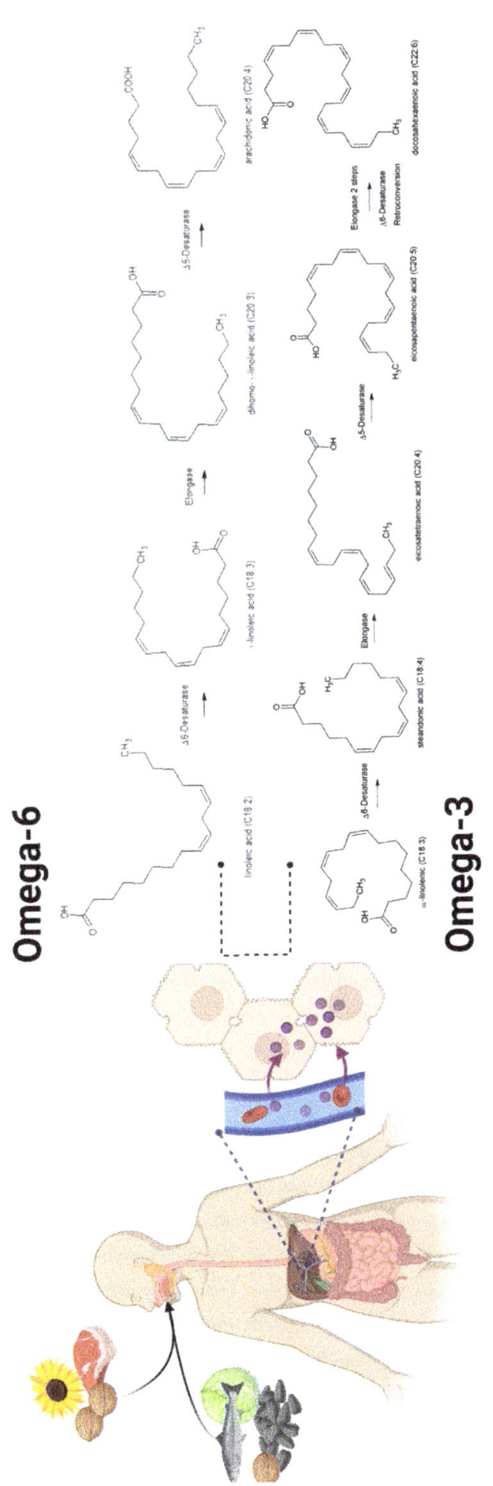

Figure 1. Biosynthetic pathways of major lipid mediators derived from dietary omega-6 **(top)** and omega-3 **(bottom)** fatty acids, highlighting the key enzymes involved in each conversion process.

https://doi.org/10.3390/biom16010139

2. Biosynthesis of Maresins

MaRs were discovered by Serhan et al. in 2009 [17]. These, along with other molecules of the same category such as PDs and Rvs, were described as SPMs, derived from EPA, and DHA [12]. The term "maresin" refers to *ma*crophage mediators in the *res*olution of *in*flammation (ma-res-in), reflecting its origin and biological function [18].

MaRs biosynthesis begins in macrophages with the 14-lipoxygenation of DHA by the human 12-lipoxygenase (12-LOX), that produces 14S-hydro(peroxy)-docosa-4Z,7Z,10Z,12E,16Z,19Z-hexaenoic acid (14S-HpDHA), a key intermediate in the MaRs pathway. Subsequently, the same enzyme promotes the formation of 13S,14S-epoxy-maresin (13,14-epoxide), an unstable intermediate that serves as precursor for MaRs metabolites [18,19]. Finally, through the sequential action of several enzymes, some of which remain to be fully characterized, the biosynthesis of the four MaRs subtypes described to date is completed. These different MaRs forms, along with their biosynthetic routes and biological activities, are discussed in detail in this review (Figures 2–5).

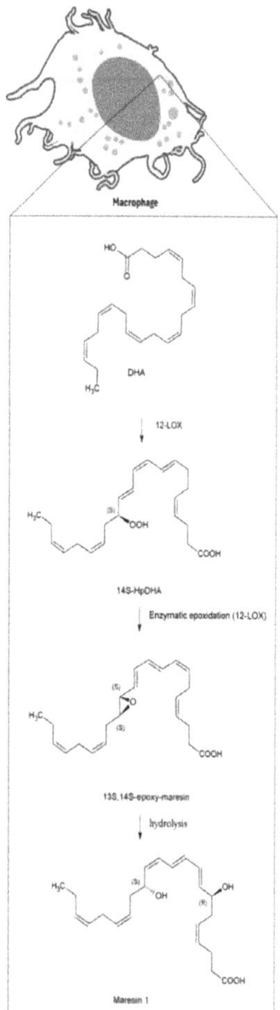

Figure 2. Biosynthesis of MaR1 from DHA in macrophages. Note: Abbreviations: Docosahexaenoic acid (DHA); 12-lipoxygenase (12-LOX); 14S-hydro (peroxy)-docosa-4Z,7Z,10Z,12E,16Z,19Z-hexaenoic acid (14S-HpDHA).

 https://doi.org/10.3390/biom16010139

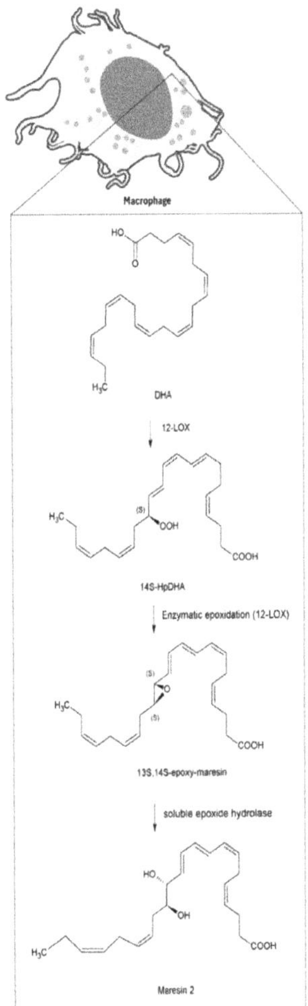

Figure 3. Biosynthesis of MaR2 from DHA in macrophages. Note: Abbreviations: Docosahexaenoic acid (DHA); 12-lipoxygenase (12-LOX); 14S-hydro(peroxy)-docosa-4Z,7Z,10Z,12E,16Z,19Z-hexaenoic acid (14S-HpDHA); soluble epoxide hydrolase (sEH).

https://doi.org/10.3390/biom16010139

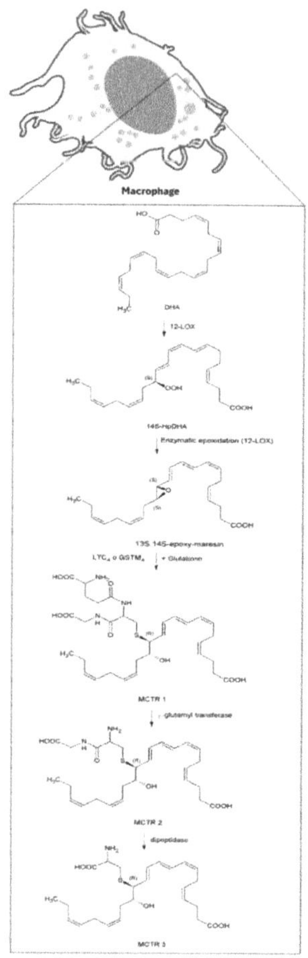

Figure 4. Biosynthesis of MCTR from DHA in macrophages. Note: Abbreviations: Docosahexaenoic acid (DHA); 12-lipoxygenase (12-LOX); 14S-hydro(peroxy)-docosa-4Z,7Z,10Z,12E,16Z,19Z-hexaenoic acid (14S-HpDHA); Leukotriene C4 synthase (LTC_4 S); glutathione S-transferase MU4 ($GSTM_4$).

https://doi.org/10.3390/biom16010139

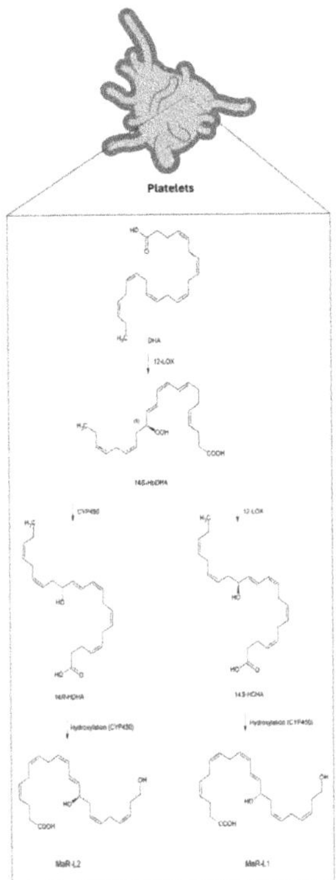

Figure 5. Biosynthesis of MaR-L from DHA in platelets. Note: Abbreviations: Docosahexaenoic acid (DHA); 12-lipoxygenase (12-LOX); 14S-hydro(peroxy)-docosa-4Z,7Z,10Z,12E,16Z,19Z-hexaenoic acid (14S-HpDHA); Cytochrome P450 (CYP450).

3. Types of Maresins

To date, four main types of MaRs have been identified: MaR1; MaR2; MCTR and MaR-L [20]. They derive from the initial intake of omega-3 fatty acids and the subsequent lipoxygenation of DHA by the enzyme 12-LOX. Detailed information on their structural characteristics is provided in Table 1.

Table 1. Types of MaRs and their specifications.

Common Name	Chemical Formula	IUPAC Designation	PubChem Number
MaR1	$C_{22}H_{32}O_4$	(7R,14S)-dihydroxy-(4Z,8E,10E,12Z,16Z,19Z)-docosahexaenoic acid	60201795
MaR2	$C_{22}H_{32}O_4$	(13R,14S)-dihydroxy-(4Z,7Z,9E,11E,16Z,19Z)-docosahexaenoic acid	101894912
MCTR1	$C_{32}H_{47}N_3O_9S$	(13R)-S-glutathionyl-(14s)-hydroxy-(4Z,7Z,9E,11E,16Z,19Z)-docosahexaenoic acid	122368871
MCTR2	$C_{27}H_{40}N_2O_6S$	(13R)-S-cysteinylglycinyl-(14S)-hydroxy-(4Z,7Z,9E,11E,16Z,19Z)-docosahexaenoic acid	122368872
MCTR3	$C_{25}H_{37}NO_5S$	(13R)-S-cysteinyl-(14S)-hydroxy-(4Z,7Z,9E,11E,16Z,19Z)-docosahexaenoic acid	122368873
MaR-L1	$C_{22}H_{31}O_4^-$	(4Z,7Z,10Z,12E,14S,16Z,19Z)-14,22-dihydroxydocosahexaenoate	126456468
MaR-L2	$C_{22}H_{31}O_4^-$	(4Z,7Z,10Z,12E,14R,16Z,19Z)-14,22-dihydroxydocosahexaenoate	126456470

Note: Abbreviations: Maresin 1 (MaR1); Maresin 2 (MaR2); Maresin-like lipid mediators (MaR-L) and Maresin conjugate in tissue regeneration (MCTR).

https://doi.org/10.3390/biom16010139

3.1. Maresin 1 (MaR1)

MaR1 is primarily synthesized by macrophages and exerts potent pro-resolving actions during inflammation. Its biosynthetic pathway begins with the formation of the intermediate 13,14-epoxy-maresin, which subsequently undergoes epoxide hydrolysis at the 13,14S-epoxy double bond. This reaction leads to the formation of a Z/E configuration, ultimately yielding 7R,14S-dihydroxydocosa-4Z,8E,10E,12Z,16Z,19Z-hexaenoic acid, commonly known as maresin 1 [18,21] (Figure 2).

3.2. Maresin 2 (MaR2)

MaR2, similarly to MaR1, limits infiltration of polymorphonuclear neutrophils and stimulates macrophages phagocytosis [22]. Its biosynthesis begins, with the 13S,14S-epoxy-maresin intermediate, then it is oxidized to 13R, 14S-dihydroxydocosahexaenoic acid (13R-14S-diHDHA), MaR2, by the action of soluble epoxide hydrolase (sEH), a key enzyme found in human macrophages [23] (Figure 3).

3.3. Maresin Conjugate in Tissue Regeneration (MCTR)

MCTRs are bioactive maresin-derived metabolites involved in tissue repair and cellular regeneration. The biosynthesis of these mediators occurs by lipoxygenation of DHA (see Figure 4), which produces 14-hydro(peroxy)-docosahexaenoic acid plus an intermediate epoxide that, through the catalysis of glutathione S-transferase MU4 (GSTM4) and leukotriene C4 synthase (LTC4 S) results in 13R-glutathione,14S-hydroxy-4Z,7Z,9E,11E,13R,14S,16Z,19Z-docosahexaenoic acid (MCTR1). Through the catalysis of γ-glutamyl transferase (GGT), 13R-cysteinylglycinyl,14S-hydroxy-4Z,7Z,9E,11E,13R,14S,16Z, 19Z-docosahexaenoic acid (MCTR2) is generated [24–26] and, finally, by dipeptidase catalysis (DPEP), 13R-cysteinyl,14S-hydroxy-4Z,7Z,9E,11E,13R,14S,16Z,19Z-docosahexaenoic acid can be obtained [26].

3.4. Maresin-like Lipid Mediators (MaR-L)

Following the conversion of DHA into 14S-hydroperoxy-DHA (14S-HpDHA) by 12-lipoxygenase (12-LOX), the biosynthesis of MaRs and related lipid mediators can proceed through two distinct hydroxylation pathways. The 14S-hydroxylation pathway is initiated by 12-LOX, while an alternative 14R-hydroxylation is catalyzed by cytochrome P450 (CYP450) (Figure 5). Subsequently, 14S-hydroxydocosahexaenoic acid (14S-HDHA) undergoes a second hydroxylation by CYP450; this step has been attributed to CYP450 activity rather than 12-LOX, as suggested by recent studies indicating that multiple enzymatic systems may contribute to Mar-L biosynthesis, leading to the formation of 14,22-dihydroxydocosahexaenoic acid (14S,22-diHDHA), also known as MaR-L1. In parallel, MaR-L2 is synthesized through CYP450-mediated hydroxylation of 14R-HDHA, yielding 14R,22-diHDHA [27].

4. Conversion of DHA to Maresins: Evidence and Knowledge Gaps

Although, the biosynthesis of MaRs from DHA has been well described at the mechanistic level, quantitative data on the conversion rate remains unavailable. In addition to conversion efficiency, the synthesis of MaRs is influenced by several factors, including enzyme expression (12-LOX for MaRs, 15-LOX for Rvs), substrate availability (DHA vs. EPA), and cell type-specific signaling. Macrophages preferentially produce MaRs during resolution phases, whereas neutrophils may favor Rvs under acute inflammation. These decisions are modulated by receptor activation, intracellular redox status, and spatial enzyme organization within lipid rafts. Indeed, this process is tightly regulated and occurs primar-

 https://doi.org/10.3390/biom16010139

ily in macrophages, where the activity of 12-LOX plays a critical role in the conversion of DHA to MaRs. Importantly, recent evidence indicates that glutathione peroxidase (GPx) activity critically regulates MaRs synthesis. The formation of the 13S,14S-epoxy-maresin intermediate occurs only when GPx activity is insufficient to reduce the hydroperoxide precursor of its hydroxy derivative, favoring the epoxide pathway under conditions of partial GPx deficiency. This dependency explains why MaRs and PD1 are often difficult to quantify in tissues, as their levels are highly sensitive to cellular antioxidant status [28]. Several studies have employed liquid chromatography–tandem mass spectrometry (LC-MS/MS) to detect MaR1 both in vitro and in vivo models [23]. In neuronal cell cultures subjected to oxidative stress, MaR1 concentrations ranging from 1 to 100 nanomolar have demonstrated neuroprotective effects that surpass those of DHA, suggesting a functionally efficient conversion at this concentration range [29,30].

Although precise conversion percentages remain unavailable, the observed relationship between administered DHA concentrations (1–10 μM) and detected MaR1 levels (1–100 nM) [21] suggests a conversion efficiency in the range of approximately 0.01–1%, depending on the cellular and physiological context.

5. Preclinical Investigations

A growing body of preclinical research has explored the biological activity of MaRs in the context of human diseases. These studies, encompassing both in vitro and in vivo models, have been instrumental in elucidating the immunomodulatory, anti-inflammatory, and tissue-protective properties of MaRs. By simulating pathological conditions, these models provide valuable insights into the mechanisms through which MaRs exert their proresolving actions and support the development of novel therapeutic strategies.

5.1. In Vitro Studies

The functional characterization of MaRs has been extensively investigated using in vitro models, particularly those employing human-derived cells. These systems have proven essential for dissecting the cellular and molecular pathways modulated by MaRs, allowing researchers to evaluate their effects on immune cell behaviors, epithelial integrity, oxidative stress, and cytokine production. The following sections summarizes the current evidence on the biological actions of different MaRs subtypes in in vitro settings.

5.1.1. Maresin 1

Central Nervous System

The central nervous system has emerged as a key target in the study of MaR1, particularly due to its relevance in neurodegenerative conditions such Alzheimer's disease. Current evidence suggests that MaR1 exerts immunomodulatory effects on microglia, the resident immune cells of the brain that play a central role in the progression of neuroinflammation and neuronal damage. In this context, two studies have explored the impact of MaR1 on microglial responses to the amyloid-beta42 (AB_{42} peptide), a neurotoxic fragment derived from amyloid precursor protein that aggregates into plaques and triggers inflammatory responses in the brain. In the first of these, MaR1 (1 and 100 nM) was able to reduce proinflammatory markers such as CD33, activated CD11b, MHC-II, and CD86 and promoted phagocytosis of AB_{42} in CHME3 cells, a human microglial cell line [31]. This functional shift was further supported by studies in the human monocytic leukemia cell line (THP-1), and in human monocyte-derived microglia, where a micromolar dose of MaR1 increased the cellular uptake of AB_{42}, reducing the secretion of inflammatory cytokines and chemokines, and reduced the CD40 marker and nuclear factor kappa-light-chain-enhancer of activated B cells (NF-κB) activity, reinforcing its role in the active resolution of microglial inflammation [32]. Beyond surface markers and cytokines profile, transcriptomic analyses in human monocyte-derived microglia showed that the treatment with MaR1 (5 μM) suppressed the expression of proinflammatory genes upregulated by AB_{42}, including tumor necrosis factor alpha (TNF-α) and NF-κB, suggesting a pro-resolving effect associated with Alzheimer's disease [33]. Interestingly, the neuroprotective potential of MaR1 extends beyond microglia. In a hybrid cell line created by fusion of mouse neuroblastoma cells with motoneurons-enriched embryonic spinal cord cells (NSC-34) subjected to oxidative stressors, MaR1 (10 nM) reduced reactive oxygen species (ROS) production and NF-κB activation. This suggested that MaR1 may also act directly on neuronal populations, contributing to cellular resilience under pathological conditions [30].

Taken together, these studies illustrate the multifaceted role of MaR1 in the central nervous system, where it modulates immune response, facilitates the clearance of neurotoxic aggregates, and protects neuronal integrity. These mechanism position MaR1 as a promising candidate for therapeutic strategies focused at mitigating neuroinflammation and neurodegeneration, particularly Alzheimer's diseases.

Respiratory System

The respiratory system has emerged as another relevant target for MaR1, particularly in the context of acute and chronic inflammatory responses. In vitro models have provided compelling evidence of its ability to regulate epithelial and immune cells behavior under stress conditions. In human neutrophils exposed to lipopolysaccharide (LPS), pretreatment with MaR1 (1–100 nM) inhibited phosphorylation of survival signaling proteins, triggering proapoptotic signals and promoting the resolution of the inflammation [34]. This suggests that MaR1 may restore the physiological turnover of neutrophils, a process often impaired in chronic lung injury. In murine lung cells (MLE-12) incubated with LPS, MaR1 showed a concentration-dependent reduction in epithelial permeability, mostly effective at 100 nM. Importantly, MaR1 upregulated the expression of claudin-1 (CLDN1) and tight junction protein (ZO-1), both essential for the integrity of tight junctions, indicating a barrier-protective role that could be critical in preventing pulmonary edema and pathogens invasion [35]. Further insights were obtained from a model of airway inflammation induced by organic dust exposure. Using BEAS-2B bronchial epithelial cells pretreated with MaR1 (0–200 nM) for 24 h, Nordgren et al. showed that MaR1 reduced the activation of protein kinase C (PKC), and consequently decreased the production of TNF-α, IL-6 and IL-8 [36]. These findings reinforce the anti-inflammatory profile of MaR1 and its capacity to modulate epithelial signaling cascades. Beyond epithelial responses, MaR1 has also been evaluated in a tuberculosis model using monocyte-derived macrophages differentiated from human peripheral blood, stimulated with LPS or *M. tuberculosis*. Treatment with MaR1 (150 nM) resulted in a significant reduction in TNF-α production in both conditions. Additionally, MaR1 enhanced the antimicrobial activity of macrophages by lowering the bacterial load and inducing the production of bactericidal/permeability-increasing protein (BPI). Importantly, MaR1 promoted nuclear translocation of NF-κB p65 and nuclear factor erythroid 2-related factor 2 (Nrf2), suggesting a dual regulatory effect on inflammatory and antioxidant pathways [37].

In the case of the pulmonary system, the studies highlight the potential of MaR1 to modulate lung inflammation through different pathways, by restoring apoptosis, protecting epithelial integrity or decreasing the production of proinflammatory cytokines.

https://doi.org/10.3390/biom16010139

Oral Diseases

In addition to its action on major organ systems, MaR1 has demonstrated relevant effects in infectious and inflammatory disease of the oral cavity. Wang et al. investigated its action in macrophages derived from patients with localized aggressive periodontitis, showing a significant restoration of their phagocytic capacity of *P. gingivalis* and *A. actinomycetemcomitans*. Even at a concentration of 1 nM, phagocytosis increased by 37% and 65%, respectively, reaching levels comparable to healthy subjects and restoring homeostasis in chronic infections [38]. Complementarily, MaR1 has been studied in human bone marrow-derived mesenchymal stem cells (hBMMSCs), to explore new endodontic strategies for bacterial infections in pulpal and periapical pathology. In this model, hBMMSCs were stimulated with LPS and the treatment with MaR1 (1 to 100 nM) resulted in enhanced cell viability, proliferation, migration, survival, and reduced expression of inflammatory cytokines. The most potent effects were obtained when MaR1 was combined with RvE1 [39].

These findings highlight the dual role of MaR1 in oral pathologies where it promotes immune resolution and supports tissue regeneration through coordinated cellular mechanism.

Endocrine and Metabolic Diseases

MaR1 has demonstrated coordinated actions across multiple metabolic and regenerative contexts, including adipose tissue, bone, and bladder epithelium. In differentiated human adipocytes (hMSCs and hSPs), incubation with MaR1 (1–10 nM) promoted the expression of key adipokines such as adiponectin (ADIPOQ), dipeptidyl peptidase 4 (DPP-4) and cardiotrophin-1 (CT-1). In TNF-α induced inflammation, MaR1 (1–200 nM) partially restored the gene expression of adipokines, ADIPOQ, leptin (LEP), and DPP-4, suggesting a role in preserving adipocyte function in inflammatory diseases [40]. Likewise, MaR1 has been implicated in the modulation of ferroptosis in the context of diabetes-induced osteoblast dysfunction. In MC3T3-E1 osteoblastic cells, incubation with MaR1 (1–10 nM) partially restored the proliferative capacity and expression of osteogenic proteins, and enhanced the expression of NRF2, a key regulator of the antioxidant response linked to the inhibition of ferroptosis. These findings suggest that MaR1 may contribute to preserving osteoblast function in altered metabolic states by regulating pathways involved in oxidative stress and cell death [41]. In a bone repair model, MaR1 was evaluated in the crosstalk between macrophages and mesenchymal stem cell (MSCs), using a demineralized bone matrix scaffold model modified with MaR1 (at 150 nM) and 19S DNA aptamer (Apt19S). In this model, MaR1 activated PPARγ in macrophages, promoting their polarization toward the M2 phenotype and reducing proinflammatory mediators, which in turn enhanced MSC proliferation, migration, and osteogenic differentiation. The effects were abolished by PPARγ inhibition, confirming the dependence of this pathway [42]. Finally, in the bladder, MaR1 significantly accelerated wound closure in vitro and reduced inflammation in vivo within 3 days of cyclophosphamide (CP) induced injury [43].

Together, these findings illustrate the integrative role of MaR1 in modulating inflammation and promoting tissue repair across metabolic vulnerable systems. The experimental conditions and outcomes of the in vitro studies using MaR1 described above are summarized in Table 2, highlighting its consistent anti-inflammatory and regenerative effects across different cell types.

https://doi.org/10.3390/biom16010139

Table 2. Summary of in vitro studies evaluating the effects of MaR1 on various cell types. The table includes stimuli, doses, exposure times, and observed responses, emphasizing MaR1's anti-inflammatory and proresolving actions.

Cell Line	Stimulus	[MaR1]	Time	Response	Reference
CHME3	1 μg/mL of Aβ_{42}	0–100 nM	0–6 h	↑ phagocytosis of Amyloid β_{42}.	[31]
Human monocyte-derived microglia and THP-1 cells	5 μM Aβ_{42}	5 μM	24 h	↑ phagocytosis of Amyloid β_{42}. ↓ CD40 and NF-κB.	[32]
Human monocyte-derived microglia	5 μM Aβ_{42}	5 μM	4 h	Inflammatory effect ↓ TNF-α and NF-κB.	[33]
NSC-34 motor cells	Oxidative stress/SOD1 G93A, TDP-43 A315T mutation	1–1000 nM	-	↓ ROS reduction. ↓ NF-κB activation. Neuronal protection.	[30]
Primary human neutrophils	Lipopolysaccharides	1–100 nM	1–3 h	Restoration of phagocytosis.	[34]
MLE-12 cells	Lipopolysaccharides (1 μg/mL)	0.1–100 nM	30 min	↓ Epithelial permeability, ↑ Claudin-1, Zo-1.	[35]
BEAS-2B cells	Organic powder 5%	0–200 nM	-	↓ PKC, ↓ TNF-α, IL-6, IL-8.	[36]
hMSC and hSP cells	TNF-α	1–200 nM	24 h	Partial reverse lof disruption in ADIPOQ, LEP, DPP-4 and FNDC5.	[40]
MC3T3-E1 cells	High glucose plus palminate	1–10 nM	-	↑ Proliferation, ↑ Osteogenic proteins, ↑ NRF2; ↓ Dysfunction induced by the diabetic environment.	[41]
Peripheral blood macrophages	*P. gingivalis* and *A. actinomycetemcomitans*	0, 1–10 nM	-	↑ phagocytosis.	[38]
Monocyte-derived macrophages	LPS/*M. tuberculosis*	150 nM	1 h	↓ TNF-α ↑ BPI induced translocation of NF-κB p65 and Nrf2 to the nucleus	[37]
hBMMSC	LPS	100 nM	-	↑ TGF-β1, IL-10, IL-4 ↓ IFN-γ, RANKL y TNF-α	[39]
Bone repair model (demineralized bone matrix scaffold + MaR1 + Apt19S)	Macrophage-MSC interaction	150 nM	-	Activation PPARγ in macrophages, ↓ Proinflammatory mediators, ↑ MSC proliferation, migration and osteogenic differentiation	[41]
Bladder model (epithelial cells and in vivo CP injury)	Cyclophosphamide		3 days	↑ Accelerated wound closure, ↓ inflammation	[42]

5.1.2. Maresin 2

The benefits of MaR2 in in vitro models have only been reported in two studies to date, both related to their role in epithelial repair. In the first, the authors tested different doses of MaR2 in an intestinal wound healing model with HT29/B6 cells, where they showed no effects on the wound closure rate. However, when pre-incubated with proinflammatory cytokines, such as TNF-α and IFN-γ, MaR2 (50–200 nM) was able to promote an increase in wound closure at 24 h′ post-injury [44]. In the second, the effect of MaR2 was evaluated on rat-derived conjunctival goblet cells, where it was able to promote an increase in

 https://doi.org/10.3390/biom16010139

intracellular calcium concentration ($[Ca^{2+}]_i$), and stimulated the secretion of high molecular weight glycoconjugates, including mucin. These responses were mediated through G protein-coupled receptors (GPCRs) that involved the activation of PKC [22].

These studies suggest that MaR2 acts as a pro-reparative mediator that enhances epithelial and secretory function under inflammatory conditions, by modulating the secretion of proinflammatory cytokines and stimulating the production of glycoconjugates. Although limited, the current evidence on MaR2 in in vitro settings is summarized in Table 3, which highlights its role in epithelial repair and secretory function.

Table 3. Summary of in vitro studies using MaR2, focusing on its effects on epithelial wound healing and mucin secretion under inflammatory conditions.

Cell Line	Stimulus	[MaR2]	Time	Response	Reference
HT29/B6 cells	Scratch injury	50–200 nM	24 h	↑ wound closure.	[44]
Globet cells	Without external stimulus	0.1 nM–10 nM	2 h	↑ $[Ca^{2+}]_i$ ↑ Secretion of glycoconjugates.	[22]

5.1.3. Maresin Conjugate in Tissue Regeneration (MCTR)

In the case of MCTRs, there are also only two studies describing their benefits to date. In one of them, using primary macrophages incubated with MCTR1 (1 pM–10 nM) concentration-dependently increased phagocytosis of *Escherichia coli* (*E. coli*), promoting efferocytosis by neutrophils, a key mechanism for tissue regeneration. In the case of MCTR3, it produced a greater increase in phagocytosis of *E. coli* compared to MCTR2 and MCTR1, highlighting its potent anti-inflammatory and pro-resolving actions [26]. On the other hand, Chiang et al. investigated how these MCTRs could counteract the vascular response stimulated by leukotriene D4 (LTD_4), using CHO cells transfected with CysLT1, an LTD_4 receptor. In this case incubation with MCTR (10–100 nM) reduced the signaling induced by LTD_4 [24].

Both approaches reinforce the potential of MCTRs as efficient mediators of tissue repair, by enhancing resolution mechanisms such as efferocytosis and phagocytosis or by inhibiting proinflammatory signaling pathways. The distinct actions of MCTR1-3 in promoting phagocytosis and modulating inflammatory signaling are illustrated in Table 4, which presents the main findings in vitro.

Table 4. Experimental parameters and out of in vitro studies using MCTR 1–3. The table highlights their role in enhancing phagocytosis, efferocytosis, and suppressing leukotriene-mediated responses.

Cell Line	Stimulus	Concentration	Time	Response	Reference
Primary leukocytes (macrophages and neutrophils)	*E. coli*	MCTR1, MCTR2, MCTR3 1 pM-10nM	-	↑ phagocytosis and efferocytosis.	[26]
CysLT1-transfected CHO cells	LTD_4	MCTR 10–100 nM	-	↓ of LTD_4-induced signals.	[24]

5.1.4. Maresin-like Mediators (MaR-L)

Regarding MaR-L lipid mediators, to date, there are only two in vitro studies examining MaR-L effects. In one, the authors used macrophages derived from diabetic mice, characterized by a delay in the resolution of inflammation. Treatment with MaR-L (10 or 50 nM) promoted the restoration of the reparative activity of macrophages, facilitating cell migration, modulating the production of growth factors, and suppressing chronic inflammation, even at concentrations as low as 10 nM [27]. The second study by Thamizhchelvan et al. evaluated the

https://doi.org/10.3390/biom16010139

effect of MaR-Ls (MaR-L1, MaR-L2 and MaR-L3) in the context of complex bacterial infections associated with traumatic burns, in combination with a bactericide on biofilms formed by strains of *Staphylococcus aureus*, *Pseudomonas aeruginosa* and *Escherichia coli*. The biofilms were treated with MaR-Ls (1–100 nM), alone or in combination with carbenicillin for 24 h. While treatment with MaR-Ls alone did not show significant effects, co-administration with the bactericide resulted in a significant reduction in bacterial cell viability [45].

These studies suggest that MaR-L is not only involved in the resolution of inflammation but may also be used as co-adjuvant of existing therapies in conditions where tissue repair or infection control is compromised. The emerging evidence on MaR-L mediators is presented in Table 5, showcasing their potential in restoring macrophage function and enhancing antimicrobial responses.

Table 5. Summary of in vitro studies evaluating MaR-Ls including their effects on macrophage repair activity and bacterial biofilm viability.

Cell line	Stimulus	Concentration	Time	Response	Reference
Macrophages	Scratch injury	MaR-L 10 o 50 nM	-	↑ repair activity of macrophages. ↓ in chronic inflammation.	[27]
(Not applicable)/ Biofilms of clinical strains	Carbenicillin (±MaRLs)	MaRL1, MaRL2, MaRL3 1–100 nM	24 h	↓ in bacterial viability.	[45]

5.2. Studies in Animal Models of Human Diseases

In this context, in vivo studies have been essential to evaluate the therapeutic potential of MaRs. These models allow for the assessment of systemic responses to inflammation and tissue injury, providing insight into how MaRs modulate disease progression and promote resolution in physiologically relevant settings.

5.2.1. Maresin 1

Liver

MaR1 has shown consistent hepatoprotective effects across a wide range of liver injury conditions, including acute inflammation, chronic fibrosis, and metabolic dysfunction. In carbon tetrachloride (CCL4)-induced acute liver injury, MaR1 administration 10–100 ng/mouse intraperitoneal (*i.p.*), reduced serum alanine aminotransferase (ALT) and aspartate aminotransferase (AST) levels, suppressed oxidative stress and lipid peroxidation, and decreased hepatic necrosis and apoptosis. These effects were associated with reduced expression of TNF-α, interleukin-6 (IL-6), interleukin-1 beta (IL-1β), monocyte chemoattractant protein-1 (MCP-1), cyclooxygenase-2 (COX-2), and inducible nitric oxide synthase (iNOS), and were mediated through inhibition of NF-κB and mitogen-activated protein kinase (MAPK) signaling pathways [46]. The relevance of this finding lies in the ability MaR1 to act early in the inflammatory cascade, potentially preventing the transition from acute damage to chronic liver dysfunction. In another acute model using lipopolysaccharide/D-galactosamine (LPS/D-GalN), MaR1 (50–100 ng/mouse, *i.p.*) not only attenuated systemic inflammation and reduced ROS, but also promoted macrophage polarization towards the M2 phenotype, and suppressed pyroptosis by decreasing the N-terminal fragment of gasdermin D (GSDMD-N), NLR family domain containing 3 (NLRP3), and IL-1β expression [47]. These results were complemented by in vitro experiments in cells stimulated with LPS where limited apoptosis was observed, and the expression of NLRP3, GSDMD-N and IL-1β was suppressed [47]. In another model of acute injury, hepatic ischemia/reperfusion (I/R), MaR1 treatment (5–20 ng/mouse, *i.p.*) significantly decreased several parameters associated with liver damage, including serum levels of ALT and AST, necrotic areas, and hepatocyte apoptosis in dose-dependent manner.

 https://doi.org/10.3390/biom16010139

Systemic inflammation was also reduced, as evidence by lower expression of IL-6 and IL-1B, suggesting that MaR1 could be exerting this effect at least in part through the ALX receptor (ALXR)/protein kinase B (Akt) signaling pathway [15].

In fibrotic liver models (chronic liver disease), such a as those induced by diethylnitrosamine (DEN), MaR1 normalized transaminase levels, improved hepatic architecture, and reduced inflammation and oxidative stress by inhibiting NF-κB translocation and activating Nrf2, TNF-α and IL-1β [48], suggesting that MaR1 may not only halts fibrotic progression but also promote tissue remodeling and recovery, a critical need in chronic liver disease management. In the context of obesity and no-alcoholic fatty liver disease, MaR1 ameliorated hepatic steatosis in both genetically obese (ob/ob) and diet induced obese (DIO) mice. MaR1 (2–10 μg/kg/day, *i.p.*, for 20 days) reduced hepatic triglyceride (TG) content, downregulated fatty acid synthase (FAS) and stearoyl-CoA desaturase-1 (SCD1), and increased phosphorylation of acetyl-CoA carboxylase (ACC), LC3-II expression and autophagic vacuole formation. In DIO mice, MaR1 (2 μg/kg/day, *i.p.*, or 50 μg/kg/day, oral gavage, for 10 days) produced similar effects, including reduced serum transaminases, and upregulated carnitine palmitoyltransferase 1A (CPT1A), acyl-CoA oxidase 1 (ACOX1), and autophagy-related genes Atg5 and Atg7. These effects were dependent on AMP-activated protein kinase (AMPK) activation, as AMPK inhibition abolished MaR1-induced gene expression changes [49]. In high-fat diet (HFD)-fed mice, MaR1 also suppressed endoplasmic reticulum (ER) stress stimulating AMPK and increasing sarco/endoplasmic reticulum Ca^{2+}-ATPase 2b (SERCA2b) expression, thereby reducing hepatic lipid synthesis and steatosis [50]. Additionally, MaR1 protected hepatocytes from palmitate-induced lipotoxicity and hypoxia-induced ER stress by activating unfolded protein response (UPR) prosurvival mechanism and modulating specific microRNA signatures involved in protein folding and apoptosis [51].

Together, these findings establish MaR1 as a multifunctional hepatoprotective mediator capable of modulating inflammation, oxidative stress, pyroptosis, ER stress, lipid metabolism, and autophagy, with strong therapeutic potential for both acute and chronic liver diseases.

Cardiovascular Diseases

Heart

MaR1 has shown cardioprotective effects across diverse models of cardiac injury, acting through antioxidants, anti-inflammatory, and anti-ferroptotic mechanisms. In neonatal cardiomyocytes, MaR1 promoted physiological hypertrophy, by activating the retinoic acid-related orphan receptor alpha (RORα), which in turn induced the production of insulin-like growth factor 1 (IGF-1), and the triggered the phosphatidylinositol 3-kinase (PI3k/Akt) pathway [16]. This mechanism suggest that MaR1 may contribute to adaptative cardiac growth rather than pathological remodeling, offering a potential strategy to support cardiac development or recovery from injury. Similarly, in a murine model of myocardial infarction (MI), the treatment with MaR1 (10 ng/g for 28 days post-MI) reduced the cardiac remodeling and the incidence of arrythmias. These effects were associated with the activation of the Nrf2 pathway and attenuation of NF-κB [52], indicating that MaR1 may protect the myocardium by enhancing antioxidant defenses while suppressing proinflammatory transcriptional programs. Importantly, the reduction in arrhythmias suggests a stabilizing effect on cardiac electrophysiology, which could be clinically relevant in infarcted patients. In a model of I/R injury, MaR1 treatment (1 ng/mouse) decreased infarcted size and cardiomyocytes cell death by suppressing both apoptosis and pryroptosis, again associated with an inhibition of the NF-κB pathway and a reduction in the Silent Information Regulator Factor 2-related enzyme 1 (SIRT1) activity, thus inhibiting the activation of the inflammasome NLPR3 [53]. In the case of sepsis-induced cardiac dysfunction in mice, another model of acute injury, the administration

https://doi.org/10.3390/biom16010139

of the MaR1 (100 ng/mouse) mitigated the cardiac impairment caused by LPS administration. This protective effect was attributed to reduced oxidative stress via activation of NRF2-HO pathway, and decreased inflammation, likely through modulation of cardiac macrophage activity [54]. More recently, MaR1 (10 ng/g) was evaluated in a model of cardiotoxicity generated by the chemotherapeutic doxorubicin, where it attenuated lipid peroxidation via the NRF2/GPX4 axis, reducing cardiomyocytes death by ferroptosis [55]. This is particularly relevant given the clinical limitations of doxorubicin due to its cardiotoxic profile.

Taken together, these findings highlight the ability of MaR1 exert positive effects multiple cardiac pathophysiological processes, offering protection through convergent molecular pathways that regulate oxidative damage, cell death, and inflammatory signaling.

Vasculature

Beyond its effects in other organs and systems, MaR1 has also been investigated in the vascular system, where it has shown a broad capacity to regulate vascular homeostasis. The first report in this context was in a model of neointimal hyperplasia by carotid ligation. Here, MaR1 (100 ng), similarly to Resolvin D2 (RvD2) (100 ng), reduced the proliferation of vascular smooth muscle cells (VSMCS), reduced neutrophils, and monocytes infiltration into the vessel wall. Notably MaR1 also promoted macrophage polarization toward the M2 phenotype [56], suggesting a shift toward a proresolving immune environment. Similarly, in a mouse model of atherosclerosis (*Apoe*$^{-/-}$ mice fed a high-fat diet), co-administration of MaR1 and RvD2 (100 ng of each SPM), prevented the progression of the atherosclerotic lesion, reduced necrotic core size, and improved plaque stability. Interestingly, while MaR1 reduced macrophage content within the plaque, it simultaneously induced VSMCs proliferation, a paradoxical but beneficial effect in this context, as VSMCs contribute to fibrous cap formation and plaque stabilization [57].

Further evidence from cardiovascular models supports the vascular protective role of MaR1. In a mouse model of systemic arterial hypertension induced by angiotensin II (AngII) infusion, administration of MaR1 (2 μg/kg for 28 days) reduced blood pressure and vascular remodeling by acting on VSMCs through activation of the LGR6 receptor. MaR1 treatment inhibited VSMC proliferation, migration, and pyroptosis, leading to the reversal of pathological vessel wall changes [58]. Similarly, in a model of aortic aneurysm, MaR1 (4 ng/g body weight), also acting via LGR6, inhibited VSMC activation and attenuated aneurysm development. Interestingly, MaR1 also enhanced efferocytosis of dying VSMCs by macrophages, contributing to inflammation resolution and tissue repair [59].

Finally, in mouse models of pulmonary hypertension, MaR1 treatment showed therapeutic potential by ameliorating disease progression. The administration of MaR1 (1 μg per mouse, followed by 100 ng every two days, for three weeks) decreased pulmonary arterial pressure and right ventricular hypertrophy. It also decreased wall thickening and luminal narrowing, evidencing an improvement in this pathology, and inhibited pulmonary vascular smooth muscle cells proliferation by inhibiting STAT, Akt, ERK, and FoxO1 phosphorylation via the activation of the LGR6 receptor [60]. Also, it was reported that MaR1 ameliorated pulmonary hypertension in mice, acting on the ALX receptor, by reducing pulmonary smooth muscle cells proliferation [61].

Kidney Diseases

The kidney is another important organ affected by prevalent pathologies. Several studies MaR1 has shown protective effects in models of acute, ischemic, and metabolic kidney injury, acting through coordinated regulation of inflammatory, oxidative, and cell death pathways. First, in sepsis-induced kidney injury, the administration of MaR1 (5 μg/kg) decreased the expression of proinflammatory factors such as TNF-α, IL-1β, Il-6

 https://doi.org/10.3390/biom16010139

and MCP-1 while it decreased the production of NOX4-derived ROS and the expression of IκBα and p65, key components of the NF-κB pathway [62]. MaR1 also induced renal protection via inhibition of the inflammasome, reducing endoplasmic reticulum stress, and hence pyroptotic cell death in a sepsis model of cecal ligation and puncture (0.04, 0.4 and 4 μg/kg) [63]. In a similar study, MaR1 (at 0.5 and 1 ng/mouse) acted via inhibition of the NF-κB/STAT3/MAPK pathways, resulting in decreased inflammatory cytokines production and increased survival rate [64]. In I/R injury, MaR1 (1.0 ng/mouse) targeted the TLR4, inhibiting the activation of NF-κB, attenuating inflammation and inducing the activation of Nrf2, improving the antioxidant defense [65]. Regarding diabetes, the potential of MaR1 to prevent the development of nephropathy was evaluated in mice treated with streptozotocin and a high fat diet (a type 2 diabetes model, DM2), at 4 μg/kg, for 14 weeks. Here, MaR1 reduced glomerular damage by decreasing the inflammation and the induction of the antioxidant response by overexpression of superoxide dismutase [66]. These findings are especially relevant given the limited therapeutic options for diabetic kidney disease and the central role of oxidative stress and inflammation in its progression.

Lung Diseases

The impact of Mar1 has been evaluated also in different models of pulmonary disease, where it exerts protective effects by modulating oxidative stress, inflammatory signaling and immune cells behavior. In lung I/R injury, MaR1 (1 ng/mouse) ameliorated pulmonary damage, inflammation and inflammatory cell infiltration associated with diminished ROS production, by enhancement of the antioxidant response of the Nrf-2 -heme oxygenase 1 (Nrf2-HO-1) pathway [67], suggesting that MaR1 activates endogenous antioxidant defenses, which are critical in limiting reperfusion-induced oxidative injury. MaR1 has also been shown to be effective in murine models of sepsis-associated lung injury. This was first described in a model of LPS-induced acute lung injury in mice. MaR1 (1 ng per mouse) reduced the damage by diminishing the inflammatory response, i.e., the surge of inflammatory cytokine and chemokines and neutrophils infiltration [68]. These observations were further reinforced in the same model by the analysis of the effects of MaR1 on the alveolar permeability, where it induced the expression of claudin-1 and ZO-1, thus reducing the pulmonary edema [35]. In addition, using the cecal ligation and puncture method of sepsis-associated acute lung injury, it was described that the treatment with MaR1 (at 0.5 and 1 ng/mouse) attenuated the inflammatory response by activation of the JAK2/STA3 and MAPK/NF-κB pathways [69]. Using the same model, but at 10 ng/mouse, MaR1 treatment reduced the pulmonary damage and induced the switch of alveolar macrophages to a M2 anti-inflammatory phenotype, through activation of peroxisome proliferator-activated receptor γ (PPAR-γ) [70]. Also, in this cecal ligation model, MaR1 (at 0.5 and 1 ng/mouse) reduced the levels of Th17 lymphocytes that favor the inflammatory response, while increasing the activation of Treg cells through the activation of the retinoid-related orphan nuclear receptor γ (ROR-γ) via the STAT3 pathway activation [71]. Finally, in this same model, at a dose of 1 ng/mouse, using single cell-sequencing, it was reported that MaR1 acts on a subpopulation of CXCL3 positive neutrophils, inhibiting the expression of the inflammatory pathway genes, reducing infiltration [72]. In addition, in a mouse model of asthma induced by ovalbumin, MaR1 (0.1, 1 and 10 ng/mouse) reduced inflammation dose-dependently by diminishing inflammatory cell recruitment, especially eosinophils and neutrophiles, associated with a reduction in NF-κB activation [14]. Finally, MaR1 was evaluated in a model of respiratory syncytial virus-induced lung inflammation. Here, MaR1 (10 ng/mouse) inhibited the viral aberrant transformation of Tregs cells and promoted interferon-β production, by activation LGR6 receptors. This blocked the production of IL-13, which is key to the transformation of Tregs lymphocytes into a prophlogistic phenotype [73].

https://doi.org/10.3390/biom16010139

Others

Beyond its effects on major organs, MaR1 has shown promising results in models of neuroinflammation, pancreatitis, and osteoarticular pain. In sepsis-associated encephalopathy (SAE) model, induced by cecal ligation and puncture, pretreatment with MaR1 (4n/g) attenuated the adverse effects of neuroinflammation by activating the SLC7A11/GPx4 ferroptosis pathway [74]. In acute (AP) and chronic (CP) pancreatitis induced by *i.p.* administration of caerulein (50 μg/kg), MaR1 treatment (2 ng) was evaluated. Here, MaR1 administration improved histopathological alterations, decreased the number of receptor-interacting protein kinase 3 (RIP3) and p-MLKL cells involved in necrosome activation, reduced macrophage infiltration and attenuated pancreatic fibrosis [75]. More recently the effects of MaR1 on osteoarthritis (OA)-like pain induced by injection with monosodium iodoacetate in animals was investigated. Eight weeks after injury, this treatment attenuated the pain-associated behavior, evidenced by increased paw withdrawal threshold and increased weight bearing. In addition, a reduction in CGRP expression, macrophage activation, and inflammatory cytokine levels were monitored. These findings suggest that MaR1 could directly modulate the functional response of neurons through the RORα mediated signaling pathway [76].

In short, MaR1 has been a focus of research in addressing various diseases affecting multiple organs. These studies position this molecule as a future therapeutic alternative for these conditions where only palliative options are available. To provide a comprehensive view of MaR1 systemic effects, Table 6 compiles the in vivo studies across various organs and disease models, emphasizing its therapeutic versatility.

5.2.2. Maresin 2

MaR2 has shown a wide range of therapeutic effects in preclinical models of pain, inflammation, epithelial injury, and metabolic dysfunction. In murine models of *Bothrops jararaca* venom-induced pain and inflammation, MaR2 administered intraperitoneally at doses of 0.3–3 ng reduced mechanical and thermal hyperalgesia, restored hindpaw weight distribution, and decreased the levels of TNF-A, IL-1B, IL-6, myeloperoxidase activity, and superoxide anion production, while enhancing total antioxidant capacity and reducing hemorrhage and leukocyte infiltration [77]. In a model of trigeminal neuropathic pain, MaR2 (1–10 ng, intrathecally) inhibited neuronal activation in the trigeminal ganglion, reduced $CGRP^+$ and c-Fos^+ neurons, and normalized phospho NF-κB, supporting neuromodulator and anti-nociceptive potential in chronic pain conditions involving central sensitization [78]. In allergic airway inflammation and asthma, MaR2 reduced the presence of inflammatory cells such as IL-4, proinflammatory cytokines such as IL-1B and IL-18. Additionally, MaR1 treatment reduced oxidative stress markers such as malondialdehyde (MDA), while increasing levels of antioxidant enzymes superoxide dismutase (SOD) and glutathione synthetase (GHS), indicating a restoration of redox balance in lung tissue [79]. In models of colonic injury, MaR2 (2 ng/g, *i.p.*) accelerated mucosal wound healing following DSS-induced colitis and biopsy-induced epithelial damage. This treatment promoted epithelial migration via activation of focal adhesion kinase (FAK), Scr, paxillin, vinculin, and talin signaling [44]. In diet-induced obese mice, cold exposure and β3-adrenergic stimulation activated brown adipose tissue (BAT) to produce MaR2, which targeted hepatic macrophages, suppressed TNF-α, IL-1β, and NLRP3 inflammasome activation, and improved insulin sensitivity, linking MaR2 to metabolic inflammation resolution [80]. Finally, in zebrafish models exposed to *Loxosceles intermedia* venom, although MaR2 was not directly tested, the study highlighted systemic inflammation, oxidative stress, and behavioral alterations, reinforcing the relevance of pro-resolving mediators like MaR2 in venom-induced pathology [81]. In conjunctival goblet cells, MaR2 (0, 1–10 nM) increased intracellular calcium

 https://doi.org/10.3390/biom16010139

and stimulated mucin secretion via GPCR- mediated pathways, suggesting a role in ocular surface homeostasis and dry eye diseases [22].

Together, these findings establish MaR2 as a multifunctional lipid mediator capable of resolving inflammation, modulating pain, promoting epithelial repair, restoring redox balance, and improving metabolic outcomes, with strong translational potential across diverse pathophysiological conditions. The anti-inflammatory and antioxidant properties of MaR2 in respiratory disease models are summarized in Table 7, highlighting its potential in allergic airway inflammation.

5.2.3. Maresin Conjugate in Tissue Regeneration (MCTR)

Another biomolecule part of the MaRs family is MCTR, where the strongest evidence focusses on the use of MCTR1. In a model of acute lung injury, MCTR1 administration (10–100 ng/mouse) led to a reduction in morphological parameters of damage, including decreased inflammatory cell infiltrate, interstitial edema, among others (ref). These effects were accompanied by lower levels of proinflammatory cytokines TNF-α and IL-1β. They also determined that the action of this pro-resolving mediator is mediated by the ALX receptor activation, a G-protein-coupled receptor that induces increase in intracellular cAMP [82]. In the kidney, MCTR1 was evaluated in a model of septic acute kidney injury, where it attenuated ferroptosis by upregulating glutathione peroxidase 4 (GPX4) levels and downregulating prostaglandin-endoperoxide synthase 2 (PTGS2).

These molecular changes correlated with improved renal morphological changes and restoration of kidney function [83]. In a model of sepsis-induced cardiac injury, MCTR1 treatment improved cardiac function by decreasing the left ventricular end-systolic volume, increased after LPS administration. In addition to improving left ventricular fractional shortening and ejection fraction, MCTR1 treatment also decreased the mRNA expression of proinflammatory factors such as IL-1β, IL-6 and TNF-α [84]. In LPS-induced lung injury, MCTR1 treatment inhibited reverse transendothelial migration of neutrophils, reducing systemic inflammation and pulmonary damage [85]. It also protected the pulmonary endothelial glycocalyx by upregulating syndecan-1 and heparan sulfate, thereby preserving vascular integrity and reducing permeability [86]. In bleomycin-induced pulmonary fibrosis, MCTR1 administered from day 7 to 21 post-injury reversed epithelial-to-mesenchymal transition (EMT), reduced collagen deposition, improved lung function, and increased survival rates [87]. In remifentanil-induced hyperalgesia, MCTR1 treatment (10–100 ng, *i.p.*) alleviated pain by regulating mitochondrial fission though suppression of dynamin-related protein 1 (DRP1) expression and restoring mitochondrial morphology in dorsal root ganglia neurons [88]. MCTR3, acting via the ALX/PINK1 signaling pathway, reduced mitochondrial dysfunction and oxidative stress in LPS-induced lung injury, improving histological outcomes and reduced inflammatory cytokines [89]. In influenza A virus–primed mice challenged with *Streptococcus pneumoniae*, administration of MCTR1–3 (10 ng, *i.p.*) restored alveolar macrophage migration, reduced CXCL1 secretion, and gene expression, thereby reducing bacterial load and lung inflammation [90]. Mechanistically, cysteinyl-MCTRs activated TRAF3 via cAMP–PKA signaling, increasing IL-10 production, and enhancing macrophage phagocytic capacity, with effects confirmed in both planarian and mammalian models [91]. Additionally, continual efferocytosis was shown to be metabolically primed by MCTR production in macrophages. Apoptotic cell breakdown upregulated 12-lipoxygenase, leading to MCTR synthesis, which enhanced Ras-related C3 botulinum toxin substrate 1 (Rac1)-mediated glycolysis and glucose uptake. Genetic ablation of MCTR synthesis impaired efferocytosis, confirming autocrine and paracrine roles for MCTR in metabolic reprogramming, sustained clearance and repair [92].

 https://doi.org/10.3390/biom16010139

Table 6. Compilation of in vivo studies using MaR1 in models of liver, heart, kidney, lung, brain, and bone diseases. The table includes experimental models, pathologies, and observed protective effects.

Year	Organ	Pathogeny	Model	Impact	Reference
2016	Liver	Acute liver injury	BALB/c mice	↓ ALT/AST, oxidative stress, necrosis, apoptosis, TNF-α, IL-6, IL-1β, MCP-1, COX-2, iNOS, via NF-κB/MAPK inhibition.	[46]
2021	Liver	Acute liver injury	C57BL/6 mice	↓ ROS, Systemic inflammation, NLRP3, GSDMD, IL-1β, ↑ M2.	[47]
2021	Liver	Hepatic I/R injury	Male C57BL/BJ	↓ ALT/AST, necrosis/apoptosis, IL-6, IL-1β via ALXR/Akt pathway.	[15]
2021	Liver	Liver fibrosis	Male rats Sprague-Dawley	↓ Inflammation and oxidative stress, ↓ NF-κB, Nrf2.	[48]
2018	Liver	Obesity and NAFLD	Ob/Ob and Diet-Induced Obese Mice	↓ TGs, FAS/SCD1, ↑ ACC phosphorylation, LC3-II, autophagy, CPT1A, ACOX1.	[49]
2018	Liver	NAFLD	Male C57BL/6J mice	↓ ER stress, ↑ SERCA2b pathway AMPK activation.	[50]
2017	Liver	Lipotoxicity	Male C57BL/6J mice	↑ UPR survival pathway, modulation of miRNAs related to protein folding and apoptosis	[51]
2021	Heart	Cardiac hypertrophy	Male mice C57BL/6J	↑ Cardiomyocyte cell surface, ↑ IGF-1 production that activates P1EK/AKT. Producing physiological hypertrophy. ↑ Overexpression of the enzyme arachidonate 12-lipoxygenase in post-myocardial infarction patients.	[16]
2022	Heart	Myocardial infarction	Male mice C57BL/6J	↓ Cardiac remodeling and arrhythmias, ↑NRF2, ↓NF-κB signaling.	[52]
2025	Heart	Cardiac I/R	Male mice C57BL/6J	↓ Heart failure size. ↑ Cardiac function. Structural recomposition of myocardial tissue. ↓ IL-1β y TNF-α. NF-κB, ↓ Pyroptosis. ↓ NLRP-3.	[53]
2021	Heart	Sepsis-induced cardiac dysfunction	Mice C57BL/6J	↓ Cardiac dysfunction, ↑ NRF2-HO pathway, ↓ Inflammation.	[54]
2025	Heart	Cardiotoxicity	Mice C57BL/6J	↓ Lipid peroxidation through NRF2/GPX4, ↓ Ferroptosis-mediated cardiomyocyte death.	[55]
2016	Vasculature	Neointimal hyperplasia	Mice	↓ VSMC proliferation, neutrophil/monocyte infiltration, ↑ M2 macrophages	[56]
2016	Vasculature	Atherosclerosis	Mice apoE $^{-/-}$	↓ Lesion progression, plaque vulnerability, VSMC proliferation, ↑ M2 macrophages.	[57]
2020	Vasculature	Hypertension	Mice C57BL/6J	↓ Blood pressure, vascular remodeling, pyroptosis via LGR6 receptor. ↓ Pyroptosis.	[58]
2022	Vasculature	Aortic aneurysm	Mice C57BL/6J	↓ VSMC activation and aneurysm development ↑ efferocytosis by macrophages	[59]
2022	Vasculature	Hypertension	Adult male mice C57BL/6J	↓ Pulmonary pressure, RV hypertrophy, wall thickening via LGR6- mediated inhibition of STAT AKT, ERK, FOxO1.	[60]

https://doi.org/10.3390/biom16010139

Table 6. *Cont.*

Year	Organ	Pathogeny	Model	Impact	Reference
2023	Vasculature	Hypertension	Mice and Rats	↓ VSMC proliferation pathway ALX activation.	[61]
2021	Kidney	Acute kidney injury associated with sepsis (LRA-S)	Male C57BL/6J mice	↓ Inflammation, IL-6, IL-1B, TNF-A, and MCP-1, ↓ BAX and caspase-3, ↑ ATP production, ↓ free oxygen radicals (ROS), ↓ Phosphorylation of IkBa and p65.	[62]
2024	Kidney	Sepsis	Male BALB/C mice	↓ Inflammasome, ER stress, pyroptosis.	[63]
2019	Kidney	Sepsis	Male C57BL/6J mice	↓ of the NF-κB/STAT3/MAPK pathways, ↓ Inflammatory cytokines, ↑ Survival.	[64]
2019	Kidney	Ischemia/reperfusion	Male C57BL/6J mice	↓ NF-κB, ↑ Nrf2 activation, antioxidant defense, ↓ Inflammation.	[65]
2022	Kidney	Diabetic nephropathy	Male C57BL/6J mice	↓ Glomerular damage, inflammation, ↑ SOD, antioxidant response	[66]
2017	Lung	I/R lung injury	BALB/c mice	↓ Pulmonary inflammation, ROS via NRf2 HO-1 pathway.	[67]
2014	Lung	Acute lung injury	BALB/c mice	↓ Cytokines and chemokines, neutrophil infiltration.	[68]
2016	Lung	Barrier function	Male BALB/c mice	↑ Claudin, ZO, ↓ Edema and permeability.	[35]
2020	Lung	Sepsis-associated acute lung injury	Male C57 mice	↓ Inflammation via JAK2/STAT3 and MAPK/NFκB	[69]
2020	Lung	Sepsis-associated acute lung injury	Male BALB/c mice	↑ M2 macrophages pathway ↑ PPARγ.	[70]
2020	Lung	Sepsis	Male C57BL/6 mice	↓ Th17, ↑ Treg via RORγ/STAT3 activation.	[71]
2022	Lung	Sepsis	Male C57BL/6	↓ CXCL3 neutrophil, ↓ Infiltration	[72]
2021	Lung	Asthma	Female BALB/c mice	↓ Eosinophil and neutrophil infiltration, ↓ NFκB.	[14]
2023	Lung	Inflammation	BALB/cByJ and FoxP3^{eGFP} mice	↓ Treg aberrant transformation, ↑ IFN-β pathway LGR6, ↓ IL-13.	[73]
2024	Brain	Sepsis-associated encephalopathy	Rat Sprague-Dawley	↓ Proinflammatory cytokines from MaR1 administration. MaR1 prevents spatial learning difficulties in mice with sepsis-associated encephalopathy.	[74]
2025	Bones	Osteoarthritis	Mice	↑ in endogenous MaR1 due to treadmill exercise. ↓ MMP13 synthesis.	[76]
2021	Pancreas	Acute and chronic pancreatitis (caerulein-induced)	ICR male mice	↓ RIP3/p-MLKL, macrophage infiltration and fibrosis	[75]

https://doi.org/10.3390/biom16010139

Table 7. Experimental conditions and therapeutic outcomes of in vivo studies using MaR2 in murine models of asthma and allergic airway inflammation.

Year	Organ	Pathogeny	Model	Impact	Reference
2025	Peripheral tissue	Pain/inflammation	Swiss Female mice	↓ Hyperalgesia ↓ TNF-α, IL-1β, IL-6, MPO activity, Hemorrhage, leukocyte infiltration ↑Total antioxidant capacity	[77]
2023	Nervous system	Neuropathic pain	Male Wistar rats and Swiss mice	↓ Neuronal activation, $CGRP^+$, and $c\text{-}FOs^+$ neurons Normalized pNF-κB	[78]
2022	Lung	Asthma	Female mice BALB/c	↓ Neutrophil infiltration. ↓ IL-4, IL-1β, IL-18, MDA ↑ SOD, GSH	[79]
2023	Colon	Epithelial injury/colitis		↑Mucosal repair ↑ FAK/Src/paxillin/vinculin/talin	[44]
2022	Liver	Obesity and metabolic inflammation	C57BL6/J mice	↓ TNF-α, IL-1β, NLRP3 Improved insulin sensitivity and glucose tolerance	[80]
2022	Whole organism	Inflammation	Zebrafish		[81]
2022	Eye	Epithelial dysfunction	Male albino Sprague-Dawley rats	↑ Intracellular Ca^{2+}, mucin secretion ↑ Stimulated mucin secretion via GPCR mediated pathways	[22]

Together, these findings establish MCTR1–3 as multifunctional lipid mediators that orchestrate resolution of inflammation, mitochondrial protection, immune reprogramming, and tissue regeneration across diverse models of sterile and infectious injury. The organ-specific protective effects of MCTR in acute inflammatory conditions are detailed in Table 8, which includes studies in lungs, kidney, and heart models.

Table 8. Summary of in vivo studies using MCTR1 in models of acute lung, kidney, and cardiac injury. The table outlines molecular and functional outcomes, including inflammation, ferroptosis, and cardiac performance.

Year	Organ	Pathogeny	Model	Impact	Reference
2020	Lung	Acute lung injury	Male rats Sprague-Dawley	Improvement in histopathological changes in lung tissue, ↓ TNF-a, IL-1B and IL-10, ↑ Alveolar fluid clearance, Na+/K+-ATPase activity, cAMP concentration in lungs, P-Nedd4-2.	[82]
2021	Kidney	Acute septic kidney injury	Male mice C57BL/6	↑ GPX4, GSH and Nrf2, ↓ PTGS2, ↓ MDA ↑ mitochondrial length. ↑ Renal function, ↓ Serum creatinine, and blood urea nitrogen. Suppression of ferroptosis. Improvement of pathogenic changes.	[83]
2021	Heart	Sepsis-induced cardiac dysfunction	Male mice C57BL/6	↓ B natriuretic peptide. ↓ IL-6, IL-1B, IL-17A, TNF-a, CCL2, CCL7, CXCL1, and G-CSF. Promotes downregulation of CXCL1 and G-CSF. ↓ Neutrophil recruitment and infiltration.	[84]

 https://doi.org/10.3390/biom16010139

Table 8. *Cont.*

Year	Organ	Pathogeny	Model	Impact	Reference
2025	Lung	Lung injury	Male mice C57BL/6	Inhibited reverse transendothelial neutrophil migration, ↓ Systemic inflammation, pulmonary damage	[85]
2020	Lung	Acute lung injury	Mice	↓ TNF-α, IL-1β, IL-6, ↑ Syndecan 1, heparin sulfate, ↓ Permeability, Preserved vascular integrity.	[86]
2021	Lung	Pulmonary fibrosis	Male mice C57BL/6	↓ Collagen deposition, ↑ Survival, Reversed EMT.	[87]
2025	Nervous system	Hyperalgesia	Rats	↓ Pain, ↓ DRP1, NR2B, ↓ MDA and ROS Restored mitochondrial morphology.	[88]
2021	Lung	Lung injury	C57BL/6 mice	↓ Mitochondrial dysfunction, ↓ Oxidative stress, ↓ Cytokines.	[89]
2022	Lung	Secondary bacterial infection	Male mice C57BL/6	↓ CXCL1, ↓ Bacterial load and inflammation.	[90]
2023	Immune system	Inflammation resolution	Mac $^{alox12/15\ KO}$ Mice	↑ IL-10, ↑ Macrophage phagocytosis.	[91]
	Macrophages	Efferocytosis and metabolic reprogramming		↑ 12-LOX, ↑ Rac-1, ↑ Glycolysis.	[92]

5.2.4. Maresin-like Lipid Mediators (MaR-L)

The last group of molecules in the MaRs family are MaR-L, which have recently gained attention for their roles in immune modulation and tissue repair (see Table 9). To date, only a few studies have evaluated their biological effects in vivo. In a transgenic 5xFAD mouse model of Alzheimer's disease, Shrivastava et al. investigated the long-term intranasal administration of MaR-L1 at 100 ng/mouse, three times per week from 1.5 to 9 months of age. MaR-L1 treatment significantly reduced amyloid-β plaque burden in the hippocampus and cortex, preserved cholinergic neurons (NauN$^+$ and ChAT$^+$), and decreased cleaved caspase-3 expression associated with apoptosis. It also suppressed M1 microglial markers (CD68$^+$), promoted M2 polarization (Iba1$^+$/Arg-1$^+$), and reduced neutrophil infiltration while enhancing claudin-5 expression, suggesting improved blood–brain barrier integrity and neuroimmune homeostasis [93]. Complementarily, Hong et al. reported that MaR-L lipid mediators are endogenously produced by leukocytes and platelets and can restore the reparative function of macrophages impaired by diabetes. In this study, macrophages derived from diabetic mice exhibited reduced efferocytosis and tissue repair capacity, which was reversed by treatment with MaR-Ls (10–100 nM). These mediators enhanced Rac1-dependent cytoskeletal remodeling and glucose uptake, reprogramming macrophages toward a pro-resolving phenotype. Lipidomic analysis confirmed the presence of MaR-L1, MaR-L2, and MaR-L3 in human leukocytes and platelets, and their biosynthesis was shown to be dependent on 12-lipoxygenase activity [27].

Together, these findings position MaR-Ls as emerging lipid mediators capable of restoring immune resolution, promoting neuronal survival, and enhancing macrophage reparative functions, with promising implications for neurodegenerative and metabolic diseases.

https://doi.org/10.3390/biom16010139

Table 9. Experimental design and outcomes of the in vivo study using MaR-L1 in a mouse model of Alzheimer's disease and diabetes (Db/Db). The table highlights its impact on neuronal survival, microglial phenotype, blood–brain barrier integrity and metabolic parameters.

Year	Organ	Pathogeny	Model	Impact	Reference
2024	Brain	Alzheimer's	Male mouse 5xFAD C57BL/6J	↑ Amyloid beta. MaR-L preserved the neuronal population. ↑ Cells ChAT$^+$, ↓ Cells caspase-3$^+$, ↓ Iba-1 ↓ Microglial cells CD68$^+$. MaR-L restored the microglial phenotype. ↑ Partially branched microgrill cells, ↑ Microgrill cells type M2, ↓ Neutrophil infiltrate, ↑ Claudin-5. MaR-L improves the integrity of the blood–brain barrier	[93]
2014	Adipose tissue	Type 2 diabetes-induced macrophage dysfunction	Diabetic mice (Db/Db)	↑ Efferocytosis, ↑ Tissue repair capacity, ↑ Glucose uptake, ↑ Rac1-dependent cytoskeletal remodeling, ↑ Pro-resolving macrophage phenotype, ↑ Expression of M2 markers, ↓ Pro-inflammatory cytokines TNF-α, IL-6.	[27]

5.3. Studies in Humans

Although there are currently no clinical trials evaluating the direct administration of MaRs in humans, several studies have investigated their endogenous presence in various tissues and diseases. Among these, plasma levels of MaR1 have been the most extensively studied, particularly in metabolic disorders, where resolution mechanisms are often impaired. One of the earliest studies by Skarke et al. (2015) [94] assessed the formation of bioactive lipids following fish oil supplementation at high (21 g/day) and standard (4 g/day) doses. While an increase in the erythrocyte membrane ratio of EPA and DHA relative to AA was observed, SPMs such as MaR1 remained below detection limits, even after inflammation was induced via LPS administration, suggesting that the mere availability of precursors may not be sufficient to trigger SPM biosynthesis in vivo, especially under controlled inflammatory conditions. Similarly, in subjects with metabolic syndrome, omega-3 fatty acids intake did not alter plasma MaR1 levels, and precursor molecules such as 18-hydroxyeicosapentaenoic acid (18-HEPE), 17-hydroxydocosahexaenoic acid (17-HDHA), and 14-HDHA were also reduced [94]. This reduction in precursors bioavailability suggests a compromised biosynthetic capacity for SPMs in metabolic syndrome, potentially limiting the endogenous production of MaR1 despite omega-3 supplementation. It also raises questions about enzymatic activity and substrate accessibility in metabolically altered tissues. In contrast, a later study in obese individuals supplemented with marine oil enriched in 14-HDHA, 17-HDHA, and 18-HEPE (2 g/day for one month), showed a 4.7-fold increase in plasma MaR1 levels [95], possibly due to persistent inflammation in the study population. In another clinical study involving T2DM and diabetic foot ulcer (DFU) patients, plasma MaR1 levels were significantly lower compared to normoglycemic individuals, with negative correlations to body mass index (BMI), systolic blood pressure (SBP), low-density lipoprotein cholesterol (LDL-C), fasting plasma glucose (FPG), glycated hemoglobin (HbA1c), and homeostasis model assessment for insulin resistance (HOMA-IR) [96]. In acute respiratory distress syndrome (ARDS), lower MaR1 levels were associated with prolonged mechanical ventilation and intensive care unit (ICU) stay, while higher levels correlated with shorter ICU stays (<7 days), suggesting that defective resolution of inflammation may contribute to ARDS pathophysiology [97]. In NAFLD, serum MaR1 levels were significantly reduced compared to healthy controls, with positive cor-

https://doi.org/10.3390/biom16010139

relations to albumin (ALB), high-density lipoprotein cholesterol (HDL-C), and negative associations with age, BMI, triglycerides, and hepatic transaminases (ALT, AST) [98]. In atherosclerotic cardiovascular disease (ASCVD), higher MaR1 levels were associated with reduced ASCVD risk, potentially mediated by LDL-C [99]. In women with coronary microvascular dysfunction (CMD), MaR1 levels were significantly lower compared to reference groups, while precursors such as EPA, DHA, and 18-HEPE were elevated, suggesting impaired enzymatic conversion of SPMs [100]. Omega-3 supplementation during pregnancy increased precursors levels at birth, but this effect was not sustained at 12 years of age, and MaR1 was undetectable despite elevated 14-HDHA. In preeclampsia (PE), although no significant differences in MaR1 levels were found between pregnant and non-pregnant women, the ratio of leukotriene B4 (LTB4) to MaR1 was markedly lower in PE patients, indicating an imbalance in inflammatory resolution [101]. Interestingly, a group of researchers investigated the effect of moderate alcohol consumption on plasma concentrations of SPMs in patients with T2DM. The study reported elevated SPM levels in T2DM subjects, suggesting a possible homeostatic response. However, no significant changes were observed in MaR1 or other SPMs following alcohol intake [102].

Although most human evidence has focused on MaR1, other members of the MaRs family, such as MCTRs, have also been detected endogenously in humans. One study examined allergic airway inflammation using human lung tissue and found that healthy tissue exhibited higher levels of MCTR1, MCTR2, and MCTR3 compared to diseased tissue. Upon incubation with MCTR1, metabolic conversion to MCTR2 and MCTR3 was observed, indicating active interconversion. MCTRs significantly blocked the action of leukotriene D4 (LTD4), revealing potential pro-resolving mechanisms in pulmonary responses that may be disrupted by disease [103]. A second study conducted in patients with rheumatoid arthritis identified all three MCTR family members in plasma. Each showed negative correlations with markers of systemic and joint disease activity, with statistically significant associations for MCTR3. These findings support the hypothesis that MCTRs may play a protective role in the resolution of chronic inflammation [104].

These studies expand the scope of MaRs research in humans, highlighting not only MaR1 but also MCTR1–3 as endogenous mediators potentially involved in the resolution of chronic inflammation and immune regulation across metabolic, respiratory, and autoimmune conditions.

6. Conclusions and Projections

MaRs, a family of SPMs derived from DHA, have emerged as potent regulators of inflammation, tissue repair, and resolution of pathological processes. Among them, MaR1 stands out due to its broad therapeutic spectrum, exhibiting anti-inflammatory, pro-resolving, and cytoprotective effects across multiple organs, including the liver, heart, lungs, kidneys, and central nervous system. Preclinical studies consistently show the ability of MaR1 to modulate key signaling pathways such as NF-κB, STAT3, and PI3K/Akt, while promoting macrophage polarization, enhancing phagocytosis, and preserving epithelial and endothelial barrier integrity.

Beyond MaR1, other members of the MaRs family including MaR2, MCTR1–3, and MaR-L1/2, have shown promising bioactivities, particularly in epithelial repair, immune modulation and resolution of chronic inflammation. However, their mechanisms of action remain less well-characterized. Emerging evidence suggests potential involvement of receptors such as LGR6, ALX, and GPCRs, as well as intracellular pathways like PPAR-γ, Nrf2, and RORγ, but further studies are needed to clarify their biosynthetic regulation, receptor specificity, and pharmacokinetics.

https://doi.org/10.3390/biom16010139

A particularly compelling aspect of MaRs is their ability to modulate specific immune cell types, including CXCL3^{+} neutrophils, Th17/Treg lymphocytes, and epithelial cells in several tissues. This cellular specificity suggests that MaRs may offer a more targeted and physiological approach to resolving inflammation compared to conventional anti-inflammatory drugs. Moreover, their influence on processes such as ferroptosis, pyroptosis, and efferocytosis opens new therapeutic avenues in degenerative, autoimmune, and metabolic diseases.

Despite the strong preclinical foundation, several challenges remain for clinical translation. These include dose optimization, molecular stability, delivery strategies, and the identification of biomarkers of resolution. The development of synthetic analogs, controlled-release formulations, and receptor-specific agonists may help overcome these barriers and accelerate the integration of MaRs into therapeutic protocols.

From a clinical perspective, MaRs offer promising strategies for managing chronic inflammatory diseases and acute conditions. Current approaches under investigation include synthetic analogs, receptor-specific agonist, and targeted delivery systems to enhance stability and bioavailability, aiming to harness their pro-resolving properties for therapeutic benefit.

Taken together, MaRs represent a novel class of bioactive lipids with the potential to reshape the treatment landscape for inflammatory and degenerative diseases. Their ability to resolve, rather than merely suppress, inflammation positions them as key candidates in the design of next-generation therapies that are both effective and aligned with the body's natural healing processes.

Author Contributions: All authors contributed significantly to the work as follows: conceptualization, F.B.-N., P.M.-S. and J.Z.-H.; design, F.B.-N. and P.M.-S. and J.Z.-H.; data acquisition, F.B.-N., P.M.-S., R.N.-G., F.H.-V., M.Q.S.M. and J.Z.-H.; data interpretation and analysis, F.B.-N., P.M.-S. and D.R.G., writing—original draft preparation, F.B.-N., P.M.-S. and R.N.-G., writing—review and editing, F.B.-N., P.M.-S., D.R.G. and J.Z.-H.; supervision, D.R.G., M.Q.S.M. and J.Z.-H.; critical revision of the manuscript, F.B.-N., P.M.-S., D.R.G., R.N.-G., M.Q.S.M. and J.Z.-H. All authors have read and agreed to the published version of the manuscript.

Funding: This study was sponsored by the National Agency for Research and Development (ANID). National Doctorate Grant N° 21212100, N° 21220031 and N° 21221512.

Institutional Review Board Statement: Not applicable.

Informed Consent Statement: Not applicable.

Data Availability Statement: No new data were created or analyzed in this review.

Acknowledgments: The authors have reviewed and edited the output and take full responsibility for the content of this publication.

Conflicts of Interest: The authors declare no conflicts of interest.

Abbreviations

PUFAs	Polyunsaturated fatty acids
SPMs	Specialized pro-resolving mediators
MaRs	Maresins
MaR1	Maresin 1
MaR2	Maresin 2
MCTR	Maresin conjugates in tissue regeneration
MaR-L	Maresin-like lipid mediators

https://doi.org/10.3390/biom16010139

LA	Linoleic acid
ALA	α-linolenic acid
AA	Arachidonic acid
EPA	Eicosapentaenoic acid
DHA	Docosahexaenoic acid
LXs	Lipoxins
RVs	Resolvins
PDs	Protectins
12-LOX	12-Lipoxygenase
14S-HpDHA	14S-hydro(peroxy)-docosa-4Z,7Z,10Z,12E,16Z,19Z-hexaenoic acid
sEH	Soluble epoxide hydrolase
GSTM4	Glutathione S-transferase MU4
LTC4S	Leukotriene C4 synthase
GGT	γ-glutamyl transferase
DPEP	Dipeptidase catalysis
CYP450	Cytochrome P450
14S,22-diHDHA	14,22-dihydroxydocosahexaenoic acid
iNOS	Inducible nitric oxide synthase
COX2	Cyclooxygenase 2
AB_{42}	Amyloid beta 42 peptide
MHC-II	Major histocompatibility complex class 2
CHME3	Human microglial cell line
THP-1	Human monocytic leukemia cell line
NSC-34	Motor neurons
LPS	Lipopolysaccharide
MLE-12	Epithelial Cells
CLDN1	Claudin 1
ZO-1	Protein zonula occludens-1
BEAS-2B	Immortalized human bronchial epithelial cell line
PKC	Protein kinase C
hMSCs	Human mesenchymal stromal/stem cells
hSPs	Heat shock Protein
Nrf2	Erythroid-derived nuclear factor 2

References

1. Bang, H.O.; Dyerberg, J.; Nielsen, A.B. Plasma Lipid and Lipoprotein Pattern in Greenlandic West-Coast Eskimos. *Lancet* **1971**, *297*, 1143–1146. [CrossRef] [PubMed]
2. Perica, M.M.; Delaš, I. Essential Fatty Acids and Psychiatric Disorders. *Nutr. Clin. Pract.* **2011**, *26*, 409–425. [CrossRef]
3. Huerta-Yépez, S.; Tirado-Rodriguez, A.B.; Hankinson, O. Role of Diets Rich in Omega-3 and Omega-6 in the Development of Cancer. *Bol. Med. Hosp. Infant. Mex.* **2016**, *73*, 446–456. [CrossRef] [PubMed]
4. Innis, S.M. Fatty Acids and Early Human Development. *Early Hum. Dev.* **2007**, *83*, 761–766. [CrossRef] [PubMed]
5. Querques, G.; Forte, R.; Souied, E.H. Retina and Omega-3. *J. Nutr. Metab.* **2011**, *2011*, 748361. [CrossRef]
6. Li, J.; Zou, B.; Yeo, Y.H.; Feng, Y.; Xie, X.; Lee, D.H.; Fujii, H.; Wu, Y.; Kam, L.Y.; Ji, F.; et al. Prevalence, Incidence, and Outcome of Non-Alcoholic Fatty Liver Disease in Asia, 1999–2019: A Systematic Review and Meta-Analysis. *Lancet Gastroenterol. Hepatol.* **2019**, *4*, 389–398. [CrossRef]
7. Zúñiga, J.; Venegas, F.; Villarreal, M.; Núñez, D.; Chandía, M.; Valenzuela, R.; Tapia, G.; Varela, P.; Videla, L.A.; Fernández, V. Protection against in Vivo Liver Ischemia-Reperfusion Injury by *n-3* Long-Chain Polyunsaturated Fatty Acids in the Rat. *Free Radic. Res.* **2010**, *44*, 854–863. [CrossRef]
8. Zárate, R.; el Jaber-Vazdekis, N.; Tejera, N.; Pérez, J.A.; Rodríguez, C. Significance of Long Chain Polyunsaturated Fatty Acids in Human Health. *Clin. Transl. Med.* **2017**, *6*, 25. [CrossRef]
9. Simopoulos, A. An Increase in the Omega-6/Omega-3 Fatty Acid Ratio Increases the Risk for Obesity. *Nutrients* **2016**, *8*, 128. [CrossRef]

 https://doi.org/10.3390/biom16010139

10. Mariamenatu, A.H.; Abdu, E.M. Overconsumption of Omega-6 Polyunsaturated Fatty Acids (PUFAs) versus Deficiency of Omega-3 PUFAs in Modern-Day Diets: The Disturbing Factor for Their "Balanced Antagonistic Metabolic Functions" in the Human Body. *J. Lipids* **2021**, *2021*, 8848161. [CrossRef]
11. Torres-Castillo, N.; Silva-Gómez, J.A.; Campos-Perez, W.; Barron-Cabrera, E.; Hernandez-Cañaveral, I.; Garcia-Cazarin, M.; Marquez-Sandoval, Y.; Gonzalez-Becerra, K.; Barron-Gallardo, C.; Martinez-Lopez, E. High Dietary ω-6:ω-3 PUFA Ratio Is Positively Associated with Excessive Adiposity and Waist Circumference. *Obes. Facts* **2018**, *11*, 344–353. [CrossRef]
12. Serhan, C.N.; Clish, C.B.; Brannon, J.; Colgan, S.P.; Chiang, N.; Gronert, K. Brief Definitive Report Novel Functional Sets of Lipid-Derived Mediators with Antiinflammatory Actions Generated from Omega-3 Fatty Acids via Cyclooxygenase 2-Nonsteroidal Antiinflammatory Drugs and Transcellular Processing. *J. Exp. Med.* **2000**, *192*, 1197–1204. [CrossRef]
13. Chatterjee, A.; Sharma, A.; Chen, M.; Toy, R.; Mottola, G.; Conte, M.S. The Pro-Resolving Lipid Mediator Maresin 1 (MaR1) Attenuates Inflammatory Signaling Pathways in Vascular Smooth Muscle and Endothelial Cells. *PLoS ONE* **2014**, *9*, 113480. [CrossRef] [PubMed]
14. Ou, G.; Liu, Q.; Yu, C.; Chen, X.; Zhang, W.; Chen, Y.; Wang, T.; Luo, Y.; Jiang, G.; Zhu, M.; et al. The Protective Effects of Maresin 1 in the OVA-Induced Asthma Mouse Model. *Mediators Inflamm.* **2021**, *2021*, 4131420. [CrossRef]
15. Tang, D.; Fu, G.; Li, W.; Sun, P.; Loughran, P.A.; Deng, M.; Scott, M.J.; Billiar, T.R. Maresin 1 Protects the Liver against Ischemia/Reperfusion Injury via the ALXR/Akt Signaling Pathway. *Mol. Med.* **2021**, *27*, 18. [CrossRef] [PubMed]
16. Wahyuni, T.; Kobayashi, A.; Tanaka, S.; Miyake, Y.; Yamamoto, A.; Bahtiar, A.; Mori, S.; Kametani, Y.; Tomimatsu, M.; Matsumoto, K.; et al. Maresin-1 Induces Cardiomyocyte Hypertrophy through IGF-1 Paracrine Pathway. *Am. J. Physiol.-Cell Physiol.* **2021**, *321*, C82–C93. [CrossRef]
17. Serhan, C.N.; Yang, R.; Martinod, K.; Kasuga, K.; Pillai, P.S.; Porter, T.F.; Oh, S.F.; Spite, M. Maresins: Novel Macrophage Mediators with Potent Antiinflammatory and Proresolving Actions. *J. Exp. Med.* **2009**, *206*, 15–23. [CrossRef]
18. Serhan, C.N.; Dalli, J.; Karamnov, S.; Choi, A.; Park, C.; Xu, Z.; Ji, R.; Zhu, M.; Petasis, N.A. Macrophage Proresolving Mediator Maresin 1 Stimulates Tissue Regeneration and Controls Pain. *FASEB J.* **2012**, *26*, 1755–1765. [CrossRef]
19. Serhan, C.N.; Dalli, J.; Colas, R.A.; Winkler, J.W.; Chiang, N. Protectins and Maresins: New pro-Resolving Families of Mediators in Acute Inflammation and Resolution Bioactive Metabolome. *Biochim. Et Biophys. Acta (BBA) Mol. Cell Biol. Lipids* **2015**, *1851*, 397–413. [CrossRef] [PubMed]
20. Liu, M.; He, H.; Chen, L. Protective Potential of Maresins in Cardiovascular Diseases. *Front. Cardiovasc. Med.* **2022**, *9*, 923413. [CrossRef]
21. Dalli, J.; Zhu, M.; Vlasenko, N.A.; Deng, B.; Haeggström, J.Z.; Petasis, N.A.; Serhan, C.N. The novel 13S, 14S-epoxy-maresin is converted by human macrophages to maresin 1 (MaR1), inhibits leukotriene A4 hydrolase (LTA4H), and shifts macrophage phenotype. *FASEB J.* **2013**, *27*, 2573–2583. [CrossRef]
22. Olsen, M.V.; Lyngstadaas, A.V.; Bair, J.A.; Hodges, R.R.; Utheim, T.P.; Serhan, C.N.; Dartt, D.A. Signaling Pathways Used by the Specialized Pro-Resolving Mediator Maresin 2 Regulate Goblet Cell Function: Comparison with Maresin 1. *Int. J. Mol. Sci.* **2022**, *23*, 6233. [CrossRef] [PubMed]
23. Deng, B.; Wang, C.-W.; Arnardottir, H.H.; Li, Y.; Cheng, C.-Y.C.; Dalli, J.; Serhan, C.N. Maresin Biosynthesis and Identification of Maresin 2, a New Anti-Inflammatory and Pro-Resolving Mediator from Human Macrophages. *PLoS ONE* **2014**, *9*, e102362. [CrossRef]
24. Chiang, N.; Riley, I.R.; Dalli, J.; Rodriguez, A.R.; Spur, B.W.; Serhan, C.N. New Maresin Conjugates in Tissue Regeneration Pathway Counters Leukotriene D_4–Stimulated Vascular Responses. *FASEB J.* **2018**, *32*, 4043–4052. [CrossRef] [PubMed]
25. Dalli, J.; Chiang, N.; Serhan, C.N. Identification of 14-Series Sulfido-Conjugated Mediators That Promote Resolution of Infection and Organ Protection. *Proc. Natl. Acad. Sci. USA* **2014**, *111*, E4753–E4761. [CrossRef]
26. Dalli, J.; Sanger, J.M.; Rodriguez, A.R.; Chiang, N.; Spur, B.W.; Serhan, C.N. Identification and Actions of a Novel Third Maresin Conjugate in Tissue Regeneration: MCTR3. *PLoS ONE* **2016**, *11*, e0149319. [CrossRef]
27. Hong, S.; Lu, Y.; Tian, H.; Alapure, B.V.; Wang, Q.; Bunnell, B.A.; Laborde, J.M. Maresin-like Lipid Mediators Are Produced by Leukocytes and Platelets and Rescue Reparative Function of Diabetes-Impaired Macrophages. *Chem. Biol.* **2014**, *21*, 1318–1329. [CrossRef]
28. O'Donnell, V.B.; Schebb, N.H.; Milne, G.L.; Murphy, M.P.; Thomas, C.P.; Steinhilber, D.; Gelhaus, S.L.; Kühn, H.; Gelb, M.H.; Jakobsson, P.-J.; et al. Failure to Apply Standard Limit-of-Detection or Limit-of-Quantitation Criteria to Specialized pro-Resolving Mediator Analysis Incorrectly Characterizes Their Presence in Biological Samples. *Nat. Commun.* **2023**, *14*, 7172. [CrossRef]
29. Yin, P.; Wang, X.; Wang, S.; Wei, Y.; Feng, J.; Zhu, M. Maresin 1 Improves Cognitive Decline and Ameliorates Inflammation in a Mouse Model of Alzheimer's Disease. *Front. Cell Neurosci.* **2019**, *13*, 466. [CrossRef]

 https://doi.org/10.3390/biom16010139

30. Ohuchi, K.; Ono, Y.; Joho, M.; Tsuruma, K.; Ogami, S.; Yamane, S.; Funato, M.; Kaneko, H.; Nakamura, S.; Hara, H.; et al. A Docosahexaenoic Acid-Derived Pro-Resolving Agent, Maresin 1, Protects Motor Neuron Cells Death. *Neurochem. Res.* **2018**, *43*, 1413–1423. [CrossRef]
31. Zhu, M.; Wang, X.; Hjorth, E.; Colas, R.A.; Schroeder, L.; Granholm, A.-C.; Serhan, C.N.; Schultzberg, M. Pro-Resolving Lipid Mediators Improve Neuronal Survival and Increase Aβ42 Phagocytosis. *Mol. Neurobiol.* **2016**, *53*, 2733–2749. [CrossRef]
32. Wang, Y.; Leppert, A.; Tan, S.; van der Gaag, B.; Li, N.; Schultzberg, M.; Hjorth, E. Maresin 1 Attenuates Pro-inflammatory Activation Induced by Β-amyloid and Stimulates Its Uptake. *J. Cell Mol. Med.* **2021**, *25*, 434–447. [CrossRef] [PubMed]
33. Wang, Y.; Zhang, X.; Biverstål, H.; Bazan, N.G.; Tan, S.; Li, N.; Ohshima, M.; Schultzberg, M.; Li, X. Pro-Resolving Lipid Mediator Reduces Amyloid-Β42–Induced Gene Expression in Human Monocyte–Derived Microglia. *Neural Regen. Res.* **2025**, *20*, 873–886. [CrossRef] [PubMed]
34. Gong, J.; Liu, H.; Wu, J.; Qi, H.; Wu, Z.; Shu, H.; Li, H.; Chen, L.; Wang, Y.; Li, B.; et al. Maresin prevents lipopolysaccharide-induced neutrophil survival and accelerates resolution of acute lung injury. *Shock* **2015**, *44*, 371–380. [CrossRef]
35. Chen, L.; Liu, H.; Wang, Y.; Xia, H.; Gong, J.; Li, B.; Yao, S.; Shang, Y. Maresin 1 Maintains the Permeability of Lung Epithelial Cells In Vitro and In Vivo. *Inflammation* **2016**, *39*, 1981–1989. [CrossRef]
36. Nordgren, T.M.; Heires, A.J.; Wyatt, T.A.; Poole, J.A.; LeVan, T.D.; Cerutis, D.R.; Romberger, D.J. Maresin-1 Reduces the pro-Inflammatory Response of Bronchial Epithelial Cells to Organic Dust. *Respir. Res.* **2013**, *14*, 51. [CrossRef]
37. Ruiz, A.; Sarabia, C.; Torres, M.; Juárez, E. Resolvin D1 (RvD1) and Maresin 1 (Mar1) Contribute to Human Macrophage Control of M. Tuberculosis Infection While Resolving Inflammation. *Int. Immunopharmacol.* **2019**, *74*, 105694. [CrossRef]
38. Wang, C.-W.; Colas, R.A.; Dalli, J.; Arnardottir, H.H.; Nguyen, D.; Hasturk, H.; Chiang, N.; Van Dyke, T.E.; Serhan, C.N. Maresin 1 Biosynthesis and Proresolving Anti-Infective Functions with Human-Localized Aggressive Periodontitis Leukocytes. *Infect. Immun.* **2016**, *84*, 658–665. [CrossRef] [PubMed]
39. AlZahrani, S.; Shinwari, Z.; Gaafar, A.; Alaiya, A.; Al-Kahtani, A. Anti-Inflammatory Effect of Specialized Proresolving Lipid Mediators on Mesenchymal Stem Cells: An In Vitro Study. *Cells* **2022**, *12*, 122. [CrossRef] [PubMed]
40. Martínez-Fernández, L.; Burgos, M.; Sáinz, N.; Laiglesia, L.M.; Arbones-Mainar, J.M.; González-Muniesa, P.; Moreno-Aliaga, M.J. Maresin 1 Exerts a Tissue-Specific Regulation of Adipo-Hepato-Myokines in Diet-Induced Obese Mice and Modulates Adipokine Expression in Cultured Human Adipocytes in Basal and Inflammatory Conditions. *Biomolecules* **2023**, *13*, 919. [CrossRef]
41. Zhang, Z.; Ji, C.; Wang, Y.-N.; Liu, S.; Wang, M.; Xu, X.; Zhang, D. Maresin1 Suppresses High-Glucose-Induced Ferroptosis in Osteoblasts via NRF2 Activation in Type 2 Diabetic Osteoporosis. *Cells* **2022**, *11*, 2560. [CrossRef]
42. Yao, D.; Zou, Y.; Lv, Y. Maresin 1 Enhances Osteogenic Potential of Mesenchymal Stem Cells by Modulating Macrophage Peroxisome Proliferator-Activated Receptor-γ-Mediated Inflammation Resolution. *Biomater. Adv.* **2022**, *141*, 213116. [CrossRef] [PubMed]
43. Hughes, F.M.; Allkanjari, A.; Odom, M.R.; Jin, H.; Purves, J.T. Specialized Pro-Resolution Mediators in the Bladder: Receptor Expression and Recovery of Bladder Function from Cystitis. *Exp. Biol. Med.* **2022**, *247*, 700–711. [CrossRef]
44. Miranda, J.; Brazil, J.C.; Morris, A.H.; Parkos, C.A.; Quiros, M.; Nusrat, A. Maresin-2 Promotes Mucosal Repair and Has Therapeutic Potential When Encapsulated in Thermostable Nanoparticles. *Proc. Natl. Acad. Sci. USA* **2023**, *120*, e2218162120. [CrossRef]
45. Thamizhchelvan, A.M.; Masoud, A.-R.; Su, S.; Lu, Y.; Peng, H.; Kobayashi, Y.; Wang, Y.; Archer, N.K.; Hong, S. Bactericidal Efficacy of the Combination of Maresin-like Proresolving Mediators and Carbenicillin Action on Biofilm-Forming Burn Trauma Infection-Related Bacteria. *Int. J. Mol. Sci.* **2024**, *25*, 2792. [CrossRef]
46. Li, R.; Wang, Y.; Zhao, E.; Wu, K.; Li, W.; Shi, L.; Wang, D.; Xie, G.; Yin, Y.; Deng, M.; et al. Maresin 1, a Proresolving Lipid Mediator, Mitigates Carbon Tetrachloride-Induced Liver Injury in Mice. *Oxid. Med. Cell Longev.* **2016**, *2016*, 9203716. [CrossRef] [PubMed]
47. Yang, W.; Tao, K.; Zhang, P.; Chen, X.; Sun, X.; Li, R. Maresin 1 Protects against Lipopolysaccharide/d-Galactosamine-Induced Acute Liver Injury by Inhibiting Macrophage Pyroptosis and Inflammatory Response. *Biochem. Pharmacol.* **2022**, *195*, 114863. [CrossRef]
48. Rodríguez, M.J.; Sabaj, M.; Tolosa, G.; Herrera Vielma, F.; Zúñiga, M.J.; González, D.R.; Zúñiga-Hernández, J. Maresin-1 Prevents Liver Fibrosis by Targeting Nrf2 and NF-ΚB, Reducing Oxidative Stress and Inflammation. *Cells* **2021**, *10*, 3406. [CrossRef] [PubMed]
49. Laiglesia, L.M.; Lorente-Cebrián, S.; Martínez-Fernández, L.; Sáinz, N.; Prieto-Hontoria, P.L.; Burrell, M.A.; Rodríguez-Ortigosa, C.M.; Martínez, J.A.; Moreno-Aliaga, M.J. Maresin 1 Mitigates Liver Steatosis in Ob/Ob and Diet-Induced Obese Mice. *Int. J. Obes.* **2018**, *42*, 572–579. [CrossRef]
50. Jung, T.W.; Kim, H.-C.; Abd El-Aty, A.M.; Jeong, J.H. Maresin 1 Attenuates NAFLD by Suppression of Endoplasmic Reticulum Stress via AMPK–SERCA2b Pathway. *J. Biol. Chem.* **2018**, *293*, 3981–3988. [CrossRef]

51. Rius, B.; Duran-Güell, M.; Flores-Costa, R.; López-Vicario, C.; Lopategi, A.; Alcaraz-Quiles, J.; Casulleras, M.; José Lozano, J.; Titos, E.; Clària, J. The Specialized Proresolving Lipid Mediator Maresin 1 Protects Hepatocytes from Lipotoxic and Hypoxia-induced Endoplasmic Reticulum Stress. *FASEB J.* **2017**, *31*, 5384–5398. [CrossRef]
52. Wang, F.; Gong, Y.; Chen, T.; Li, B.; Zhang, W.; Yin, L.; Zhao, H.; Tang, Y.; Wang, X.; Huang, C. Maresin1 Ameliorates Ventricular Remodelling and Arrhythmia in Mice Models of Myocardial Infarction via NRF2/HO-1 and TLR4/NF-KB Signalling. *Int. Immunopharmacol.* **2022**, *113*, 109369. [CrossRef]
53. Ding, H.-S.; Qu, J.-F.; Luo, X.-J.; Luo, Z.-H.; Huang, Z.-Y.; Zhong, P.; Li, X.; Liu, X.-X. Maresin 1 Attenuates Myocardium Ischemia/Reperfusion Injury via SIRT1/HMGB-1/NLRP-3-Related Mechanisms. *Eur. J. Pharmacol.* **2025**, *998*, 177456. [CrossRef]
54. Li, D.; Wang, M.; Ye, J.; Zhang, J.; Xu, Y.; Wang, Z.; Zhao, M.; Ye, D.; Wan, J. Maresin 1 Alleviates the Inflammatory Response, Reduces Oxidative Stress and Protects against Cardiac Injury in LPS-Induced Mice. *Life Sci.* **2021**, *277*, 119467. [CrossRef] [PubMed]
55. Wang, F.; Zhang, W.; Yin, L.; Huang, C.; Zhao, H. Exogenous Maresin1 Attenuates Doxorubicin-Induced Cardiomyocyte Ferroptosis and Mitochondrial Impairment via NRF2/GPX4 Axis. *Free Radic. Biol. Med.* **2025**, *235*, 335–346. [CrossRef] [PubMed]
56. Akagi, D.; Chen, M.; Toy, R.; Chatterjee, A.; Conte, M.S. Systemic Delivery of Proresolving Lipid Mediators Resolvin D_2 and Maresin 1 Attenuates Intimal Hyperplasia in Mice. *FASEB J.* **2015**, *29*, 2504–2513. [CrossRef]
57. Viola, J.R.; Lemnitzer, P.; Jansen, Y.; Csaba, G.; Winter, C.; Neideck, C.; Silvestre-Roig, C.; Dittmar, G.; Döring, Y.; Drechsler, M.; et al. Resolving Lipid Mediators Maresin 1 and Resolvin D2 Prevent Atheroprogression in Mice. *Circ. Res.* **2016**, *119*, 1030–1038. [CrossRef]
58. Yin, Z.; Zhang, J.; Zhao, M.; Peng, S.; Ye, J.; Liu, J.; Xu, Y.; Xu, S.; Pan, W.; Wei, C.; et al. Maresin-1 Ameliorates Hypertensive Vascular Remodeling through Its Receptor LGR6. *MedComm* **2024**, *5*, e491. [CrossRef]
59. Elder, C.T.; Filiberto, A.C.; Su, G.; Ladd, Z.; Leroy, V.; Pruitt, E.Y.; Lu, G.; Jiang, Z.; Sharma, A.K.; Upchurch, G.R. Maresin 1 Activates LGR6 Signaling to Inhibit Smooth Muscle Cell Activation and Attenuate Murine Abdominal Aortic Aneurysm Formation. *FASEB J.* **2021**, *35*, e21780. [CrossRef] [PubMed]
60. Li, H.; Li, X.; Hao, Y.; Wu, C.; Fu, Y.; Su, N.; Chen, H.; Ying, B.; Wang, H.; Su, L.; et al. Maresin 1 Intervention Reverses Experimental Pulmonary Arterial Hypertension in Mice. *Br. J. Pharmacol.* **2022**, *179*, 5132–5147. [CrossRef]
61. Liu, M.; He, H.; Fan, F.; Qiu, L.; Zheng, F.; Guan, Y.; Yang, G.; Chen, L. Maresin-1 Protects against Pulmonary Arterial Hypertension by Improving Mitochondrial Homeostasis through ALXR/HSP90α Axis. *J. Mol. Cell Cardiol.* **2023**, *181*, 15–30. [CrossRef]
62. Li, J.; Zhang, Z.; Wang, L.; Jiang, L.; Qin, Z.; Zhao, Y.; Su, B. Maresin 1 Attenuates Lipopolysaccharide-Induced Acute Kidney Injury via Inhibiting NOX4/ROS/NF-KB Pathway. *Front. Pharmacol.* **2021**, *12*, 782660. [CrossRef]
63. Sun, M.; Wang, F.; Li, H.; Li, M.; Wang, Y.; Wang, C.; Zhang, Y.; Zhang, D.; Li, J.; Yao, S. Maresin-1 Attenuates Sepsis-Associated Acute Kidney Injury via Suppressing Inflammation, Endoplasmic Reticulum Stress and Pyroptosis by Activating the AMPK/SIRT3 Pathway. *J. Inflamm. Res.* **2024**, *17*, 1349–1364. [CrossRef]
64. Sun, S.; Wang, J.; Wang, J.; Wang, F.; Yao, S.; Xia, H. Maresin 1 Mitigates Sepsis-Associated Acute Kidney Injury in Mice via Inhibition of the NF-KB/STAT3/MAPK Pathways. *Front. Pharmacol.* **2019**, *10*, 1323. [CrossRef]
65. Qiu, Y.; Wu, Y.; Zhao, H.; Sun, H.; Gao, S. Maresin 1 Mitigates Renal Ischemia/Reperfusion Injury in Mice via Inhibition of the TLR4/MAPK/NF-κB Pathways and Activation of the Nrf2 Pathway. *Drug Des. Dev. Ther.* **2019**, *13*, 739–745. [CrossRef] [PubMed]
66. Li, X.; Xu, B.; Wu, J.; Pu, Y.; Wan, S.; Zeng, Y.; Wang, M.; Luo, L.; Zhang, F.; Jiang, Z.; et al. Maresin 1 Alleviates Diabetic Kidney Disease via LGR6-Mediated CAMP-SOD2-ROS Pathway. *Oxid. Med. Cell Longev.* **2022**, *2022*, 7177889. [CrossRef] [PubMed]
67. Sun, Q.; Wu, Y.; Zhao, F.; Wang, J. Maresin 1 Ameliorates Lung Ischemia/Reperfusion Injury by Suppressing Oxidative Stress via Activation of the Nrf-2-Mediated HO-1 Signaling Pathway. *Oxid. Med. Cell Longev.* **2017**, *2017*, 9634803. [CrossRef]
68. Gong, J.; Wu, Z.; Qi, H.; Chen, L.; Li, H.; Li, B.; Yao, C.; Wang, Y.; Wu, J.; Yuan, S.; et al. Maresin 1 Mitigates LPS-induced Acute Lung Injury in Mice. *Br. J. Pharmacol.* **2014**, *171*, 3539–3550. [CrossRef]
69. Wang, F.; Wang, M.; Wang, J.; Chen, M.; Sun, S.; Yao, S.; Xia, H. Maresin1 Ameliorates Sepsis-Associated Lung Injury by Inhibiting the Activation of the JAK2/STAT3 and MAPK/NF-KB Signaling Pathways. *Microb. Pathog.* **2020**, *148*, 104468. [CrossRef] [PubMed]
70. Qiao, N.; Lin, Y.; Wang, Z.; Chen, J.-Y.; Ge, Y.-Y.; Yao, S.-L.; Gong, J. Maresin1 Promotes M2 Macrophage Polarization Through Peroxisome Proliferator–Activated Receptor-γ Activation to Expedite Resolution of Acute Lung Injury. *J. Surg. Res.* **2020**, *256*, 584–594. [CrossRef]
71. Xia, H.; Wang, F.; Wang, M.; Wang, J.; Sun, S.; Chen, M.; Huang, S.; Chen, X.; Yao, S. Maresin1 Ameliorates Acute Lung Injury Induced by Sepsis through Regulating Th17/Treg Balance. *Life Sci.* **2020**, *254*, 117773. [CrossRef] [PubMed]
72. Wang, F.; Chen, M.; Wang, C.; Xia, H.; Zhang, D.; Yao, S. Single-Cell Sequencing Reveals the Regulatory Role of Maresin1 on Neutrophils during Septic Lung Injury. *Cells* **2022**, *11*, 3733. [CrossRef]

73. Krishnamoorthy, N.; Walker, K.H.; Brüggemann, T.R.; Tavares, L.P.; Smith, E.W.; Nijmeh, J.; Bai, Y.; Ai, X.; Cagnina, R.E.; Duvall, M.G.; et al. The Maresin 1–LGR6 Axis Decreases Respiratory Syncytial Virus-Induced Lung Inflammation. *Proc. Natl. Acad. Sci. USA* **2023**, *120*, e2206480120. [CrossRef]
74. Wu, H.; Li, N.; Peng, S.; Fu, H.; Hu, Z.; Su, L. Maresin1 Improves Hippocampal Neuroinflammation and Cognitive Function in Septic Rats by Activating the SLC7A11/GPX4 Ferroptosis Signaling Pathway. *Int. Immunopharmacol.* **2024**, *131*, 111792. [CrossRef]
75. Lu, Y.; Lu, G.; Gao, L.; Zhu, Q.; Xue, J.; Zhang, J.; Ma, X.; Ma, N.; Yang, Q.; Dong, J.; et al. The Proresolving Lipid Mediator Maresin1 Alleviates Experimental Pancreatitis via Switching Macrophage Polarization. *Mediat. Inflamm.* **2021**, *2021*, 6680456. [CrossRef]
76. Shih, Y.-R.V.; Tao, H.; Gilpin, A.; Lee, Y.-W.; Perikamana, S.M.; Varghese, S. Specialized Pro-Resolving Mediator Maresin 1 Attenuates Pain in a Mouse Model of Osteoarthritis. *Osteoarthr. Cartil.* **2025**, *33*, 341–350. [CrossRef] [PubMed]
77. Dantas, K.L.S.; Bianchini, B.H.S.; da Silva, M.D.V.; Piva, M.; da Cunha, J.M.; Zanoveli, J.M.; Cardoso, F.C.; Vicentini, F.T.M.C.; Ferraz, C.R.; Clissa, P.B.; et al. Maresin 2, a Specialized Pro-Resolution Lipid Mediator, Reduces Pain and Inflammation Induced by Bothrops Jararaca Venom in Mice. *Toxins* **2025**, *17*, 367. [CrossRef]
78. Lopes, R.V.; Baggio, D.F.; Ferraz, C.R.; Bertozzi, M.M.; Saraiva-Santos, T.; Verri Junior, W.A.; Chichorro, J.G. Maresin-2 Inhibits Inflammatory and Neuropathic Trigeminal Pain and Reduces Neuronal Activation in the Trigeminal Ganglion. *Curr. Res. Neurobiol.* **2023**, *4*, 100093. [CrossRef] [PubMed]
79. Yu, C.; Shi, Z.; Ou, G.; Chen, X.; Liu, Q.; Zeng, D.; Nie, X.; Chen, J. Maresin-2 Alleviates Allergic Airway Inflammation in Mice by Inhibiting the Activation of NLRP3 Inflammasome, Th2 Type Immune Response and Oxidative Stress. *Mol. Immunol.* **2022**, *146*, 78–86. [CrossRef]
80. Sugimoto, S.; Mena, H.A.; Sansbury, B.E.; Kobayashi, S.; Tsuji, T.; Wang, C.-H.; Yin, X.; Huang, T.L.; Kusuyama, J.; Kodani, S.D.; et al. Brown Adipose Tissue-Derived MaR2 Contributes to Cold-Induced Resolution of Inflammation. *Nat. Metab.* **2022**, *4*, 775–790. [CrossRef] [PubMed]
81. Tozzi, O.N.; Jiacomini, I.G.; Bastos, T.S.B.; Nicolazzi, L.H.C.N.; dos Santos Luz, R.B.; Paredes, L.C.; Gonçalves, L.E.; Lima, M.H.S.; Verri, W.A.; Camara, N.O.S.; et al. Evaluation of the Effects of Loxosceles Intermedia's Venom in Zebrafish. *Toxicol. Rep.* **2022**, *9*, 1410–1418. [CrossRef]
82. Han, J.; Li, H.; Bhandari, S.; Cao, F.; Wang, X.; Tian, C.; Li, X.; Zhang, P.; Liu, Y.; Wu, C.; et al. Maresin Conjugates in Tissue Regeneration 1 Improves Alveolar Fluid Clearance by Up-regulating Alveolar ENaC, Na, K-ATPase in Lipopolysaccharide-induced Acute Lung Injury. *J. Cell Mol. Med.* **2020**, *24*, 4736–4747. [CrossRef] [PubMed]
83. Xiao, J.; Yang, Q.; Zhang, Y.; Xu, H.; Ye, Y.; Li, L.; Yang, Y.; Jin, S. Maresin Conjugates in Tissue Regeneration-1 Suppresses Ferroptosis in Septic Acute Kidney Injury. *Cell Biosci.* **2021**, *11*, 221. [CrossRef] [PubMed]
84. Yang, Y.; Li, X.-Y.; Li, L.-C.; Xiao, J.; Zhu, Y.-M.; Tian, Y.; Sheng, Y.-M.; Chen, Y.; Wang, J.-G.; Jin, S.-W. Γδ T/Interleukin-17A Contributes to the Effect of Maresin Conjugates in Tissue Regeneration 1 on Lipopolysaccharide-Induced Cardiac Injury. *Front. Immunol.* **2021**, *12*, 674542. [CrossRef]
85. Cheng, Y.; Xiao, Y.; Wang, D.-M.; Liu, K.; Yang, X.; Zheng, C.-X.; He, Z.-B.; Guo, Z.-Y.; Yang, Y. MCTR1 Ameliorates LPS-Induced Lung Injury by Inhibiting Neutrophil Reverse Transendothelial Migration. *Int. Immunopharmacol.* **2025**, *157*, 114777. [CrossRef]
86. Li, H.; Hao, Y.; Yang, L.; Wang, X.; Li, X.; Bhandari, S.; Han, J.; Liu, Y.; Gong, Y.; Scott, A.; et al. MCTR1 Alleviates Lipopolysaccharide-induced Acute Lung Injury by Protecting Lung Endothelial Glycocalyx. *J. Cell Physiol.* **2020**, *235*, 7283–7294. [CrossRef]
87. Pan, J.; Li, X.; Wang, X.; Yang, L.; Chen, H.; Su, N.; Wu, C.; Hao, Y.; Jin, S.; Li, H. MCTR1 Intervention Reverses Experimental Lung Fibrosis in Mice. *J. Inflamm. Res.* **2021**, *14*, 1873–1881. [CrossRef]
88. Gu, L.; Chen, L.; Wang, Y.; Gao, L.; Ye, J.; Xu, Z.; Zhou, Z.; Sun, J. MCTR1 Alleviates Remifentanil-Induced Hyperalgesia by Regulating Mitochondrial Fission Protein DRP1 in Rats. *Neuroreport* **2025**, *36*, 760–766. [CrossRef]
89. Zhuang, R.; Yang, X.; Cai, W.; Xu, R.; Lv, L.; Sun, Y.; Guo, Y.; Ni, J.; Zhao, G.; Lu, Z. MCTR3 Reduces LPS-Induced Acute Lung Injury in Mice via the ALX/PINK1 Signaling Pathway. *Int. Immunopharmacol.* **2021**, *90*, 107142. [CrossRef]
90. Tavares, L.P.; Brüggemann, T.R.; Rezende, R.M.; Machado, M.G.; Cagnina, R.E.; Shay, A.E.; Garcia, C.C.; Nijmeh, J.; Teixeira, M.M.; Levy, B.D. Cysteinyl Maresins Reprogram Macrophages to Protect Mice from Streptococcus Pneumoniae after Influenza A Virus Infection. *mBio* **2022**, *13*, e01267-22. [CrossRef] [PubMed]
91. Chiang, N.; de la Rosa, X.; Libreros, S.; Pan, H.; Dreyfuss, J.M.; Serhan, C.N. Cysteinyl-Specialized Proresolving Mediators Link Resolution of Infectious Inflammation and Tissue Regeneration via TRAF3 Activation. *Proc. Natl. Acad. Sci. USA* **2021**, *118*, e2013374118. [CrossRef]
92. Koenis, D.S.; de Matteis, R.; Rajeeve, V.; Cutillas, P.; Dalli, J. Efferocyte-Derived MCTRs Metabolically Prime Macrophages for Continual Efferocytosis via Rac1-Mediated Activation of Glycolysis. *Adv. Sci.* **2024**, *11*, e2304690. [CrossRef] [PubMed]

93. Shrivastava, P.; Lu, Y.; Su, S.; Kobayashi, Y.; Zhao, Y.; Lien, N.; Masoud, A.-R.; Lukiw, W.J.; Hong, S. Maresin-like 1 Ameliorates Neuropathology of Alzheimer's Disease in Brains of a Transgenic Mouse Model. *Biomedicines* **2024**, *12*, 2865. [CrossRef] [PubMed]
94. Skarke, C.; Alamuddin, N.; Lawson, J.A.; Li, X.; Ferguson, J.F.; Reilly, M.P.; FitzGerald, G.A. Bioactive Products Formed in Humans from Fish Oils. *J. Lipid Res.* **2015**, *56*, 1808–1820. [CrossRef]
95. Barden, A.E.; Mas, E.; Croft, K.D.; Phillips, M.; Mori, T.A. Specialized Proresolving Lipid Mediators in Humans with the Metabolic Syndrome after n–3 Fatty Acids and Aspirin. *Am. J. Clin. Nutr.* **2015**, *102*, 1357–1364. [CrossRef]
96. Miao, T.; Huang, B.; He, N.; Sun, L.; Du, G.; Gong, X.; Xu, Y.; Zheng, Y.; Zheng, H.; Qu, H. Decreased Plasma Maresin 1 Concentration Is Associated with Diabetic Foot Ulcer. *Mediators. Inflamm.* **2020**, *2020*, 4539035. [CrossRef] [PubMed]
97. Tejera, P.; Abdulnour, R.-E.E.; Zhu, Z.; Su, L.; Levy, B.D.; Christiani, D.C. Plasma Levels of Proresolving and Prophlogistic Lipid Mediators: Association with Severity of Respiratory Failure and Mortality in Acute Respiratory Distress Syndrome. *Crit. Care Explor.* **2020**, *2*, e0241. [CrossRef]
98. Fang, X.; Wang, H.; Ye, T.; Fu, X.; Tan, X.; Zeng, Y.; Fan, J.; Xu, Y. Low Serum Maresin-1 Levels Are Associated with Non-Alcoholic Fatty Liver Disease: A Cross Sectional Study. *Lipids Health Dis.* **2021**, *20*, 96. [CrossRef]
99. Li, M.; Sun, Y.; Liu, B.; Xue, Y.; Zhu, M.; Zhang, K.; Jing, Y.; Ding, H.; Liang, Y.; Zhou, H.; et al. Association between Plasma Maresin 1 and the Risk of Atherosclerotic Cardiovascular Disease in Chinese Adults: A Community-Based Cohort Study. *Nutr. Metab. Cardiovasc. Dis.* **2024**, *34*, 1631–1638. [CrossRef] [PubMed]
100. Keeley, E.C.; Li, H.J.; Cogle, C.R.; Handberg, E.M.; Merz, C.N.B.; Pepine, C.J. Specialized Proresolving Mediators in Symptomatic Women with Coronary Microvascular Dysfunction (from the Women's Ischemia Trial to Reduce Events in Nonobstructive CAD [WARRIOR] Trial). *Am. J. Cardiol.* **2022**, *162*, 1–5. [CrossRef]
101. Oliveira Perucci, L.; Pereira Santos, T.A.; Campi Santos, P.; Ribeiro Teixeira, L.C.; Nessralla Alpoim, P.; Braga Gomes, K.; Pires Sousa, L.; Sant'Ana Dusse, L.M.; Talvani, A. Pre-eclampsia Is Associated with Reduced Resolvin D1 and Maresin 1 to Leukotriene B4 Ratios in the Plasma. *Am. J. Reprod. Immunol.* **2020**, *83*, e13206. [CrossRef] [PubMed]
102. Barden, A.; Shinde, S.; Phillips, M.; Beilin, L.; Mas, E.; Hodgson, J.M.; Puddey, I.; Mori, T.A. The Effects of Alcohol on Plasma Lipid Mediators of Inflammation Resolution in Patients with Type 2 Diabetes Mellitus. *Prostaglandins Leukot. Essent. Fatty Acids* **2018**, *133*, 29–34. [CrossRef]
103. Levy, B.D.; Abdulnour, R.-E.E.; Tavares, A.; Brüggemann, T.R.; Norris, P.C.; Bai, Y.; Ai, X.; Serhan, C.N. Cysteinyl Maresins Regulate the Prophlogistic Lung Actions of Cysteinyl Leukotrienes. *J. Allergy Clin. Immunol.* **2020**, *145*, 335–344. [CrossRef]
104. Pistorius, K.; Ly, L.; Souza, P.R.; Gomez, E.A.; Koenis, D.S.; Rodriguez, A.R.; Foster, J.; Sosabowski, J.; Hopkinson, M.; Rajeeve, V.; et al. MCTR3 Reprograms Arthritic Monocytes to Upregulate Arginase-1 and Exert pro-Resolving and Tissue-Protective Functions in Experimental Arthritis. *EBioMedicine* **2022**, *79*, 103974. [CrossRef]

https://doi.org/10.3390/biom16010139

MDPI AG
Grosspeteranlage 5
4052 Basel
Switzerland
Tel.: +41 61 683 77 34

Biomolecules Editorial Office
E-mail: biomolecules@mdpi.com
www.mdpi.com/journal/biomolecules

www.ingramcontent.com/pod-product-compliance
Lightning Source LLC
Chambersburg PA
CBHW061602210926
51850CB00009B/2184
* 9 7 8 3 7 2 5 8 6 8 8 4 1 *